Chemotaxis: Its Biology and Biochemistry

Antibiotics and Chemotherapy

Vol. 19

Editors

H. Schönfeld, Grenzach; R. W. Brockman, Birmingham, Ala.;
F. E. Hahn, Washington, D.C.

S. Karger · Basel · München · Paris · London · New York · Sydney

Chemotaxis:
Its Biology and Biochemistry

Editor

E. Sorkin, Davos Platz

98 figures and 77 tables

S. Karger · Basel · München · Paris · London · New York · Sydney 1974

Antibiotics and Chemotherapy

Vol. 12: Pharmakokinetik und Arzneimitteldosierung/Pharmacokinetics and Drug Dosage. Kolloquium im Forschungsinstitut Borstel 1962. Ed.: ENNO FREERKSEN (Borstel), unter Mitarbeit von/with the collaboration of L. DETTLI (Basel), E. KRÜGER-THIEMER (Borstel), E. NELSON (Buffalo, N.Y.). XVI + 455 p., 161 fig., 46 tab., 1964. ISBN 3-8055-0477-2

Vol. 13: VIII + 316 p., 25 fig., 20 tab., 1965. ISBN 3-8055-0478-0

Vol. 14: VIII + 280 p., 75 fig., 41 tab., 1968. ISBN 3-8055-0479-9

Vol. 15: The Immune Response and its Suppression. An International Symposium held at the Forschungsinstitut, Davos 1968. VIII + 424 p., 148 fig., 64 tab., 1969. ISBN 3-8055-0480-2

Vol. 16: Experimental and Clinical Evaluation of the Tuberculostatics Capreomycin, Isoxyl, Myambutol, Rifampicin. International Colloquy, Borstel 1968. Ed.: H.-J. THUMIM (Borstel). X + 536 p., 150 fig., 251 tab., 1970. ISBN 3-8055-0481-0

Vol. 17: Mode of Action. Ed.: H. SCHÖNFELD (Grenzach) and A. DE WECK (Bern). XII + 170 p., 19 fig., 8 tab., 1971. ISBN 3-8055-1225-2

Vol. 18: Chemotherapy under Special Conditions. Ed.: H. SCHÖNFELD (Grenzach), R. W. BROCKMAN (Birmingham, Ala.) and F. E. HAHN (Washington, D. C.). VIII + 210 p., 32 fig., 41 tab., 1974. ISBN 3-8055-1646-0

S. Karger · Basel · München · Paris · London · New York · Sydney
Arnold-Böcklin-Strasse 25, CH-4011 Basel (Switzerland)

List of Contributors

Adler, Julius, Departments of Biochemistry and Genetics, University of Wisconsin, Madison, WI 53706 (USA)

Allison, A. C., Clinical Research Centre, Watford Road, Harrow, Middlesex (England)

Austen, K. Frank, Department of Medicine, Robert B. Brigham Hospital, Boston, MA 02120 (USA)

Baisero, M. H., Département d'Immunologie et d'Allergie, Service de Médecine, Hôpital Cantonal Universitaire, CH-1011 Lausanne (Switzerland)

Becker, Elmer L., Department of Pathology, University of Connecticut Health Center, Farmington, CT 06032 (USA)

Berg, Howard C., Department of Molecular, Cellular and Developmental Biology, University of Colorado, Boulder, CO 80302 (USA)

Bessis, Marcel, Institut de Pathologie Cellulaire (INSERM, U. 48), Hôpital de Bicêtre, F-94270 Le Kremlin- Bicêtre (France)

Boos, Winfried, Department of Biological Chemistry, Harvard Medical School and the Biochemical Research Laboratory, Massachusetts General Hospital, Boston, MA 02114 (USA)

Borel, J. F., Biological and Medical Research Division, Sandoz Ltd., CH-4002 Basel (Switzerland)

Brown, Douglas A., Department of Molecular, Cellular and Developmental Biology, University of Colorado, Boulder, CO 80302 (USA)

Cottier, H., Department of Pathology, University of Bern, CH-3008 Bern (Switzerland)

Feurer, Camille, Biological and Medical Research Division, Sandoz Ltd., CH-4002 Basel (Switzerland)

Frei, P. C., Département d'Immunologie et d'Allergie, Service de Médecine, Hôpital Cantonal Universitaire, CH-1011 Lausanne (Switzerland)

Gallin, John I., Laboratory of Clinical Investigation, National Institute of Allergy and Infectious Diseases, National Institutes of Health, Bethesda, MD 20014 (USA)

Goetzl, Edward J., Department of Medicine, Harvard Medical School, Boston, MA 02120 (USA)

Hayashi, H., Department of Pathology, Kumamoto University Medical School, Kumamoto 860 (Japan)

Hess, M. W., Department of Pathology, University of Bern, CH-3008 Bern (Switzerland)

Kay, A. B., Department of Respiratory Diseases, University of Edinburgh at the City Hospital, Greenbank Drive, Edinburgh EH10 5SB (Scotland)

Keller, Evelyn Fox, Division of Natural Science, State University of New York, College at Purchase, Purchase, NY 10577 (USA)

Keller, H. U., Department of Pathology, University of Bern, CH-3008 Bern (Switzerland)

Konijn, Theo M., Cell Biology and Morphogenesis Unit, University of Leiden, Kaiserstraat 63, Leiden (The Netherlands)

McKay, I. C., Department of Bacteriology and Immunology, University of Glasgow, Glasgow (Scotland)

Miller, Michael E., Department of Pediatrics, Martin Luther King, jr. General Hospital, 1620 E. 119th Street, Los Angeles, CA 90059 (USA)

Ochsner, Michèle, Département d'Immunologie et d'Allergie, Service de Médecine, Hôpital Cantonal Universitaire, CH-1011 Lausanne (Switzerland)

Parish, W. E., Lister Institute of Preventive Medicine, Elstree, Hertfordshire (England)

Ramsey, W. Scott, Sullivan Park, Corning Glass Works, Corning, NY 14830 (USA)

Snyderman, R., Division of Rheumatic and Genetic Diseases, Departments of Medicine and Immunology, Duke University Medical Center, Durham, NC 27710 (USA)

Sorkin, E., Schweizerisches Forschungsinstitut, Medizinische Abteilung, CH-7270 Davos Platz (Switzerland)

Stahl, C. E., Division of Rheumatic and Genetic Diseases, Departments of Medicine and Immunology, Duke University Medical Center, Durham, NC 27710 (USA)

Stecher, V. J., Research Division, Ciba–Geigy Corporation, Ardsley, NY 10502 (USA)

Ward, Peter A., Department of Pathology, University of Connecticut Health Center, Farmington, CT 06032 (USA)

Wilkinson, P. C., Department of Bacteriology and Immunology, University of Glasgow, Glasgow (Scotland)

White, R. G., University Department of Bacteriology and Immunology, Western Infirmary, Glasgow G11 6NT (Scotland)

Wissler, J. H., Schweizerisches Forschungsinstitut, Medizinische Abteilung, CH-7270 Davos Platz (Switzerland)

Yamamoto, S., Department of Pathology, Kumamoto University Medical School, Kumamoto 860 (Japan)

Yoshinaga, M., Department of Pathology, Kumamoto University Medical School, Kumamoto 860 (Japan)

Zigmond, Sally H., Section of Cell Biology, Yale Medical School, 333 Cedar Street, New Haven, CT 06510 (USA)

Contents

Contents

Foreword

Chemotaxis is generally understood to be concerned with directional locomotion of cells towards chemical substances in their environment. It is the purpose of this written symposium to bring together for the first time in one volume current research on chemotaxis. It is evident that this rapidly expanding field has much to offer for our understanding of several important biological and medical problems, such as cell and molecule recognition, differentiation, growth, organ development, discrimination by scavenger cells of normal and altered self-components, e.g. cancer cells, and accumulation of specific cell types in the inflammatory process.

It is clear that this field, which owes so much to its discoverers such as ENGELMANN, PFEFFER, METCHNIKOFF and LEBER, about a century ago, has really come into its stride. The necessary techniques have been developed, not the least the one for leucotaxis by BOYDEN, so that we now have rational means for dealing with many of the questions about chemotaxis. It can already be perceived that chemotherapy of inflammation will eventually profit from this knowledge.

Extensive as is the coverage of this multifaceted subject there are still certain omissions, some because of heavy commitment on the part of the specialists selected for the particular task and others that reflect the lack of awareness on the part of the editor.

I want especially to thank the contributors for their dedicated efforts, HELENE KREUZER, my secretary, and Miss D. GREDER of the Karger Press, Basel, for her fine cooperation.

E. SORKIN

Antibiotics and Chemotherapy, vol. 19, pp. 1–9
(Karger, Basel 1974)

Introduction

Chemotaxis: Impressions and Problems

E. SORKIN

Schweizerisches Forschungsinstitut, Medizinische Abteilung, Davos Platz

There is a number of points that seem to me especially noteworthy in the rapidly expanding field of chemotaxis. Having emerged from an originally mere curiosity, the problem of chemotaxis impinges today on many aspects of modern biology and medicine. Some of the more fascinating questions relate to the following issues:

(1) How is the multiplicity of signals explained?

(2) How are signals recognized?

(3) How are signals translated into locomotion?

(4) Implications for problems of differentiation, development of organs and orderly growth?

(5) What are the medical implications?

The very character of these issues makes it amply clear that chemotaxis has merged with the mainstream of biology. The increasing knowledge in this field will assuredly be applicable and will rapidly prove helpful in understanding and for meaningful intervention in various clinical disorders.

Methodological Aspects

In the critically important matter of quantitating bacterial chemotaxis, enormous technical progress has been made as described here by ADLER and by BERG and BROWN. The latter authors designed a microscope which automatically follows the motion of individual cells and thus offers an excellent way of analysing the directional changes in response to various chemicals and changing concentrations of chemotactic agents. Their method will assuredly help us to understand mechanisms of how cells 'sense' chemoattractants and how concentration gradients are perceived by them.

Much of the work in the area of leucotaxis has been beset and hampered by controversy on just how quantitative the test methods really are. Recently, further developments of the Boyden technique in leucocyte chemotaxis, while being less sophisticated than the assay for bacterial chemotaxis, should soon prove of great help so that hopefully data from various laboratories can be compared. Some methods presently still in use should perhaps be abandoned since they are likely to yield misleading results. Especially in the evaluation of certain clinical disorders of the leucotactic system, only the most critical attitude towards test methods can forestall the 'discovery' of too many new syndromes, when in fact appropriate controls (cave concentration) are omitted. The critical evaluation of methodology by KELLER and his colleagues, by ZIGMOND and by FREI and his colleagues should help us towards this goal. The assay with radioactively labelled cells, developed by GALLIN and described in this volume, should prove of considerable value, since it allows evaluation of 200 samples per day and it eliminates staining, variability, and individual bias.

The refinement of a traditional method by JUNGI which prevents detachment of cells from the filter by reversing the direction of cell migration has been successfully applied for measuring more reliably neutrophil, eosinophil and macrophage chemotaxis. Eosinophils, for example, detach easily, and counting errors are frequent. Such losses are obviated by reversing the filter during the experiments. The method described by ZIGMOND which measures the advancing front of the chemotactically activated cells also seems very reliable and eliminates the need for laborious counting.

Chemotaxis of neutrophils, eosinophils, basophils, and macrophages *in vitro* has been amply demonstrated. Nearly all workers agree that the classical Boyden technique using Millipore filters is unsuitable for measuring directional migration of lymphocytes (KELLER *et al.;* but see HAYASHI *et al.*).

The microbeam technique with Laser or ultraviolet was introduced into the field of chemotaxis by BESSIS and promises to become a powerful method for tracking scavenging phagocytes.

Leucocyte chemotaxis *in vivo* has notoriously been difficult to demonstrate and in fact its very existence has been denied. However, earlier work by BUCKLEY, HURLEY, and analysis made by GRANT as well as clinical and experimental correlations with *in vitro* chemotaxis data (e.g., FEURER and BOREL) support the view that leucotaxis is a phenomenon that occurs *in vivo*.

The Recognition Problem

Whether one deals with individual cells or with the intact mamalian host, it is crucial for survival to discriminate between cells and molecules of self and nonself. It is clear that for reproduction, orderly growth, and development to occur, cells must be able to recognize each other. Locomotion of cells, the related phenomenon of cell trapping and contact inhibition probably play an important role in these phenomena as well as in tumor biology. Cell surfaces with receptors and markers of a highly organized and specific form must play an important part in the performance of many cell tasks. Chemotaxis of leucocytes is one special form of migration in which recognition is an obvious necessity.

In a multicellular organism it is essential that nonself components, altered or aged self-cells are constantly screened for 'histoincompatibility' by a recognition mechanism. We can safely assume that the phagocytic system of multicellular organisms has this function. In the normal host then must be a constant unremitting surveillance particularly of erythrocytes and leucocytes by mononuclear phagocytes, seeking out all cells with significantly altered surface characteristics. As biologists we tacitly assume the efficient scanning and elimination of such effete cells – when in fact this process is a far more extraordinary and indeed marvellous event than the ability of the mononuclear phagocytic system to cope with infection. On evolutionary grounds it is now considered most likely that the phagocytes were later supplemented by an additional helper cell system with more specialized functions, the lymphocyte-plasma cell system, which produces specific antibodies and is capable of recognizing most foreign components and maybe even the host's own modified self-components in another, antibody-specific manner.

In my view, recognition by phagocytic cells stands in contrast to this type of 'immune' recognition by lymphocytes, based on stereospecific antibody or like receptors. The phagocytic cells recognize other cells and molecules as well in the absence of antibodies or like structures, although the additional possibility of recognition by the large set of passively absorbed cytophilic antibodies with different combining sites is a further consideration. However, analogies exist in slime molds which recognize cyclic AMP (KONIJN) or bacteria which recognize oxygen or sugars (ADLER). This obviously occurs without antibody, making it likely that phagocyte cell membranes recognize chemotactic signals also without the help of antibody. Indeed, phagocytes must have recognized molecules or cells long before antibody was invented. This does not exclude, of course, the possibility of chemotaxis in cells induced by an antigen-antibody reaction on cell membranes or by generation of plasma mediators from precursors.

What is the molecular basis of the primary recognition event, how then is the attractant signal translated over intermediate steps into locomotion of cells? We have no convincing answers to these important questions. It seems that in bacteria, e.g. in *Escherichia coli*, there exist specific receptor proteins which can bind the chemotaxis-inducing sugars (ADLER, BOOS). What steps follow this event is not clear. Interestingly enough, bacteria can also respond chemotactically to certain sugars, not necessarily ones they are capable of metabolizing, and thus change them to chemotactically active molecules. Extensive metabolism of chemicals is neither required nor sufficient for attraction of bacteria by the chemicals. ADLER has reported on 40 mutants which are generally nonchemotactic to amino acids, sugars, and oxygen – though the bacteria are motile. He argues that since it is unlikely that a single mutation would lead to a loss of all the kinds of chemoreceptors, these mutants are probably defective at some stage beyond the receptors. The defect could be in a transmitting system through which information from all the receptors is channeled.

As regards phagocyte chemotaxis, we are nearly completely ignorant about the character of their cell surface receptors capable of recognizing the signals necessary for directional locomotion. We do not know in what way the signal is translated and transmitted into the intracellular compartments. Some data might be obtained by enzyme pretreatment of locomotory cells, others by specific chemical modifications of the receptor structures. Treatment with proteolytic enzymes might be especially useful, as it would at the same time allow exclusion of antibodies as receptors. Alterations in the chemotactic signal molecule may also provide information on the relation of its structure to its functions. This approach is now being explored by several laboratories (WILKINSON and McKAY; WISSLER *et al.*) and has already led to important conclusions.

While providing no definite clues as regard the receptor structure, these studies provide some indication that at least some types of leucocytes seem to possess no identifiable stereospecific receptors for chemotactic signals. WILKINSON investigated the chemotactic activity of proteins such as human serum albumin denatured by various methods, and found that HSA then became chemotactic for neutrophils and macrophages. An especially intriguing example for discrimination by leucocytes of native versus degraded proteins is given by dissociation of hemoglobin and myoglobin. It was found that on removal of hem from these molecules, the resultant globin preparations had good chemotactic activity. On readdition of hemin to globin the protein regains its original native conformation with concomitant loss of chemotactic activity.

When nonpolar side groups, e.g. tosyl, proprionyl or butyryl groups were conjugated with nonchemotactic proteins under mild nondenaturing conditions, the conjugates proved to be chemotactic. Interestingly enough, the chemotactic activity of conjugated proteins was not inhibited by the presence of the conjugate in free solution. WILKINSON and MCKAY therefore concluded that chemotactic recognition does not depend on a close stereospecific bonding between a cell surface receptor and the active site of the protein.

A novel concept on the mechanism of recognition by neutrophil leucocyte is that these cells are stimulated by secondary quantum energy transduction (WISSLER). The proposed concept is derived largely from experimental data obtained with the anaphylatoxin-related leucotactic binary peptide system and emphasizes the similarities between the recognition process in phagocytes and the stimulation of animals by specially differentiated sensory organs.

It is apparent that the cell response has to be separated from the recognition phenomenon. It seems that allosteric conformational alterations which serve as regulatory principle in many biological reactions and which are 'low energy signals' do not attract neutrophil leucocytes. These cells recognize neither a special primary, secondary, tertiary, nor quarternary structure in the protein signal. It appears that the neutrophil leucocyte recognizes the energy difference of the transition state; in other words, the recognition signal is finally identified as an energy signal. The registration of this energy signal can only occur by secondary energy transduction as the first step of the recognition reaction sequence. WISSLER suggests that, in order for a product to be chemotactic for neutrophils, the diffusion-controlled conformational transitions in natural proteins or synthetic peptides should be characterized thermodynamically by a Gibbs free energy change of $-\Delta G = >10$ kcal/mol.

Concerning the sensitivity of the signal-transducing elements in the cell membrane to a certain quantity of energy, and in view of the failure of phagocytes to register allosteric transitions in proteins, it follows that there is a lower limit of recognition of a quantity of energy and presumably an upper limit as well. Such a recognition in a distinct range of energy quanta for chemotactic stimulation would automatically limit the number of possible signals, yet account for their multitude of possibilities. It would moreover also offer an explanation for the cell-specific response of different kinds of cells to one and the same energy quantum which have most likely genetically determined ranges of energy quantum transduction.

The Chemotactic Signals

This problem is well described here for bacteria (ADLER), amoebae (KONIJN), and for leucocytes (WILKINSON and McKAY; WISSLER *et al.*). *Escherichia coli* have the capacity to respond to numerous different sugars, for which there exist seemingly at least eight different chemoreceptors.

For neutrophils there has been described a great number of chemotactically active factors, most of which have not been isolated in pure form so far or only in a denatured state (there are some exceptions; see WISSLER *et al.*, this volume). This has doubtless confused the issue of chemotactic mediators. In view of the obvious high level discriminatory power of leucocytes it is not surprising that seemingly new leucotactic agents are discovered with great frequency. It is in fact the task of these cells to recognize even small changes in molecules or cells and, when found, to eliminate them.

In the field of leucotaxis the main emphasis has so far been centered around complement components. This is now rapidly changing and giving way to a broader view of mediators of leucotaxis. It is established that complement split products can be chemotactic, but such components are probably only of minor significance in the light of the entire issue of chemotaxis in multicellular organisms. Several C components such as C3a or denatured C5a or even doubtless impure complexes (such as C567) which have been described as chemotactic should be viewed with reservations. Their chemotactic activity, as reported, is not the issue; one wonders rather whether these factors are in fact the natural chemotactic factors in, for example, immunologically induced activation of serum or plasma. Thus, the C5a anaphylatoxin has only recently been obtained in pure, crystalline form, and it was then soon established unequivocally that this product was inactive for neutrophils. However, when the pure C5a was treated at pH 4, a procedure used in some laboratories for its isolation or storage, it then acquired highly neutrophilotactic properties. Thus, the potential for artifacts and premature interpretations is very considerable.

The recently discovered split products of fibrinogen, collagen, and of immunoglobulins (HAYASHI *et al.*) as signals in inflammatory processes are of particular interest. It will be worthwhile to explore the significance of some of the fibrinogen split products, fibrinopeptides, also in formation of thrombi.

A controversial problem concerns the activity of cyclic AMP in leucotaxis. The fascinating studies of BONNER, KONJIN and others on the chemotactic activity of AMP in Amoebae led to similar studies of its activity in leucocytes. Nearly all workers agree now that pure cyclic AMP is no leucotactic signal for neutrophils. This fact does not exclude some other regulatory role of cyclic AMP (or ATP) in locomotion of leucocytes.

Chemotaxis in Its Relation to Migration Inhibition

It has been convincingly demonstrated that chemotactic stimulation results in directional migration of cells. The significance of a concentration gradient is likewise now established. It is of great interest that by reversal of the gradient, i.e., by placing the cells in a highly chemotactic environment, directional migration is prevented whereas random migration may be increased. The finding that leucocytes can be prevented from migrating away from a chemotactic center is of particular significance with respect to *in vivo* phenomena. Active factors may exert their effect *in vivo* in such a way that leucocytes migrate towards the site of generation of chemotactic material. In the area where cells have accumulated they are exposed to the highest concentration of the factors and become 'trapped'. Since the cells at the center can no longer sense a gradient they move at random and are unable to migrate away from the higher concentration of the chemotactic factor, thus becoming in a way 'immobilized' (migration inhibition, retention, trapping). Directional locomotion and migration inhibition thus represent a dual mobility behavior of cells which in effect counteract one another. However, since it now appears that random motion can be changed or remain unchanged in both behavioral phenomena, it should be considered that apparent migration inhibitory effects represent chemoattractive trapping by a single recognizable protein. The trapping of cells by a chemotactic agent may eventually prove to be of greater significance, e.g. in leucocyte accumulation in inflammatory reactions, than the actual directional migration induced by it (see also PARISH). Such trapping effects may also be critical for cellular cooperation in antibody production, now the subject of such intense study by immunologists. Phagocytophilic and lymphocytophilic agents are presumably released from thymus-derived lymphocytes, which not only attract, for example, macrophages and B cells to the site, but through their trapping effects bring and hold them in close proximity.

Failure to trap and activate scavenger cells may well be one of the reasons for uninhibited tumor growth. In order to understand cellular immunity and allergic responses, further knowledge on the leucotactic factors released by lymphocytes is therefore most important. It will be especially important to ascertain whether these factors are indeed of lymphocytic origin or whether they are really products of plasma components which were split by cell-released enzymes.

Cell Specificity of Chemotaxis

During the last century PFEFFER demonstrated cell specificity of chemotaxis in plants. In regard to leucocytes, FLOREY stated as recently as 1962 that there is no striking difference in the chemotactic response *in vitro* of monocytes and polymorphs; otherwise the development of various stimuli at different stages of inflammation could not explain why the wandering cells at the beginning are chiefly polymorphs and thereafter of the mononuclear type. The monocyte was assumed to react *in vitro* chemotactically in the same way and to the same extent as the granulocyte and apparently to the same stimuli.

The first demonstration of cell-*specific* chemotaxis of leucocytes *in vitro* was achieved only recently by KELLER and SORKIN who found no correlation between the capacity of certain cytotaxins to attract polymorphs or mononuclear cells.

One of the most remarkable findings concerning the newly detected anaphylatoxin-related leucotactic peptide system is that it exerts a cell-specific chemotactic effect on neutrophil *and* eosinophil leucocytes which is dependent on the absolute concentration and the molar ratio of anaphylatoxin and cocytotaxin. No such effects were observed on macrophages. The negative result for macrophages is in contrast to the findings of other authors that C5a anaphylatoxin is a chemotactic factor for macrophages (SNYDERMAN and STAHL). It is likely that this latter discrepancy will soon be clarified.

Perfusates of guinea pig anaphylactic tissue and human anaphylactic basophils and mast cells release a substance which selectively attracts eosinophils *in vitro* (PARISH). A seemingly similar product (ECF-A) is described by KAY, but it was also found to be chemotactic for neutrophils.

Clinical Considerations

Several authors have drawn attention to defective chemotaxis of polymorphs in human diseases. While there is no doubt about the importance of these studies, their value has been limited by the fact that most authors did not mention the lowest normal values below which chemotaxis can be considered as pathological. It is quite possible that some of the published defective chemotaxis cases were not the cause of infection but rather a transitory consequence of them (FREI *et al.*).

It remains a clinically important question whether chemotaxis is involved in immunologically and nonimmunologically induced tissue injury such as occurs in the Arthus reaction, acute and chronic human glomerulonephritis, polyarteritis nodosa, and rheumatoid arthritis. There are experimental data which suggest that chemotactic factors may be important in such inflammatory processes. In designing therapeutic agents, one should therfore aim to interfere with leucotactic mediators either by inhibiting their formation or their chemotactic action on inflammatory cells. It follows that further knowledge about the natural inhibitors of chemotaxis will be of aid in making these attempts more rational.

Author's address: E. SORKIN, Schweizerisches Forschungsinstitut, Medizinische Abteilung, *CH-7270 Davos Platz* (Switzerland)

Chemotaxis in Bacteria

Antibiotics and Chemotherapy, vol. 19, pp. 12–20
(Karger, Basel 1974)

Chemoreception in Bacteria

J. ADLER

Departments of Biochemistry and Genetics, University of Wisconsin, Madison, Wisc.

I. Introduction

Motile bacteria are attracted to a variety of chemicals – a phenomenon called chemotaxis [for a review, see ref. 1]. Although chemotaxis by bacteria has been recognized since the end of the 19th century, thanks to the pioneering work of ENGELMANN, PFEFFER and other biologists, the mechanisms involved are still almost entirely unknown. How do bacteria detect the attractants? How is this sensed information translated into action; that is, how are the flagella directed?

To learn about the detection mechanism that bacteria use in chemotaxis, it is important first to know *what* is being detected. One possibility is that the attractants themselves are detected. In that case, extensive metabolism of the attractants would not be necessary for chemotaxis. There is another possibility: the attractants themselves are not detected but, instead, some metabolite of the attractants is detected (for example, the pyruvate inside the cell); or the energy produced from the attractants, perhaps in the form of adenosine triphosphate, is detected. In these cases, metabolism of the attractants would be necessary for chemotaxis. The idea that bacteria sense the energy produced from the attractants has, in fact, gained wide acceptance for explaining chemotaxis (and also phototaxis) [2, 3].

To try to determine which of these possibilities is correct, experiments were carried out with *Escherichia coli* bacteria which, we had previously shown [4], exhibit chemotaxis toward various organic nutrients. The results show that extensive metabolism of the attractants is not required, or sufficient, for chemotaxis. Instead, the attractants themselves are detected.

The systems that bacteria use to detect chemicals without metabolizing them are here called 'chemoreceptors'. Efforts to identify the chemoreceptors are described.

II. A Quantitative Method for Studying Chemotaxis

In the 1880s PFEFFER [5, 6] demonstrated chemotaxis by pushing a capillary tube containing a solution of attractant into a suspension of motile bacteria on a slide and then observing microscopically that the bacteria accumulated first near the mouth of the capillary and later inside. A modification [7, 8] of this method, which permits quantitative study of chemotaxis, follows.

After incubation at 30°C for 60 min, the capillary is taken out of the bacterial suspension and the number of bacteria inside the capillary is measured by plating the contents of the capillary and counting colonies the next day. Reproducibility of the method is ±15%. An attractant is tested over a range of concentrations usually between 10^{-8} and 10^{-1} M in 10-fold intervals. From such an experiment one can construct a concentration-response curve and estimate a threshold concentration for accumulation inside the capillary. Some of the best attractants are galactose, glucose, ribose, aspartate and serine. (All sugars mentioned in this paper have the D configuration, and all amino acids the L configuration.)

III. Evidence that the Attractants Themselves Are Detected

The following 5 approaches lead to the conclusion that chemotaxis is not a consequence of the metabolism of the attractants but, rather, that the attractants themselves are detected. A more complete documentation of the data can be found elsewhere [8].

1. Some chemicals that are extensively metabolized fail to attract bacteria. This includes galactonate, gluconate, glucuronate, glycerol, α-keto-glutarate, succinate, fumarate, malate and pyruvate. This result makes it clear that metabolism of a chemical and energy production from it are not sufficient to make a chemical an attractant.

2. Some chemicals that are essentially nonmetabolizable attract bacteria.

(a) Mutant bacteria that have lost the ability to metabolize a chemical are attracted to it. Mutants which lack three enzymatic activities essential for the metabolism of galactose are attracted to galactose as well as wild-type

bacteria. Evidence has been presented [8] that these mutants are 99.5% or more blocked in their metabolism of galactose, relative to a wild-type strain. A similar result for glucose taxis was obtained with a mutant defective in its ability to metabolize glucose.

(b) Some essentially nonmetabolizable analogs of metabolizable chemicals attract bacteria. D-fucose (6-deoxy-D-galactose) is a galactose analog that is not metabolized by *E. coli* [8, 9]. Nevertheless, the bacteria are attracted to it very well, although its threshold concentration for chemotaxis is higher than that of D-galactose, as might be expected for an analog. The D-fucose had been purified to remove metabolizable impurities such as galactose or glucose.

Three glucose analogs which were known to be nonmetabolizable are also attractants: 2-deoxyglucose, α-methyl glucoside and L-sorbose. The three analogs were purified before use in order to remove glucose or other contaminants.

3. Chemicals attract bacteria even in the presence of a metabolizable chemical. If bacteria detect metabolites of an attractant, or energy produced from it, then the addition of a metabolizable chemical should stop chemotaxis by flooding the cells with metabolites and energy. This was not found to be the case for either metabolizable or nonmetabolizable attractants [8].

4. Attractants that are closely related in structure compete with each other but not with structurally unrelated compounds. This finding supports the conclusion that it is the attractants themselves that are detected, and that there exists a variety of specific receptors.

In these experiments one attractant, usually at 0.01 M, is put into the capillary tube, and another attractant at a concentration of 0.01 M is put into both the capillary and the bacterial suspension. If the two attractants use the same chemoreceptor the response should be inhibited; if they do not the response should not be affected. Only two examples will be presented here [see ref. 8, 10 and 11 for more results].

Chemotaxis toward fucose was completely inhibited by the presence of galactose, and in the reciprocal experiment there was nearly complete inhibition. This suggests that fucose and galactose use the same chemoreceptor (the 'galactose receptor').

Glucose completely eliminated taxis toward galactose, but in the reciprocal experiment the inhibition was only about 60–70%, no matter how high the concentration of galactose was. This suggests that the receptor which detects galactose also detects glucose but that, in addition, there is another receptor that detects glucose but not galactose (the 'glucose receptor').

5. There are mutants which fail to carry out chemotaxis to certain attractants but are still able to metabolize them. If there are chemoreceptors in bacteria and if they are specific, there should be mutants that are defective in their response to some attractants but not to others, because of a defect in a single receptor. Such mutants of *E. coli* have now been found [12].

One mutant, defective in the 'serine receptor', fails to be attracted to serine and shows much-reduced taxis toward glycine, alanine and cysteine. (These residual responses result from the 'aspartate receptor'.) The mutant is attracted normally to aspartate and glutamate, and to galactose, glucose and ribose. It oxidizes and takes up L-serine at the same rate that its parent does.

Another mutant, lacking the 'aspartate receptor', shows no chemotaxis toward aspartate and glutamate, nearly normal taxis toward serine, alanine, glycine and cysteine, and normal taxis toward galactose, glucose and ribose. The rate of oxidation and uptake of aspartate is the same for the mutant and its parent.

A third type of mutant, missing the 'galactose receptor', is not attracted to galactose and fucose and is attracted to glucose at a higher-than-normal threshold. (This residual response to glucose results from the 'glucose receptor'.) These mutants are attracted normally to fructose, ribose, serine and aspartate. Metabolism of galactose is normal in these mutants. In some of the mutants there is a defect in the uptake of galactose, which will be discussed below.

The existence of these three types of specifically nonchemotactic mutants argues for specific receptors and provides additional support for the idea that detection of the attractants is independent of their metabolism.

IV. How Many Chemoreceptors?

To determine how many kinds of chemoreceptors there are, three approaches are being used. The first is to ask whether a given attractant is still effective when another attractant is present. The second is to try to isolate mutants defective in individual receptors. A third approach is to study the inducibility of specific taxes (presumably the inducibility of specific receptors). For example, taxis toward galactose and fucose is inducible by galactose.

The conclusion from results obtained so far [8, 10–12] is that there are at least the 8 chemoreceptors shown in table I. Oxygen is known to be an attractant for *E. coli* [8], so there could be a receptor for it, but this question

Table I. Partial list of chemoreceptors in *Escherichia coli*

Attractant	Threshold[1] molarity
Fructose receptor	
D-fructose	1×10^{-5}
Galactose receptor	
D-galactose	4×10^{-7}
D-glucose	4×10^{-7}
D-fucose	3×10^{-5}
Glucose receptor	
D-glucose	3×10^{-5}
Maltose receptor	
maltose	3×10^{-6}
Mannitol receptor	
D-mannitol	7×10^{-6}
Ribose receptor	
D-ribose	3×10^{-7}
Aspartate receptor	
L-aspartate	6×10^{-8}
L-glutamate	1×10^{-5}
Serine receptor	
L-serine	2×10^{-7}
L-cysteine	5×10^{-6}
L-alanine	5×10^{-5}
glycine	5×10^{-5}

1 The threshold values are lower in mutants unable to take up or metabolize a chemical.

has not been investigated so far. A survey of possible other attractants or of repellents has not been completed.

It is conceivable that, besides chemoreceptors, at least some bacteria might have receptors specialized to detect light, gravity or temperature, since all these stimuli are known to elicit tactic responses in some bacteria [1].

V. What Is the Nature of the Chemoreceptors?

One possibility is that the chemoreceptors are the first enzymes in the metabolism of the chemicals. This possibility has been excluded in the case of galactose, because mutants that lack galactokinase still respond perfectly well. Another possibility is that the chemoreceptors are the permeases and related components essential for transport of substances into the cell. To find out, mutants defective with respect to transport have been investigated from the standpoint of chemotaxis.

An *E. coli* mutant, 20SOK⁻, that is defective in the uptake of galactose [9, 13] to the extent of a 99.5-percent block [8] is attracted to galactose perfectly well [8]. Thus, transport is not required for chemotaxis, and the chemoreceptors therefore appear to be located somewhere on the 'outside' of the cell.

The galactose binding protein [14], a component needed for the transport of galactose [15], is present in this mutant [16]; it must be missing an additional component needed for galactose transport. We [16] have now found that the galactose binding protein is also needed for galactose taxis, and that this is the component of the galactose receptor that recognizes galactose, glucose and a number of structurally related chemicals. The evidence [16] follows.

1. The specificity of the galactose binding protein is the same as the specificity of the galactose receptor.

2. Most of the galactose taxis mutants are defective in the transport of galactose, as mentioned above, and these all have very low levels of galactose binding protein.

3. Galactose taxis can be eliminated by a mild osmotic shock [14] which releases the galactose binding protein from the cells.

4. The galactose chemoreceptor is saturated at concentrations above 10^{-6}M, and so is the galactose binding protein.

There must be one or more additional components of the galactose chemoreceptor, since at least one of the galactose taxis mutants has normal galactose binding protein [16]. We are trying to find out what this additional component is.

The role of the binding protein and the relationship between transport and taxis are summarized in figure 1.

We have also found binding proteins for maltose and ribose that appear to serve the corresponding chemoreceptors [16].

VI. How Do Chemoreceptors Work?

As to the mechanism of chemoreceptors in bacteria, this still remains unknown. Boos has shown that the galactose binding protein undergoes a configurational change when it binds galactose [17]. Somehow this change

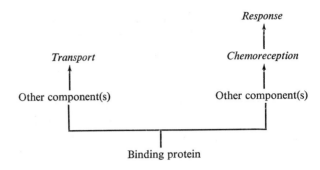

Fig. 1. Relationship between transport and chemoreception.

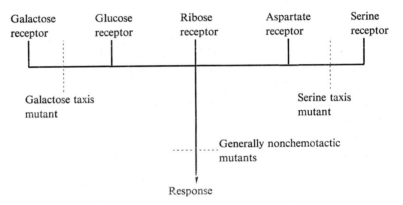

Fig. 2. The scheme for chemotaxis suggested in this article, showing location of defects in the various mutants.

may be 'felt' by the additional component(s). Then this information is transmitted to the flagella, either by a diffusible substance or by a configurational change of macromolecules perhaps in the cell membrane. The latter mechanism could involve something like a receptor potential or an action potential. The flagella then respond by changing their orientation in such a way to bring about a change in the frequency of tumbling.

We have isolated and reported on mutants which are generally non-chemotactic, i.e. they fail to carry out chemotaxis toward any of the attractants – amino acids, sugars or oxygen – though the bacteria are perfectly motile [18]. Since it is unlikely that a single mutation would lead to a loss of all the kinds of chemoreceptors, these mutants are probably defective at some stage beyond the receptors, as shown diagrammatically in figure 2.

The defect could be in a transmitting system through which information from all the receptors is channeled to the flagella, or in the responding mechanism itself. Genetic analyses of the mutants have shown that three genes are involved [19, 20]. Further studies of these mutants may lead to an understanding of the way in which the chemoreceptors direct the flagella.

Summary

Extensive metabolism of chemicals is neither required, nor sufficient, for attraction of bacteria to the chemicals. Instead, the bacteria detect the attractants themselves. The systems that carry out this detection are called 'chemoreceptors'. There are mutants that fail to be attracted to one particular chemical or to a group of closely related chemicals but still metabolize these chemicals normally. These mutants are regarded as being defective in specific chemoreceptors. Data obtained so far indicate that there are at least 8 different chemoreceptors in *E. coli*. The chemoreceptors are not the enzymes that catalyze the metabolism of the attractants; nor are they certain parts of the permeases and related transport systems, and uptake itself is not required or sufficient for chemotaxis. In the case of the galactose receptor, the galactose binding protein is the component that recognizes the galactose.

Acknowledgements

The research discussed was supported by a grant from the US National Institutes of Health. I thank MARGARET DAHL for having carried out many of the experiments mentioned here, and G. L. HAZELBAUER and R. E. MESIBOV for major contributions.

Much of this paper is reproduced from an article published in Science [8] by permission of the American Association for the Advancement of Science (copyright 1969) and by permission of the Wissenschaftliche Verlagsgesellschaft GmbH, Stuttgart.

Addendum

Negative chemotaxis has also been studied. A large number of repellents have been identified and classified into chemoreceptor groups [W.-W. Tso and J. ADLER, J. Bacteriol. *118:* 560–576 (1974)].

References

1   WEIBULL, C.: Movement; in GUNSALUS and STANIER The bacteria, vol. 1, pp. 153–205 (Academic Press, New York 1960).

2   CLAYTON, R. K.: Phototaxis in microorganisms; in GIESE Photophysiology, vol. 2, pp. 51–77 (Academic Press, New York 1964).

3   LINKS, J.: Onderzoekingen Met Polytoma Uvella; thesis, Leiden (in Dutch, summary in English) (1955).

4   ADLER, J.: Chemotaxis in bacteria. Science 153: 708–716 (1966).

5   PFEFFER, W.: Locomotorische Richtungsbewegungen durch chemische Reize. Untersuch. botan. Inst., Tübingen 1: 363–482 (1884).

6   PFEFFER, W.: Ueber chemotaktische Bewegungen von Bacterien, Flagellaten und Volvocineen. Untersuch. botan. Inst., Tübingen 2: 582–661 (1888).

7   ADLER, J.: A method for measuring chemotaxis and use of the method to determine optimum conditions for chemotaxis by Escherichia coli. J. gen. Microbiol. 74: 77–91 (1973).

8   ADLER, J.: Chemoreceptors in bacteria. Science 166: 1588–1597 (1969).

9   BUTTIN, G.: Mécanismes régulateurs dans la biosynthèse des enzymes du métabolisme du galactose chez Escherichia coli K12. I. La biosynthèse induite de la galactokinase et l'induction simultanée de la séquence enzymatique. J. molec. Biol. 7: 164–182 (1963).

10 ADLER, J.; HAZELBAUER, G. L., and DAHL, M. M.: Chemotaxis toward sugars in Escherichia coli. J. Bact. 115: 824–847 (1973).

11 MESIBOV, R. E. and ADLER, J.: Chemotaxis toward amino acids in Escherichia coli. J. Bact. 112: 315–326 (1972).

12 HAZELBAUER, G. L.; MESIBOV, R. E., and ADLER, J.: Escherichia coli mutants defective in chemotaxis toward specific chemicals. Proc. nat. Acad. Sci., Wash. 64: 1300–1307 (1969).

13 ROTMAN, B.; GANESAN, A. K., and GUZMAN, R.: Transport systems for galactose and galactosides in Escherichia coli. II. Substrate and inducer specificities. J. molec. Biol. 36: 247–260 (1968).

14 ANRAKU, Y.: Transport of sugars and amino acids in bacteria. II. Properties of galactose- and leucine-binding proteins. J. biol. Chem. 243: 3123–3127 (1968).

15 BOOS, W.: The galactose binding protein and its relationship to the β-methylgalactoside permease from Escherichia coli. Europ. J. Biochem. 10: 66–73 (1969).

16 HAZELBAUER, G. L. and ADLER, J.: Role of the galactose binding protein in chemotaxis of Escherichia coli toward galactose. Nature New Biol. 230: 101–104 (1971).

17 BOOS, W.: Substrate induced conformational change of the galactose-binding protein of E. coli. Fed. Proc. 30: 1062 (1971).

18 ARMSTRONG, J. B.; ADLER, J., and DAHL, M.: Non-chemotactic mutants of Escherichia coli. J. Bact. 93: 390–398 (1967).

19 ARMSTRONG, J. B. and ADLER, J.: Location of genes for motility and chemotaxis on the Escherichia coli genetic map. J. Bact. 97: 156–161 (1969).

20 ARMSTRONG, J. B. and ADLER, J.: Complementation of non-chemotactic mutants of Escherichia coli. Genetics 61: 61–66 (1969).

Author's address: Dr. JULIUS ADLER, Departments of Biochemistry and Genetics, University of Wisconsin, Madison, WI 53706 (USA)

Antibiotics and Chemotherapy, vol. 19, pp. 21–54
(Karger, Basel 1974)

The Properties of the Galactose-Binding Protein, the Possible Chemoreceptor for Galactose Chemotaxis in *Escherichia coli*

W. Boos[1]

Department of Biological Chemistry, Harvard Medical School, and the Biochemical Research Laboratory, Massachusetts General Hospital, Boston, Mass.

I. Introduction

Chemotaxis in bacteria has been observed for a long time [PFEFFER, 1884, 1886]. But only recently progress has been made to quantitatively approach this phenomenon. J. ADLER and his collaborators [see this volume] have introduced a semiquantitative assay which allowed them to distinguish between different chemoreceptors and to determine the relative specificity of the particular chemoreceptors. Methods were developed which made it possible to select and characterize mutants defective in general chemotaxis as well as mutants defective only in the chemotaxis of attractants belonging to a single chemoreceptor. Using both defined and stable attractant gradients and Laser-light scattering to determine the change in the concentration of chemotactic bacteria along the gradient, DAHLQUIST *et al.* [1971] showed elegantly that bacteria respond to a relative change in concentration rather than to an absolute change. Most recently BERG and BROWN [1972] have studied the behavior of single bacterium under different conditions of attractants by use of a microscope which can automatically track the 3-dimensional movements of a single bacterium. The conclusion of these studies was that bacteria were able to recognize when they moved up the gradient, but not when they moved down the gradient. However, MACNAB and KOSHLAND [1972] found that bacteria were able to alter their frequency

1   The author thanks the European Journal of Biochemistry, the American Society of Biological Chemists Inc. and Nature, Lond. for kind permission to reproduce published material.

of twiddle and run phase, depending upon whether they are experiencing a change to higher or lower attractant concentration. The time-dependent decline of the response to the altered environment led these authors to propose a time-dependent mechanism of chemotaxis which requires the existence of a simple memory mechanism in the bacteria.

Very little is known about the actual mechanism of chemotaxis, and efforts were concentrated on the elucidation of the first step in chemotaxis, the recognition of the attractant by the chemoreceptor. It was H. M. KALCKAR who suggested a possible role of the galactose-binding protein as the chemoreceptor for galactose chemotaxis [KALCKAR, 1971]. This protein was first isolated by ANRAKU [1967, 1968] and we could subsequently demonstrate that it was an essential component of a specific galactose transport system of *Escherichia coli*, the β-methylgalactoside transport system [BOOS and SARVAS, 1970; LENGELER *et al.*, 1971; BOOS, 1972]. Subsequently, HAZELBAUER and ADLER [1971] could indeed demonstrate a striking similarity between the properties of the galactose chemoreceptor *in vivo* and the properties of the isolated galactose binding protein *in vitro* [for details, see ADLER, this volume]. As striking as the similarities of the galactose binding protein and the galactose chemoreceptor in *E. coli* appeared to be, this correlation cannot be generalized. A search for the possible role of other periplasmic binding proteins in chemotaxis has shown that only two, a ribose binding protein of *Salmonella typhimurium* [AKSAMIT and KOSHLAND, 1972] and a maltose binding protein from *E. coli* [ADLER *et al.*, in press; ADLER, this volume] might be involved in the chemotaxis of the respective substrates. In contrast, the serine and asparte chemoreceptors of *E. coli* [MESIBOV and ADLER, 1972] have no counterparts in soluble binding proteins and, *vice versa*, no chemotaxis has been observed for instance with histidine and lysine as attractants [MESIBOV and ADLER, 1972], even though binding proteins for these amino acids have been isolated [LEVER, 1972; ROSEN, 1971].

In the framework of this book it was desirable to discuss in detail the properties of the galactose binding protein in lieu of its possible role as the chemoreceptor for galactose chemotaxis. It was not attempted to elaborate the numerous arguments for, and the few arguments against, the role of the galactose binding protein, either in galactose chemotaxis or in galactose transport, but simply to discuss the properties of the protein which might be relevant to its biological function. Since other chapters in this book cover bacterial chemotaxis, and since the author has a somewhat biased view of transport, reference will be made more often to the presumed function of the protein in transport than to its function in chemotaxis.

II. Properties of the Galactose Binding Protein

A. Location in the Cell Envelope of *E. coli*

The galactose binding protein is a so-called periplasmic protein because, as the name suggests, it seems to be located in the periplasm [HEPPEL, 1967], between the cell wall and the cytoplasmic membrane of gram-negative bacteria. The name for this location is purely operational and its existence is indicated by several experiments.

1. A mild cold osmotic shock after plasmolysis in hypertonic sucrose-Tris-EDTA solution releases the 'periplasmic proteins'. The same group of proteins are removed from the cell surface during spheroplast formation by EDTA lysozyme (with the exception of ribonuclease) but not by the penicillin technique. The osmotic shock treatment shows indeed an astonishing discrimination between 'periplasmic' and 'cytoplasmic' proteins. Thus, 5'-nucleotidase, cyclic phosphodiesterase and acid phosphatase are released to the extent of 98, 105 and 76%, respectively, while only 0.8% of the β-galactosidase is released [NEU and HEPPEL, 1965]. The extent of this release of the periplasmic proteins is somewhat dependent on the lipid composition of the cell envelope [ROSEN and HACKETTE, 1972]. Also, different periplasmic proteins are released by the osmotic shock procedure to a varying degree when the shock conditions are changed [NOSSAL and HEPPEL, 1966]. Similarly, mutants have been isolated which appear to leak periplasmic proteins into the growth medium, yet not all periplasmic proteins are lost to the same extent [LOPES *et al.*, 1972; HEPPEL *et al.*, 1972].

2. Alkaline phosphatase and other periplasmic proteins can be assayed *in vivo* without penetration of the substrate through the cytoplasmic membrane [BROCKMAN and HEPPEL, 1968].

3. The application of a dye, diazo-7-amino-1,3-naphthalene-disulfonate, which is not able to penetrate the cytoplasmic membrane, reacts with and inactivates a periplasmic sulfate binding protein but not the cytoplasmic β-galactosidase [PARDEE and WATANABE, 1968].

These and other experiments [as reviewed by HEPPEL, 1969, 1971] argue that periplasmic proteins are indeed located outside the cytoplasmic membrane.

Another question of considerable importance with regard to the mechanism of chemotaxis is the location of the galactose binding protein molecules within the cell envelope of the sensing bacteria. Are they uniformly distributed over the surface of the entire cell, or are they more concentrated on opposite ends of the rod-shaped cell, i.e. in the polar caps, or are they even concentrated preferentially on one end of the cell? WETZEL *et al.* [1970] could demonstrate by reaction product staining of the alkaline phosphatase that the polar caps of the cells exhibit high density in the electron microscope, indicating a preferential concentration of alkaline phosphatase. This suggests by extension that other periplasmic proteins may also concentrate on the opposite ends of the cell. However, this line of argument has been challenged by McALLISTER *et al.* [1972]. By using reaction product staining as well as ferritin-coupled antibody precipitation, these authors conclude that an even distribution of alkaline phosphatase along the cell envelope is a more likely picture.

Recently [SHEN and BOOS, 1973], we could demonstrate that the synthesis or *in vivo* assembly of the galactose binding protein occurs only immediately after cell division and not during elongation of the cell. Furthermore, this synthesis does not occur at the nonpermissive temperature in a temperature-sensitive cell division mutant. These observations suggest that the synthesis or assembly of the galactose binding protein is closely associated with septum formation. Possibly this synthesis occurs in a compartment which becomes extramembranal after the septum has formed. Depending on the mobility of the protein in the periplasmic space, the cell would obtain, for a certain length of time, an asymmetric distribution of galactose binding protein molecules on its respective ends. Furthermore, if the periplasmic proteins are more or less immobilized at their location of assembly one should find the galactose binding protein preferentially on both polar caps of the cell, and not uniformly distributed over the entire cell envelope.

Also of interest is the concentration of the galactose binding protein molecules in the periplasmic space. From binding studies with intact cells made unable to transport galactose by *p*-hydroxy-mercuribenzoate or un-couplers of oxidative phosphorylation (neither of which affect binding of galactose to the galactose binding protein) one can estimate that each cell contains 50,000–80,000 molecules of galactose binding protein. Estimating a width of 50 Å for the periplasmic space and an average cell size of 0.5-μm diameter and 1-μm length, the concentration of galactose binding protein is in the order of 10^{-2} M! This high concentration might explain why restoration of galactose transport by addition of purified galactose binding protein to shocked cells or membrane vesicles cannot be satisfactorily demonstrated. Furthermore, it can always be argued that the restoration of either transport or chemotaxis may be the result of *de novo* synthesis or assembly from pre-existing precursors of the binding protein and not the result of the addition

Table I. Amino acid composition of galactose and leucine binding protein

Amino acid	Relative composition, galactose binding protein, residues per 100 residues
Lysine	9.98
Histidine	1.11
Arginine	2.08
Aspartic acid	15.9
Threonine	4.47
Serine	4.04
Glutamic acid	9.24
Proline	2.92
Glycine	7.12
Alanine	13.8
Half-cystine[1]	
Valine	9.55
Methionine	1.90
Isoleucine	4.85
Leucine	7.78
Tyrosine	1.83
Phenylalanine	2.12
Tryptophan	1.28

1 Half-cystine was not determined. The values represent an average of a sample that was hydrolyzed for 24 and 48 h. [From ANRAKU, 1968.]

of exogenous binding protein. Therefore, restoration studies should be done in mutants which are negative in transport or chemotaxis due to a defective binding protein.

B. Amino Acid Composition and Stability

Table I shows the amino acid composition of the galactose binding protein as reported by ANRAKU [1968]. The protein contains 53% nonpolar amino acids, and this does not seem to be abnormally high (an extreme lipophilic enzyme, the membrane-bound phosphokinase of *Staphylococcus aureus* has 60.4%), but rather is comparable with typical membrane proteins which range from 47 to 55% nonpolar amino acid residues [GUIDOTTI, 1972]. Since typical soluble enzymes also show amino acid compositions of similarly high percentage of nonpolar amino acids it is not possible to demonstrate any preferential affinity of the galactose binding protein for the cytoplasmic membrane of *E. coli*.

Cysteine appears to be absent in the galactose binding protein [MC-GOWAN, unpublished results] even though its presence has been reported in the arabinose [PARSONS and HOGG, submitted for publication], the leucine [ANRAKU, 1968] and the ribose [AKSAMIT and KOSHLAND, 1972] binding proteins. The presence or absence of cysteine in binding proteins is of interest, since sulfhydryl (S-H) groups may play a crucial role in the energy coupling of active transport. The function of S-H groups in the energy coupling of transport has recently been proposed for a variety of membrane-bound sugar and amino acid 'carriers' in *E. coli* membrane vesicles [KABACK and BARNES, 1971]. Indeed, membrane-bound binding proteins solubilized by the nonionic detergent Brij 36-T have been shown to be sensitive to sulf-hydryl reagents such as N-ethylmaleimide and *p*-chloromercuribenzoate [GORDON *et al.*, 1972]. In contrast, the binding activity of the soluble peri-plasmic binding proteins are not inhibited *in vitro* by these reagents even though they strongly interfere with the *in vivo* transport activity of the respective transport system [for the galactose binding protein, see ANRAKU, 1968; PARNES and BOOS, 1973].

A general phenomenon observed with periplasmic binding proteins including the galactose binding protein is a considerable heat stability. Incubation for 10 min at 80°C reduced the binding activity for galactose by only 10% [ANRAKU, 1968]. Possibly connected to this phenomenon is the high percentage of saturation of ammonium sulfate (>60%) required to precipitate the galactose binding protein as well as other periplasmic binding proteins.

The galactose binding protein is quite insensitive to changes in pH. From pH 5 to 9 the binding activity changes less than 15% [ANRAKU, 1968]. This insensitivity against pH changes can also be seen by ultraviolet difference spectroscopy in the presence of substrate. Between pH 5.3 and 9.3 no changes of the spectrum occur. The reduction of the difference spectrum that occurs below pH 5 seems to correspond to protein denaturation. However, at a pH higher than 9.5 one sees a qualitative change in the difference spectra prior to denaturation. This seems to suggest that at a pH near 10 substrate bind-ing still occurs, but it occurs in a different manner [MCGOWAN *et al.*, 1974].

C. Molecular Weight

The recent discovery that the galactose binding protein can be separated into two forms [Boos and Gordon, 1971] made a thorough study of the molecular weight imperative. Early experiments by Anraku [1968] using sedimentation velocity in the analytical ultracentrifuge gave a molecular weight of 35,000 assuming a partial specific volume of 0.73 of the protein. Experiments in our laboratory with the analytical ultracentrifuge using sedimentation equilibrium and molecular sieve chromatography through Biogel P-150 (fig. 1) gave molecular weights of 34,000 and 36,000, respectively, for the native protein [Boos *et al.*, 1972]. These values did not change upon addition of galactose. A single band corresponding to a molecular weight of 36,500 was obtained [Boos *et al.*, 1972] by analytical polyacrylamide gel electrophoresis in the presence of sodium dodecylsulfate (SDS), a method which gives the molecular weight of the denatured polypeptide chain [Weber and Osborn, 1969].

D. Structural Feature

Little is known about the actual structure and conformation of the protein. The optical rotary dispersion (ORD) and circular dichroism (CD) spectra, which are frequently used to interpret protein conformation [Beychok, 1968], are shown in figure 2. The salient features are: (1) an ORD trough near 230 nm and a cross over below 220 nm, and (2) a CD band at 219 nm, a shoulder near 208 nm and a crossover at 203 nm. These spectral properties are indicative of a rather unusually high content of β-structure. Since no typical α-helix band can be observed the content of α-helix is predicted to be less than 10%. The high content of β-structure can also be demonstrated by infrared spectroscopy (fig. 3) which shows a strong band at 1,635 cm^{-1} which has been attributed to β-conformation [Susi *et al.*, 1967].

E. Measurement of Activity

The galactose binding protein has been implicated in two cellular functions: galactose transport and chemotaxis towards galactose. Neither phenomena can be measured in an *in vitro* system using the isolated and

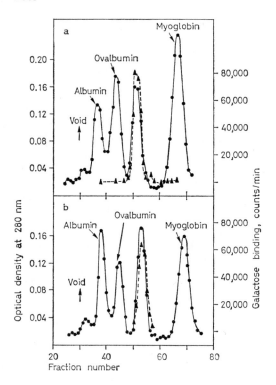

Fig. 1. Bio-gel P-150 chromatography of galactose binding protein in the presence and absence of galactose. The packed column (90 × 1.5 cm) had a flow rate of 0.06 ml/min at a pressure head of 25 cm. One milligram of galactose binding protein was applied in 1 ml of 0.01 M Tris-HCl, pH 7.3. The column was eluted with the same buffer at room temperature and 1 ml fractions were collected. •—• = Absorption at 280 nm; ▲—▲ = binding activity as determined by equilibrium dialysis at an initial [1-^{14}C]galactose concentration of 10^{-6} M. Galactose binding is given in excess counts per min found in the protein-containing chamber over the amount present in the protein-free chamber. This test is not linear since the free galactose concentration decreases up to 10-fold with increasing protein concentration. The marker proteins, bovine serum albumin, ovalbumin and myoglobin are run together with the galactose binding protein and their positions were identified in separate runs. *a* Chromatography in the absence of galactose. *b* Chromatography in the presence of 10^{-4} M galactose. Before measuring binding activity in the galactose-containing eluate, the fractions were dialysed against 0.01 M Tris, pH 7.3, overnight. [From Boos *et al.*, 1972.]

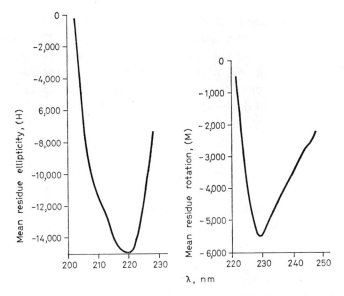

Fig. 2. Optical rotatory dispersion and circular dichroism of the galactose binding protein. The spectra were measured at room temperature with a Cary 60 spectropolarimeter with circular dichroism (CD) attachment. *a* CD spectrum at a protein concentration of 0.55 mg/ml in the presence and absence of 10^{-4} M galactose (the two curves are superimposable). Path length of the cell was 1 mm. *b* Optical rotary dispersion (ORD) at a protein concentration of 0.4 mg/ml in 0.01 M Tris-HCl, pH 7.3, in the presence and absence of 10^{-4} M galactose (the two curves are superimposable). Path length was 5 mm. Mean residue weight of the galactose binding protein was calculated according to the amino acid composition to be 126. [From Boos *et al.*, 1972.]

purified protein. Thus, measurements of activity are restricted either to assays using the binding affinity of the protein towards galactose or to assays of the intact structure of the protein. Three different kinds of assays are commonly in use: (1) the direct binding assay using radioactively labeled ligand; (2) measurements of spectral changes which occur in response to binding of ligand, and (3) measurements based on the immunological properties of the protein.

1. Binding Assay
a) Equilibrium Dialysis
The classical test for binding activity of any macromolecule is equilibrium dialysis. The equilibrium dialysis cell contains two chambers divided by a semipermeable membrane. The protein solution is added to one chamber,

and the radioactively labeled ligand is added to the other. After sufficient time, aliquots of each chamber are counted for radioactivity. The excess radioactivity in the protein containing chamber represents the concentration of ligand bound to the protein, whereas the counts in the ligand chamber represent the concentration of free ligand. Presently there are apparatuses available which contain 24 double chambers of a minimal volume of 100 μl to be operated under temperature control. Since the plastic material of the dialysis chambers adsorbs protein unspecifically the binding protein concentration has to be sufficiently high (>0.2 mg/ml) or a nonbinding active protein has to be added in excess to saturate the unspecific 'binding sites' of the plastic material. In a typical experiment using 0.4 mg/ml galactose binding protein (11 μM, based on a molecular weight of 36,000) one obtains excellent values by using 10 nM to 1 μM initial free galactose concentration.

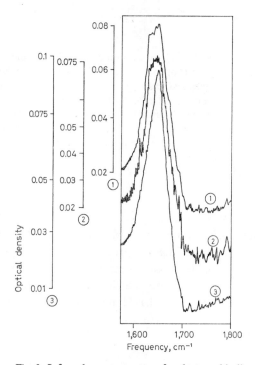

Fig. 3. Infrared spectroscopy of galactose binding protein in the presence and absence of galactose. 25 μl of galactose binding protein solution of 1.5 mg/ml in 0.01 M Tris-HCl, pH 7.3, in the presence and absence of 10^{-4} M galactose was frozen in liquid nitrogen on silver chloride windows and freeze-dried. 1 = No galactose; 2 = 10^{-4} M galactose; 3 = bovine serum albumin. [From Boos *et al.*, 1972.]

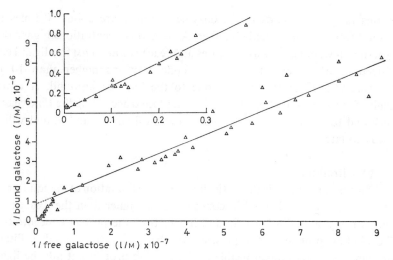

Fig. 4. Galactose binding activity of the galactose binding protein as a function of galactose concentration measured by equilibrium dialysis. Double chambers of 100 μl volume separated by dialysis tubing were filled with 90 μl of galactose binding protein (0.4 mg/ml) in 0.01 M Tris-HCl, pH 7.3, and 90 μl of [1-^{14}C]-galactose or [1-^{3}H]galactose in the same buffer. The dialysis was performed at 4 °C for at least 12 h. 50 μl from each chamber were counted for radioactivity. [From Boos *et al.*, 1972.]

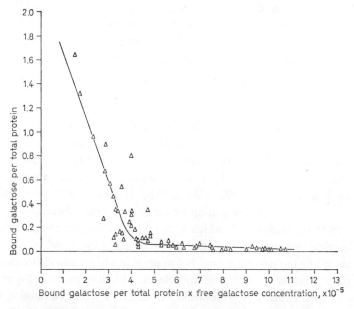

Fig. 5. Scatchard plot of the binding data of figure 4. [From Boos *et al.*, 1972.]

In this range the counts in the enzyme chamber are about 10 times the amount found in the ligand chamber. At ligand concentrations higher than 1 μM (initial value) this ratio decreases and reaches 2 at 10 μM free ligand concentration. Measurements at ligand concentrations higher than 10 μM therefore reflect increasing error due to the high background. A typical result of the above-mentioned experiment, plotted according the Lineweaver-Burk and to SCATCHARD [1949] are shown in figures 4 and 5, and will be discussed later.

b) Ultrafiltration

This binding assay is theoretically free of the limitation of the equilibrium dialysis when ligand is used in concentrations higher than the dissociation constant of the binding protein and higher than the protein concentration. The binding protein in the presence of the radioactive ligand is filtered through membranes impermeable to the binding protein but not the ligand [PAULUS, 1969]. The filter is subsequently measured for radioactivity. Using this method, 10 μg of galactose binding protein can be easily detected [BOOS and GORDON, 1971]. Yet quantitative binding assays for the galactose binding protein with this method are not possible: during the filtration assay the protein solution is necessarily being concentrated and independent studies by equilibrium dialysis have shown that the binding affinity is dependent on the protein concentration. This appears also to be true for the arabinose binding protein of *E. coli* B/r [PARSONS and HOGG, submitted for publication]. However, the ultrafiltration assay is a convenient and sensitive method to qualitatively detect galactose binding protein in different fractions during the purification of the protein.

c) Preparative Polyacrylamide Gel Electrophoresis

Another excellent method to qualitatively detect galactose binding in crude extracts of bacterial 'shock fluid' is preparative polyacrylamide gel electrophoresis through gels which were polymerized in the presence of radioactive ligand [BOOS, 1969]. This method is based on the particular elution technique in which the eluting buffer does not flow through the gel, but removes all the compounds which have migrated through the gel forced by the electrical potential and which appear on the anodic end of the gel. The uncharged ligand galactose does not migrate in the gel and radioactivity appears only together with the macromolecular binding protein. Figure 6 shows two experiments, one with crude extracts of a strain containing the galactose binding protein, the other without. A potential hazard of this

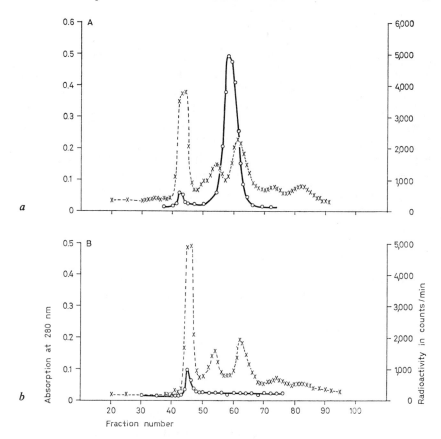

Fig. 6. Polyacrylamide electrophoresis of the shock protein preparation of W3092i and W4345 in the presence of labeled galactose. The elution rate was 0.3 ml/min; 5.0 ml were fractionated. The gels contained 1.2 μM [^{3}H]galactose (2.5×10^6 counts/min). x---x = Absorbance at 280 m; o—o = radioactivity of 0.1 ml aliquots. *a* Preparation from strain W3092i. *b* Preparation from strain W4345. [From Boos, 1969.]

method is that the peak of the galactose binding protein does not necessarily coincide with the peak of radioactivity. This can be seen when pure galactose binding protein is subjected to the electrophoresis under these conditions. The leading edge of the migrating protein may bind all the galactose, depleting the trailing edge of ligand. Therefore, the lower the concentration of ligand is in the gel, the larger is the distance between protein and radioactivity peak.

d) Precipitation with Ammoniumsulfate in the Presence of Ligand

Once ligand is bound to the galactose binding protein, it is not removed during precipitation with 80% saturated ammonium sulfate solution. This phenomenon has been used as a test for galactose binding activity [BERN-HARD-BENTABOULET and KEPES, 1973]. The ammonium sulfate precipitation is done in the presence of labeled galactose. The precipitate is filtered through a millipore filter and measured for radioactivity. Even though the amount of galactose bound to the protein under these conditions is only about one-half the amount found with the same protein preparation by equilibrium dialysis, the assay represents a reproducible and fast qualitative binding assay, most useful to determine relative amounts of galactose binding protein in crude protein preparations.

2. Spectral Changes in Response to Binding of Ligand

As will be discussed in the next paragraph, the interaction of substrate and the galactose binding protein results in the alteration of a variety of parameters. A convenient and very sensitive parameter to follow is the tryptophan fluorescence [BOOS *et al.*, 1972]. Figure 7 shows the excitation and emission spectrum of the pure galactose binding protein in the presence and absence of galactose and glucose. The excitation spectrum shows a peak at 288 nm which is increased in intensity by galactose and glucose whereas the peak position remains unchanged. At more narrow excitation slits the excitation peak is observed to split into a peak at 290 nm and a shoulder at 285 nm. These presumably correspond either to two different tryptophan residues or to tryptophan and tyrosine, respectively. The emission spectra remain unchanged in shape when protein is excited at 280 or 295 nm, showing that the emission is only due to tryptophan [UDENFRIEND, 1962]. The emission spectra (fig. 7b) show a broad maximum at 340 nm. Galactose increases the intensity and gives a 2-nm blue shift in the emission maximum. The maximum percentage increase for galactose is observed at 330 nm. Glucose also increases the intensity but gives no shift in the emission maximum. This indicates, and will be discussed in detail later, that the active site contains at least one tryptophanyl residue. The increase in the emission spectrum can conveniently be used as an activity test of the pure protein. Figure 8 shows the percent increase in the emission at 330 nm at an excitation of 290 nm upon the addition of varying concentrations of galactose and glucose. Both sugars result in a half maximal increase at a concentration of 1 μM, while the maximal percent increase is larger with galactose than with glucose. This method to determine the binding affinity of the galactose

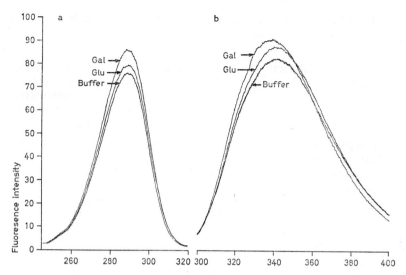

Fig. 7. Fluorescence spectra (uncorrected) of galactose binding protein in the presence and absence of 10^{-4} M galactose (Gal) and glucose (Glu). Galactose binding protein (16.7 µg/ml) in 0.01 M Tris-HCl, pH 7.3; excitation slit, 10 nm; emission slit, 8 nm. *a* Excitation spectra. Emission wave length was 330 nm. *b* Emission spectra. Excitation wave length was 290 nm. Temperature was 24°C. [From Boos *et al.*, 1972.]

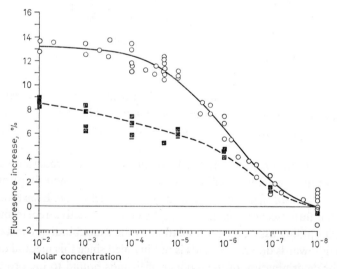

Fig. 8. Percent increase in fluorescence of galactose binding protein *versus* total sugar concentration. o—o = galactose; ■---■ = glucose. Excitation: 290; emission: 330. Excitation slit was 4 nm; emission slit was 10 nm. Protein concentration was 16.7 µg/ml. Temperature was 24°C. [From Boos *et al.*, 1972.]

Table II. The effect of different sugars on entry of galactose *in vivo* and on the fluorescence of the purified galactose binding protein

Concentration of sugar 0.1 μM	Percent inhibition of galactose uptake at an initial concentration of 0.5 μM [1-^{14}C]galactose	Percent of maximal fluorescence increase of purified galactose binding protein
(1-D-glyceryl)-β-D-galactopyranoside	96	88
L-Arabinose	54	59
D-Fucose	24	48
Methyl-β-D-galactopyranoside	22	29
D-Xylose	22	41
Methyl-1-thio-β-D-galactopyranoside	1	1
(2-glyceryl)-β-D-galactopyranoside	⟨1	⟨1
β-D-galactosyl-1-thio-β-D-galactopyranoside	⟨1	⟨1
Melibiose	⟨1	⟨1
Lactose	⟨1	⟨1
Isopropyl-1-thio-β-D-galactopyranoside	⟨1	⟨1

From PARNES and BOOS, manuscript submitted for publication.

binding protein is free of the limitations of the equilibrium dialysis method when high substrate concentrations have to be used and has proven useful to demonstrate the low binding affinity of a mutant galactose binding protein for which no binding affinity could be detected with either equilibrium dialysis or ultrafiltration [BOOS, 1972]. Unfortunately, only preparations of somewhat purified galactose binding protein (<50% of protein impurities with no affinity towards galactose) can be used in this technique. Even though measuring the fluorescence increase seems to be an attractive replacement for the direct binding assay, an extrapolation to the exact binding constants of the protein towards different ligands is not justified since this method does not allow for determination of the number of ligands bound to the protein. It also would not account for the binding of substrate to a second site not containing tryptophan. However, measurements of the increase in fluorescence provides an excellent tool to compare the relative ability of different

Table III. Specificity of the galactose chemoreceptor and the galactose binding protein as determined by inhibition studies

	Concentration (μM) required for 50% inhibition of:		
	(A) taxis towards 5 μM galactose	(B) binding of 5 μM galactose	ratio B:A
D-Glucose[1]	0.005	1.0	200
D-Galactose	0.036	7.0	190
(1-D-Glycerol)-β-D-galactoside	0.15	25	170
D-Fucose	6.2	1,100	180
Methyl-β-D-galactoside	30	3,500	120
L-Arabinose	95	17,000	180
D-Xylose	120	18,000	150

1 Most recently J. ADLER found that 6-deoxyglucose also is an excellent attractant for the galactose chemoreceptor. [From HAZELBAUER and ADLER, 1971.]

sugars to act as a substrate of the galactose binding protein. This then enables one to compare the observed specificity with the known substrate specificity of the β-methylgalactoside transport system or chemotaxis mediated by the galactose chemoreceptor. Table II shows such a comparison [PARNES and BOOS, 1973]. For comparison, table III shows the specificity of the galactose chemoreceptor evaluated by studies of inhibition of chemotaxis with galactose as attractant. The sugars listed in table II, for which the galactose binding protein shows affinity, exhibit either a 'glucose' or a 'galactose' emission spectrum, depending on the configuration of the hydroxyl (OH) group in the 4 position of the glycon ring. All active galactosides tested so far have a β-glycosidic linkage to the aglycon. It was therefore of interest to see whether the fluorescence increase with galactose was due only to the β-form of the sugar. When crystalline α-D-galactopyranose is equilibrated in 0.01 M Tris-HCl, pH 7.3, an equilibrium is reached in which 70% of the sugar has mutarotated to the β-form. The addition of such an equilibrated solution to the galactose binding protein results in an immediate change in fluorescence with no further subsequent increase. However, when α-D-galactopyranose solutions were prepared and immediately added to the protein, there was an initial fast increase followed by a slower increase which reached the value of the equilibrated galactose after about 30 min

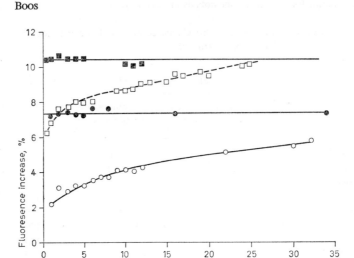

Fig. 9. Time-course of fluorescence increase following addition of galactose. Closed symbols indicate equilibrated galactose; open symbols are α-D-galactopyranose mixed at $t = 0$. Squares = galactose concentration of 2×10^{-5} M; circles represent galactose concentrations of 3×10^{-6} M. Protein concentration and instrument settings were as in figure 7. [From Boos *et al.*, 1972.]

(fig. 9). The time-course of this change is of the same order as the mutarotation required to reach equilibration of the α- and β-forms. Thus, the fluorescence increase is either caused preferentially by the β-form of galactose, or it is caused by both with the α-form being less effective. Subsequent studies using ultraviolet difference spectroscopy (discussed below) indicate the latter possibility.

The fluorescence increase by equilibrated galactose is too fast to be measured without a stopped-flow technique. This fast method has been applied for fluorescence measurement with the glutamine binding protein [WEINER and HEPPEL, 1971] and will be discussed here for comparison. Similar to the ligand-dependent increase of fluorescence of the galactose binding protein, the glutamine binding protein also shows such an increase in the presence of glutamine. From stopped-flow experiments a value of 9.8×10^7 [M^{-1} sec^{-1}] for the second-order rate constant of association and a value of 16 [sec^{-1}] for the first-order rate constant of dissociation has been determined. One is tempted to use these values to get an estimate for the percent change of binding protein molecules into the substrate bound state when bacteria are moving up a typical gradient of 0.1 mM/1 cm (10 nM/1 μM).

The calculation becomes simple if we place the bacterium on the beginning of the gradient and let it move one body length. Using the rate constant for association of the glutamine binding protein (10^8 M^{-1} sec^{-1}), the estimated molarity of the galactose binding protein in the periplasmic space (10^{-2} M), the measured rate of migration of a motile bacteria in the running phase (20 μm/sec) [BERG and BROWN, 1972] and the length of a single bacteria (1 μm) then the rate of complex formation is given by $10^8 \times 10^{-2} \times 10^{-8} = 1 \times 10^{-2}$ M/sec. For the distance of one body length the bacterium needs 0.05 sec, so that during that time 0.5 mM or 5% of the binding protein molecules should have changed their state from free to bound. Of course this approximation is based on several assumptions; one which certainly is not correct is that the concentration of galactose at any point remains constant, i.e. that the supply of ligand is unlimited. Nevertheless, from this consideration it seems not unlikely that the single bacterium would be capable to sense a gradient in a spacial way, as opposed to a memory-type mechanism proposed by MACNAB and KOSHLAND [1972].

3. Assays Based on the Immunological Properties

The purified galactose binding protein is an excellent antigen and specific antibodies have been isolated from rabbits and goats. Qualitative tests for the presence, absence or the inducibility of the galactose binding protein in mutants of the β-methylgalactoside transport system or of galactose chemotaxis have been done by Ouchterlony immunodiffusion test [BOOS and SARVAS, 1970; LENGELER et al., 1971]. This method also is useful for following the presence of a mutant galactose binding protein during purification. This protein exhibits a strongly reduced binding affinity, so that the usual binding assays could not be applied [BOOS, 1972].

A more direct immunological binding test based on the precipitation of galactose binding protein, even in crude extracts, with specific antibodies in the presence of radioactively labeled substrate [ROTMAN and ELLIS, 1972] has been reported. The antiserum has to be dialysed extensively to remove glucose which otherwise will strongly interfere with galactose binding. The forming complex of antibody, galactose binding protein and substrate is filtered through millipore filters and the radioactivity remaining on the filter can be taken as a measure of binding activity. Yet, as will be discussed in the next paragraph, the complex formation appears to change the binding affinity of the galactose binding protein [ROTMAN and ELLIS, 1972]. Therefore, the binding properties of the native protein cannot be measured quantitatively by this technique.

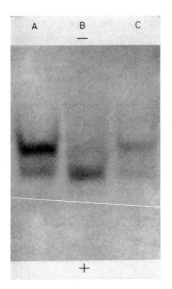

Fig. 10. Polyacrylamide gel electrophoresis of purified galactose binding protein in the presence and absence of urea. The gel slab (4 × 120 × 160 mm) contained 7.5% acrylamide, 0.2% bisacrylamide, 0.1 M Tris-borate, pH 8.4, and 0.002 M EDTA. A = untreated sample of galactose binding protein; B = sample incubated with 8 M urea in 0.01 M Tris-HCl, pH 7.4, for 1 h; C = sample incubated with 8 M urea in 0.01 M Tris-HCl, pH 7.3, for 1 h and subsequently passed through Sephadex G-100. The electrophoresis was run for 4 h at 300 V and 50 mA. [From Boos and GORDON, 1971.]

F. Conformational Change

1. Change of Electrophoretic Mobility upon Binding with Ligand

We observed recently that highly purified galactose binding protein shows two bands on polyacrylamide gels at pH 8.4 when the protein was incubated with 20% sucrose[2] prior to the electrophoresis (fig. 10). Direct binding assays on the gel by running the electrophoresis with gels which were equilibrated with 0.1 μM [^{3}H]galactose demonstrated that both components were capable of binding galactose. Also, both components consisted of the same material since reelectrophoresis of either component resulted again in the appearance of the same two bands [Boos and GORDON, 1971]. Subsequently, we discovered that the simultaneous appearance of the two

2   Sucrose was added to increase the density of the protein solution in order to facilitate the application of the protein to the gel.

bands on these gels was caused by the presence of sucrose in the incubation mixture prior to the electrophoresis. Replacing the sucrose by 20% non-polymerized acrylamide we obtained only one band. Yet the position of the band changed when substrate was present in the gel as well as the electrode buffers. The position of the band in the presence and absence of substrates corresponds to the position of the two bands of the protein when it was run after incubation with sucrose. The change in electrophoretic mobility is highly specific; it occurs with galactose, glucose and glycerolgalactoside, but not with lactose, isopropyl-thio-galactoside and other β-galactosides which are not substrates of the galactose binding protein [Boos *et al.*, 1972]. The concentration of galactose to give the half maximal effect (confluence of the two bands) is in the vicinity of 1 μM. This value represents the initial free galactose concentration in the acrylamide gel. The actual free galactose concentration on the band position, however, might be considerably lower since part of the galactose is removed by binding.

A plausible explanation for the cause of the substrate-dependent change in the migration through acrylamide gels was that the molecular weight of the galactose binding protein changes, and both forms would represent monomer and dimer of the same polypeptide chain. The subsequent detailed study of the molecular weight by different methods in the presence and absence of the substrate galactose, as was discussed in a previous paragraph, showed that the protein exhibits under all conditions a molecular weight of about 36,000. Therefore this interpretation seems unlikely. The occurrence of the two bands might be explained in several other ways: (1) substrate binding alters the protein conformation, thus changing its retardation coefficient [BANKER and COTMAN, 1972]; (2) substrate binding is accompanied by an alteration of surface charge, and (3) a combination of both of the above.

Evidence indicating that the two forms differed in charge was obtained by the separation during electrofocussing of the two species with a pI of 5.3 and 5.4, respectively[3]. The more negatively charged form, i.e. the form with higher electrophoretic mobility at pH 8.4 during electrophoresis, represents the form complexed with galactose. Assuming that the change in electrophoretic mobility is due to a change in surface charge and electrophoretic mobility increases linearly with the increase in negative charges one can estimate that the surface charge at pH 8.4 increases by 12–15%. That would mean that binding of galactose uncovers glutamic or aspartic

3   T. SILHAVY and W. BOOS, as referred to by BOOS [1972].

acid residues or buries the corresponding numbers of positively charged amino acid residues. One would expect to observe such a change in the arrangement of amino acid residues by the classical methods of CD, ORD and infrared spectroscopy. Yet, as discussed before, neither of these methods showed any change in the spectrum of the galactose binding protein in response to the presence of galactose at the wavelengths where chromophores of the polypeptide backbone absorb. The change in electrophoretic mobility does not occur in a mutant galactose binding protein (isolated from strain EH3039) even at galactose concentrations above the dissociation constant of the protein for galactose (>1 mM) [Boos, 1972]. However, electrofocussing still reveals two species with different pI. Therefore, the mutant protein must still be able to occur in two different forms and the difference in electrophoretic mobility at pH 8.4 of the wild type is not the cause for the difference in pI of the two forms.

2. Increase in Fluorescence upon Binding with Ligand

As discussed before, the tryptophan fluorescence of the galactose binding protein (fig. 7) increases up to 13.5% at 330 nm when excited at 290 nm [Boos et al., 1972]. This increase of fluorescence can be interpreted in several ways:

1. One or more tryptophanyl residues are part of the active site and the microenvironment of this tryptophanyl residue(s) is changed upon the interaction with the substrate itself.

2. The tryptophanyl residue in question is not part of the active site itself but changes its microenvironment due to a conformational change induced by the binding of galactose.

3. Both effects might occur at the same time.

As discussed before, the fluorescence spectra of the galactose binding protein in the presence of glucose and galactose are not identical. This seems to indicate that the change in the fluorescence is caused by the direct interaction of the carbohydrate with a tryptophanyl residue in the active site rather than solely by a change of microenvironment of an outside tryptophanyl residue due to a conformational change.

The tryptophan fluorescence can be quenched in the presence of potassium iodide. The effect of quenching in the wild-type protein with this agent in the presence and absence of galactose is shown in figure 11. As can be seen, galactose protects at all concentrations of iodide tested [McGowan et al., 1974]. Other substrates, such as glucose or glycerolgalactoside have the same effect. In contrast, the quenching effect of potassium iodide in a mutant

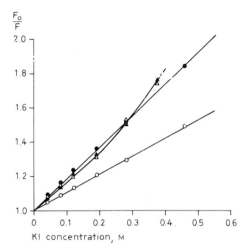

Fig. 11. Fluorescence quenching of the galactose binding protein in the presence and absence of galactose. Protein concentration was 16 μg/ml in 10 mM Tris-HCl, pH 7.3. KI was added from 6 M stock solutions. The control sample was adjusted with the corresponding amount of 10 mM Tris-HCl, pH 7.3. F_0 = fluorescence in the absence of KI; F = fluorescence in the presence of KI at the indicated concentrations; o—o = wild-type galactose binding protein; △—△ = mutant galactose binding protein (strain EH3039); open symbols = measurements in the absence of galactose; closed symbols = measurements in the presence of galactose; 0.1 mM in case of the wild-type and 20 mM in case of the mutant protein. [From McGowan *et al.*, 1974.]

galactose binding protein (EH3039) is unaffected by galactose concentrations above its K$_{DISS}$ of binding (0.01 M). As will be discussed later, the bulk solvent cannot have access to the tryptophanyl residue of the active site. Therefore, it is very likely that neither the iodide anion has access to this tryptophanyl residue. Therefore, we conclude that one of the tryptophanyl residues affected by iodide is located on the outside of the protein in close proximity to a charged group which magnifies the quenching effect. Burying of this group without burying the tryptophanyl residue due to the conformational change caused by galactose binding will decrease the quenching effect of potassium iodide (KI) and it appears as if galactose 'protects' against quenching. In agreement with this explanation is the observation that galactose does not protect against quenching in the mutant binding protein which also does not show the difference in electrophoretic mobility at pH 8.4 in the presence and absence of galactose so characteristic for the wild-type protein. The data obtained with fluorescence quenching by KI, however, do not distinguish between the two possibi-

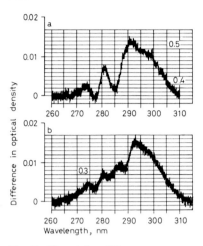

Fig. 12. Ultraviolet difference spectrum of the galactose binding protein in the presence and absence of D-galactose. *a* Wild-type protein, 0.9 mg/ml in 10 mM Tris-HCl, pH 7.3. Galactose concentration was 0.1 mM. *b* Mutant protein (strain EH3039), 1.1 mg/ml in 10 mM Tris-HCl, pH 7.3. Galactose concentration was 15 mM. [From McGowan *et al.*, 1974].

lities: a positive charge enhancing the quenching in the absence of galactose or a negative charge reducing quenching in the presence of galactose, since opposite charges also have opposite effects on the ability of KI to quench tryptophan fluorescence [Lehrer, 1971].

3. Change of the Ultraviolet Spectrum upon Binding with Ligand

The addition of galactose to the wild-type galactose binding protein results in the production of an ultraviolet difference spectrum that closely resembles the solvent perturbation difference spectrum of N-acetyl-trypto-phanyl-ethyl ester (a model compound which resembles tryptophanyl residues in a polypeptide chain). This difference spectra is shown in figure 12. As can be seen, the mutant galactose binding protein (strain EH3039) also exhibits a difference spectrum slightly distinct from that of the wild type. It should be noted that the addition of glucose to both proteins results in the production of a similar but *not* identical difference spectra (not shown).

The addition of freshly prepared α-galactose to the wild-type galactose binding protein results in a difference spectrum with the time dependent increase of two absorption maxima; this time dependence is caused by the mutarotation of the α-form to the equilibrium mixture of α- and β-forms. The third peak of the difference spectrum is independent of mutarotation.

The observation that glucose and galactose as well as α- and β-galactose give rise to similar but not identical ultraviolet difference spectra strongly suggests that at least part of the difference spectra must be caused by the direct interaction of substrate with a tryptophanyl residue in the active site of the galactose binding protein.

4. Solvent Perturbation in the Presence and Absence of Ligand

One way of investigating the relative exposure of chromophoric groups such as tryptophanyl or tyrosyl residues is by solvent perturbation [HERSKOVITS, 1967]. If a perturbant such as dimethylsulfoxide, ethylenglycol or methanol is added to a protein solution, chromophores exposed to this agent will experience an alteration in their microenvironment. This alteration of environment will be reflected by an altered ultraviolet absorption spectrum. By comparing solvent perturbation difference spectra with available model compound data [HERSKOVITS and SORENSEN, 1968a, b] one can calculate the number of exposed chromophores present in a given protein. The validity of the method can be demonstrated when proteins denatured by 8 M urea are used in the solvent perturbation difference spectroscopy, where all tryptophanyl and tyrosyl residues should be accessible to the solvent. In the case of the galactose binding protein, 5 tryptophanyl and 7–8 tyrosyl residues can be calculated for the protein denatured by 8 M urea [McGOWAN *et al.*, 1974] which is consistent with the amino acid composition reported by ANRAKU [1968].

Studies with the native galactose binding protein and dimethylsulfoxide, methanol and ethylenglycol as perturbants indicated that 2 tryptophanyl and 4–5 tyrosyl residues were accessible. The smaller perturbant D_2O resulted in the perturbation of two additional tryptophanyl residues.

If galactose produces a conformational change, then it may well alter the number of exposed chromophoric residues. Indeed, the addition of galactose to galactose binding protein did result in an altered difference spectrum but again this alteration is not conclusive for a conformational change only. As discussed before, we concluded that the substrate must interact with a tryptophanyl residue in the active site. If in addition the interaction of the galactose binding protein changes the microenvironment of the chromophoric groups outside the active site, then the solvent perturbation spectrum in the presence of galactose should be different from that in the absence of galactose. When this was done, both spectra were identical. This indicates that the ultraviolet difference spectrum observed in the presence of substrate is due solely to an interaction between substrate and tryptophanyl residue in the active site rather than caused by a change of the environment of an 'outside' tryptophanyl residue due to a conformational change. It should be noted, however, that these results do not exclude a conformational change of the galactose binding protein in the presence of substrate.

The identity of the solvent perturbation spectra in the presence and absence of galactose suggests that the tryptophanyl residue in the active site is never accessible to the perturbing solvent. This must be the case, since it is exceedingly unlikely that the tryptophanyl residue in the active site would be perturbed by any solvent in an identical way, regardless of the presence or absence of substrate.

It is interesting that the solvent perturbation difference spectra of the mutant galactose binding protein (strain EH3039) in the presence or absence of substrate are not identical. The difference between these two spectra are slight but reproducible. This might be explained in the following way.

1. Substrate interacts with the same tryptophanyl residue in the active site, but the fit is not as tight, and some contact between the tryptophanyl residue and the bulk solvent is possible.

2. In the absence of substrate some interaction between the tryptophanyl residue in the active site and the added perturbant is possible.

3. The substrate induced conformational change in the mutant galactose-binding protein *does* alter, to a small degree, the exposure of a chromophoric group located outside the active site.

To date, we have been unable to devise any experiments which would help clarify this situation.

5. Change of Binding Affinity upon Binding of Ligand

Examination of the binding activity at different free galactose concentrations by equilibrium dialysis showed that the binding behavior is biphasic [Boos *et al.*, 1972]. Figure 4 gives the results plotted according to the Lineweaver-Burk method. Extrapolation of the points obtained between 0.3 μM and 10 nM free galactose concentration yielded an apparent dissociation constant of 0.1 μM, whereas extrapolation of the values obtained between 0.3 μM and 10 μM yielded an apparent K_{DISS} of 10 μM. As discussed in the previous paragraph, the experimental data are still meaningful up to a concentration of 10 μM (protein concentration was 11 μM). When the data were plotted according to SCATCHARD [1949] the heterogeneous behavior was even more pronounced (fig. 5). Surprisingly, the extrapolation of the curve to high free galactose concentrations indicated that 2 moles galactose were bound per mole of galactose binding protein of 36,000 molecular weight.

All periplasmic binding proteins studied so far, including the galactose binding protein [ANRAKU, 1968], have been reported as having only one binding site per polypeptide chain; however, the existence of a second binding site with strongly reduced binding affinity might have been overlooked due to the increasing error in equilibrium dialysis measurements at high substrate concentrations. Indeed, the differential ability of substrate to protect against the inhibition of the M-protein by N-ethylmaleimide, the membrane-bound transport protein of the lactose transport system in *E. coli*, suggests two binding sites per polypeptide chain [CARTER *et al.*, 1968]. Thus, the existence of two different binding sites in the fully saturated galactose binding protein is not a completely isolated phenomenon. Our interpretation of the binding behavior at the time was that binding of galactose to the protein of the high affinity form induces a conformational change resulting in the formation of a protein structure with one additional binding site, but lower affinity. The extrapolation of the number of binding sites of the high affinity form (present at low galactose concentrations) gave, at the most, only 0.1 moles galactose per mole binding protein (fig. 5), an observation explainable by the continuous conversion into the low affinity form by the increasing galactose concentration. Another possibility to interpret the observed binding plot would be the existence of two different forms of the binding protein at any one time: form I, with a very low binding affinity site, and form II, with both a high ($K_{DISS} = 0.1 \ \mu M$) and low ($K_{DISS} = 10 \ \mu M$) affinity site. Both forms would be in equilibrium, favoring form I in the absence of galactose. The contribution of form II in the equilibrium would be too small to be detected beside form I during polyacrylamide gel electrophoresis, but would be present in amounts large enough to exhibit the high affinity binding activity during equilibrium dialysis (one can extrapolate a contribution of 5–10% of form II in the absence of galactose). Binding of galactose would stabilize form II and shift the equilibrium in favor of form II. Therefore, only one form (II) can be detected at high galactose concentration during polyacrylamide gel electrophoresis.

G. A Working Model

Models of a reaction sequence in any one system usually function to conceal our lack of knowledge about the actual mechanism. The model presented

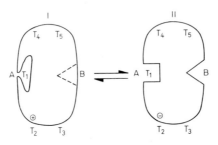

Fig. 13. Working model for the galactose binding protein. The protein can occur in two conformations (I and II), which are in equilibrium with each other. In the absence of galactose, the equilibrium is predominantly shifted to the left. The protein has two binding sites for galactose, A and B. A has a high affinity for galactose when the protein is in state II and is buried in state I. B has a low affinity for galactose and may or may not be present in state I. Binding of galactose to site A shifts the equilibrium in favor of state II. The protein contains 5 tryptophanyl residues (T_{1-5}) two of which (T_{2-3}) are accessible to bulk solvent. T_2 is located in the close proximity of a positively charged group in state I which is not present in state II, or near a negatively charged group in state II which is not present in state I. [From McGOWAN *et al.*, 1974.]

in figure 13 does not claim any resemblance to reality. Instead it represents the author's current opinion extracted from the biased interpretation of available results.

The galactose binding protein occurs in two different conformational states, I and II, which are in equilibrium with each other. In the absence of substrate, state II comprises less than 10% of the total protein. The protein has two binding sites, A and B. A has different affinities in states I and II: KDISS very high in I and 0.1 μM in II with galactose as substrate. B has a KDISS of 10μ M in II and might also be present in I with the same affinity. Due to the presence of the high active binding site in the absence of galactose, the presence of B in state I cannot be demonstrated experimentally. The only evidence for its presence in state II is the number of two binding sites found at high galactose concentrations.

Binding of substrate to A in II shifts the equilibrium to the right. I and II are different by their electrophoretic mobility in acrylamide gels, while the fluorescence increase and the ultraviolet difference spectrum are caused by interaction of the substrate with a tryptophanyl residue in the active site of A and in state II. The charge differences in the two forms are the reason of the protection by galactose against KI quenching. The position of this charge must be in the vicinity of a tryptophanyl residue on the outside of the protein not related to the active site.

The binding data are in agreement with the model as discussed before. Starting from very low galactose concentration binding would be accomplished only by site A of state II, which shifts the equilibrium of the two forms to the right (the equilibrium would require that 5–10% of the molecules found in state I to be in state II with a free binding site A). Thus, form II complexed with galactose is the species which gives rise to the fluorescence increase while both the bound and nonbound form of II show the increased electrophoretic mobility on polyacrylamide gels.

Increase in 'binding activity' has been reported when specific antibodies were present in the binding assay [ROTMAN and ELLIS, 1972]. The order of additions was important for this increase. Galactose had to be added first to react with the galactose binding protein before addition of the antibodies. An explanation, other than a trivial trapping of bound galactose by formation of the antigen-antibody precipitate can be provided by the model. The antibodies precipitate either form I or form II depending on whether or not substrate was present during the precipitation. The antibody interaction will hold the galactose binding protein in its respective state and therefore in different binding affinities.

From the results obtained from the ultraviolet difference spectra caused by glucose and galactose it is clear that at least one tryptophanyl residue occupies the active site. From the solvent perturbation studies with ethylene glycol, methanol and dimethylsulfoxide one can conclude that 2 of 5 tryptophanyl residues are accessible to the bulk solvent no matter if galactose is present or not. With D_2O as perturbing solvent 4 tryptophanyl residues are accessible again with no difference whether or not galactose is present. Therefore, none of the perturbing solvents can have access to the tryptophanyl residue of the active site.

The mutant protein does not change its electrophoretic mobility even at very high galactose concentrations, sufficient to maximally increase its fluorescence and to give rise to its ultraviolet difference spectrum. However, by electrofocussing one can still detect two different forms of the protein. We therefore conclude that the mutant binding protein can still occur in two states even though the mutation has abolished the charge differences at pH 8.4, characteristic for the wild type. Incidentally, the mutant protein migrates under these conditions identical to the wild-type protein in the absence of substrate, the form with the higher positive charge.

As in the wild type, two tryptophanyl residues from a total of 5 are accessible to bulk solvent but the solvent perturbation difference spectrum is not entirely independent of the substrate. In addition, the effect of KI to quench the tryptophan fluorescence is not counteracted by the substrate. In terms of the model, one would explain the mutant protein by saying that the mutation has led to 'loosening' of site A. The positive charge is not 'buried' or the negative charge is not 'revealed' any more when the protein is in state II; hence, galactose does not protect fluorescence quenching by KI. In addition, the 'loosening' of the active site which results in a 1,000-fold reduction of the binding affinity allows limited access of bulk solvent to the tryptophanyl residue in the active site in the presence of galactose. The mutation also might have affected the equilibrium of the two forms and must have abolished binding site B.

Despite the restriction implicated by the inaccuracy of the method of solvent perturbation, the tentative distribution of the different tryptophanyl residues (T_x) would be as follows. T_1 is part of the active site and is not accessible to the bulk solvent when the protein is in state I. When the protein is in state II, T_1 interacts with galactose. T_1 in this state must be accessible to bulk solvent, since galactose can interact. However, the amount of protein found at any time in state II, not complexed with galactose, is too small to be detected by solvent perturbation difference spectroscopy. Two tryptophanyl residues (T_2, T_3) are located on the 'outside' of the protein unrelated to the active site. At least one of these residues must be located close to the positive charge of state I or close to the negative charge in II. The remaining two residues (T_4, T_5) are 'half-buried', i.e. they are accessible to D_2O but not to larger solvents such as dimethylsulfoxide. These tryptophanyl residues also have no relation to the active site.

Summary and Comment

The galactose binding protein is localized in the periplasmic space external to the cytoplasmic membrane of E. *coli*. The substrate galactose has relatively easy access to the binding protein and active transport is not required for this binding reaction. The isolated and purified protein can exist in two different states, which are interconvertible by association and dissociation of substrate. The two different states can be distinguished from each other by their different binding affinities towards the substrate and by their electrical surface charge. Due to the formation of a new binding site of lower affinity, the protein is able to 'respond' to a wide range of galactose concentrations from 50 nM to 50 μM.

As to the actual function of the galactose binding protein in galactose transport and galactose chemotaxis, there are at present only speculations. The binding protein must respond in some way to the presence of the attractant galactose. This necessary response might be seen by the conformational change or the change in the surface charge of the binding protein. In any case the response must be transmitted more or less directly to the flagella which provides the bacterium with its motility. One of the next biochemical approaches would be the elucidation of the possible interaction of the galactose binding protein with the basal bodies of the flagella, the isolation and characterization of which have been demonstrated [DePAMPHILIS and ADLER, 1971].

References

ADLER, J.; HAZELBAUER, J. L., and DAHL, M. M.: Chemotaxis toward sugars in *Escherichia coli*. J. Bact. (in press).

AKSAMIT, R. and KOSHLAND, D. E.: A ribose binding protein of *Salmonella typhimurium*. Biochem. biophys. Res. Commun. *48:* 1348 (1972).

ANRAKU, Y.: The reduction and restoration of galactose transport in osmotically shocked cells of *Escherichia coli*. J. biol. Chem. *242:* 793 (1967).

ANRAKU, Y.: Transport of sugars and amino acids in bacteria. II. Properties of galactose- and leucine-binding proteins. J. biol. Chem. *243:* 3123 (1968).

BANKER, G. A. and COTMAN, C. W.: Measurement of free electrophoretic mobility and retardation coefficient of protein-sodium dodecyl complexes by gel electrophoresis. A method to validate molecular weight estimates. J. biol. Chem. *247:* 5856 (1972).

BERG, H. C. and BROWN, D. A.: Chemotaxis in *Escherichia coli* analysed by three-dimensional tracking. Nature, Lond. *239:* 500 (1972).

BERNHARD-BENTABOULET, M. and KEPES, A.: Energy dependent masking of substrate binding sites of the lactose permease of *Escherichia coli*. Biochim. biophys. Acta *307:* 197 (1973).

BEYCHOK, S.: Rotary dispersion and circular dichroism. Annu. Rev. Biochem. *37:* 437 (1968).

BOOS, W.: The galactose-binding protein and its relationship to the β-methylgalactoside permease from *Escherichia coli*. Europ. J. Biochem. *10:* 66 (1969).

BOOS, W.: Structurally defective galactose-binding protein isolated from a mutant negative in the β-methylgalactoside transport system of *Escherichia coli*. J. biol. Chem. *247:* 5414 (1972).

BOOS, W. and GORDON, A. S.: Transport properties of the galactose-binding protein of *Escherichia coli*. Occurrence of two conformational states. J. biol. Chem. *246:* 621 (1971).

BOOS, W.; GORDON, A. S.; HALL, R. E., and PRICE, H. D.: Transport properties of the galactose-binding protein of *Escherichia coli*. Substrate-induced conformational change. J. biol. Chem. *247:* 917 (1972).

Boos, W. and Sarvas, M. O.: Close linkage between a galactose-binding protein and the β-methylgalactoside permease in *Escherichia coli*. Europ. J. Biochem. *13:* 526 (1970).

Brockman, R. W. and Heppel, L. A.: On the localization of alkaline phosphatase and cyclic phosphodiesterase in *Escherichia coli*. Biochemistry *7:* 2554 (1968).

Carter, J. R.; Fox, C. F., and Kennedy, E. P.: Interaction with sugars with the membrane protein component of the lactose transport system of *Escherichia coli*. Proc. nat. Acad. Sci., Wash. *60:* 725 (1968).

Dahlquist, F. W.; Lovely, P., and Koshland, jr., D. E.: A quantitative analysis of bacterial migration in chemotaxis. Nature new Biol. *236:* 120 (1972).

DePamphilis, M. L. and Adler, J.: Fine structure and isolation of the hook-basal body complex of flagella from *Escherichia coli* and *Bacillus subtilis*. J. Bact. *105:* 384 (1971).

Gordon, A. S.; Lombardi, F. J., and Kaback, H. R.: Solubilization and partial purification of amino acid-specific components of the D-lactate dehydrogenase-coupled amino acid-transport systems. Proc. nat. Acad. Sci., Wash. *69:* 358 (1972).

Guidotti, G.: Membrane proteins. Annu. Rev. Biochem. *41:* 731 (1972).

Hazelbauer, G. J. and Adler, J.: Role of the galactose-binding protein in chemotaxis of *Escherichia coli* toward galactose. Nature New Biol. *230:* 101 (1971).

Heppel, L. A.: Selective release of enzymes from bacteria. Science *156:* 1451 (1967).

Heppel, L. A.: The effect of osmotic shock on release of bacterial proteins and on active transport. J. gen. Physiol. *54:* 95s (1969).

Heppel, L. A.: The concept of periplasmic enzymes; in Rothfield Structure and function of biological membranes, p. 223 (Academic Press, New York 1971).

Heppel, L. A.; Rosen, B. P.; Friedberg, I.; Berger, E. A., and Weiner, J. H.: Studies on binding proteins, periplasmic enzymes and active transport in *Escherichia coli;* in Woessner and Huijing The molecular basis of biological transport, p. 133 (Academic Press, New York 1972).

Herskovits, T. T.: Difference spectroscopy. Methods in enzymology, vol. 11, p. 748 (Academic Press, New York 1967).

Herskovits, T. T. and Sorensen, M.: Studies of the location of tyrosyl and tryptophyl residues in proteins. I. Solvent perturbation data of model compounds. Biochemistry *7:* 2523 (1968a).

Herskovits, T. T. and Sorensen, M.: Studies of the location of tyrosyl and tryptophyl residues in protein. II. Application of model data to solvent perturbation studies of proteins rich in both tyrosine and tryptophan. Biochemistry *7:* 2533 (1968b).

Kaback, H. R. and Barnes, E. M.: Mechanisms of active transport in isolated membrane vesicles. II. The mechanism of energy coupling between D-lactic dehydrogenase and β-galactoside transport in membrane preparations from *Escherichia coli*. J. biol. Chem. *246:* 5523 (1971).

Kalckar, H. M.: The periplasmic galactose binding protein of *Escherichia coli*. Science *174:* 557 (1971).

Lehrer, S. S.: Solute perturbation of protein fluorescence. The quenching of the tryptophyl fluorescence of model compounds and of lysozyme by iodide ion. Biochemistry *10:* 3254 (1971).

Lengeler, J.; Hermann, K. O.; Unsöld, H. J., and Boos, W.: The regulation of the β-methylgalactoside transport system and of the galactose-binding protein of *Escherichia coli* K12. Europ. J. Biochem. *19:* 457 (1971).

LEVER, J. E.: Purification and properties of a component of histidine transport in *Salmonella typhimurium*. J. biol. Chem. *247:* 4317 (1972).

LOPES, J.; GOTTFRIED, S., and ROTHFIELD, L.: Leakage of periplasmic enzymes by mutants of *Escherichia coli* and *Salmonella typhimurium:* Isolation of 'periplasmic leaky' mutants. J. Bact. *109:* 520 (1972).

MACNAB, R. M. and KOSHLAND, D. E.: The gradient-sensing mechanism in bacterial chemotaxis. Proc. nat. Acad. Sci., Wash. *69:* 2509 (1972).

MCALLISTER, T. J.; COSTERTON, J. W.; THOMPSON, L.; THOMPSON, J., and INGRAM, J. M.: Distribution of alkaline phosphatase within the periplasmic space of gram-negative bacteria. J. Bact. *111:* 827 (1972).

MCGOWAN, E.; SILHAVY, T., and BOOS, W.: Involvement of a tryptophan residue in the binding site of *Escherichia coli* galactose-binding protein. Biochemistry (in press, 1974).

MESIBOV, R. and ADLER, J.: Chemotaxis toward amino acids in *Escherichia coli*. J. Bact. *112:* 315 (1972).

NEU, H. C. and HEPPEL, L. A.: The release of enzymes from *Escherichia coli* by osmotic shock and during the formation of spheroplasts. J. biol. Chem. *240:* 3685 (1965).

NOSSAL, N. G. and HEPPEL, L. A.: The release of enzymes by osmotic shock from *Escherichia coli* in exponential phase. J. biol. Chem. *241:* 3055 (1966).

PARDEE, A. B. and WATANABE, K.: Location of sulfate-binding protein in *Salmonella typhimurium*. J. Bact. *96:* 1049 (1968).

PARNES, J. R. and BOOS, W.: Unidirectional transport activity mediated by the galactose-binding protein of *Escherichia coli*. J. biol. Chem. *248:* 4436 (1973).

PARSONS, R. G. and HOGG, R. W.: A comparison of L-arabinose and D-galactose-binding protein from *Escherichia coli* (submitted for publication).

PAULUS, H.: A rapid and sensitive method for measuring the binding of radioactive ligands to proteins. Analyt. Biochem. *32:* 91 (1969).

PFEFFER, W.: Locomotorische Richtungsbewegungen durch chemische Reize. Untersuch. botan. Inst., Tübingen *1:* 363 (1884).

PFEFFER, W.: Chemotaktische Bewegung von Bacterien, Flagellaten und Volvucineen. Untersuch. botan. Inst., Tübingen *2:* 582 (1886).

ROSEN, B. P.: Basic amino acid transport in *Escherichia coli*. J. biol. Chem. *246:* 3653 (1971).

ROSEN, B. P. and HACKETTE, S. L.: Effects of fatty acid substitution on the release of enzymes by osmotic shock. J. Bact. *110:* 1181 (1972).

ROTMAN, B. and ELLIS, J. H.: Antibody-mediated modification of the binding properties of a protein related to galactose transport. J. Bact. *111:* 791 (1972).

SCATCHARD, G.: The attractions of proteins for small molecules and ions. Ann. N.Y. Acad. Sci. *51:* 660 (1949).

SHEN, B. H. P. and BOOS, W.: Regulation of the β-methylgalactoside transport system and the galactose-binding protein by the cell cycle of *Escherichia coli*. Proc. nat. Acad. Sci., Wash. *70:* 1481 (1973).

SUSI, H.; TIMASHEFFS, S. N., and STEVENS, L.: Infrared spectra and protein conformations in aqueous solutions. I. The amide band in H_2O and D_2O solutions. J. biol. Chem. *242:* 5460 (1967).

UDENFRIEND, S.: Fluorescence assay in biology and medicine (Academic Press, New York 1962).

WEBER, K. and OSBORN, M.: The reliability of molecular weight determinations by dodecyl sulfate-polyacrylamide gel electrophoresis. J. biol. Chem. *244:* 4406 (1969).

WEINER, J. H. and HEPPEL, L. A.: A binding protein for glutamine and its relation to active transport in *Escherichia coli.* J. biol. Chem. *246:* 6933 (1971).

WETZEL, B. K.; SPICER, S. S.; DVORAK, H. F., and HEPPEL, L. A.: Cytochemical localization of certain phosphatases in *Escherichia coli.* J. Bact. *104:* 529 (1970).

Author's address: Dr. WINFRIED BOOS, Department of Biological Chemistry, Harvard Medical School, and the Biochemical Research Laboratory, Massachusetts General Hospital, *Boston, MA 02114* (USA)

Antibiotics and Chemotherapy, vol. 19, pp. 55–78
(Karger, Basel 1974)

Chemotaxis in *Escherichia coli* Analyzed by Three-Dimensional Tracking

H. C. BERG and D. A. BROWN

Department of Molecular, Cellular and Developmental Biology, University of Colorado, Boulder, Colo.

If a capillary tube containing an attractant is inserted into a suspension of motile bacteria, the bacteria accumulate near the mouth where the concentration of attractant is relatively high. PFEFFER [14, 15] introduced this technique as an assay for chemotaxis in 1884. In 1901, ROTHERT [16] and JENNINGS and CROSBY [10] noted that bacteria often swam past the capillary but returned after failing to enter regions of much lower concentration. It is now generally thought that chemotactic bacteria actively avoid regions of lower concentration by backing up or by choosing new directions at random [1, 17]. This is not obvious in *Escherichia coli*, since these bacteria repeatedly change their directions even in the absence of an applied stimulus.

To study this motion in detail, we built a microscope which automatically follows individual cells [5]. We have used this on mutants of *E. coli* K12 [2]: a wild type [4], a nonchemotactic mutant [4], an uncoordinated mutant [3], a mutant defective in taxis toward serine [9, 11] and a mutant defective in taxis toward aspartate [11]. The results which we describe here demonstrate that the response to serine and aspartate at concentrations of order 10^{-5} M is not an avoidance response; when cells swim down gradients of these amino acids their motion is indistinguishable from that in isotropic solutions; when they swim up the gradients they change direction less frequently.

In another communication [8] we discuss a solution to the diffusion equation which allows us to compute the concentration of attractant outside the mouth of a capillary and, thus, to follow cells in defined gradients.

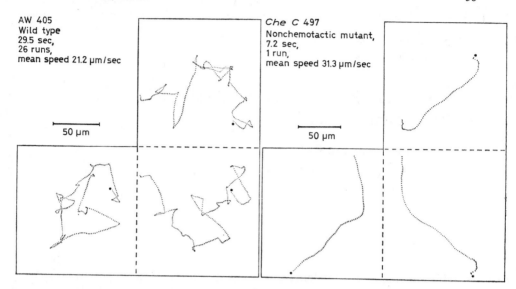

AW 405
Wild type
29.5 sec,
26 runs,
mean speed 21.2 μm/sec

50 μm

Che C 497
Nonchemotactic mutant,
7.2 sec,
1 run,
mean speed 31.3 μm/sec

50 μm

Fig. 1. Digital plots of the displacement of a wild-type bacterium, AW405, and a generally nonchemotactic mutant, *cheC* 497, at the rate of 12.6 words (data points) per second. Tracking began at the points indicated by the large dots. The plots are planar projections of 3-dimensional paths. If the left and upper panels of each figure are folded out of the page along the dashed lines, the projections appear in proper orientation on three adjacent faces of a cube. The cultures were grown in a minimal salts medium on glycerol, threonine, leucine and histidine, as described by HAZELBAUER *et al.* [9]. They were washed twice at 4 °C with a solution containing 10^{-2} M potassium phosphate (pH 7.0), 10^{-4} M ethylenediamine tetraacetate (EDTA) and 10^{-3} M magnesium sulfate and diluted at room temperature to an optical density of 0.01 (590 nm) in a solution containing 10^{-2} M potassium phosphate (pH 7.0), 10^{-4} M EDTA, and 0.18% (w/v) hydroxypropyl methyl cellulose (Dow Methocel 90 HG). They were tracked as such at 32.0° (viscosity 2.7 cp) in a tantalum and glass chamber 2 mm in diameter and 2 mm high.

Motion in Isotropic Solutions

The motion appears as an alternating sequence of intervals during which changes in direction are gradual or abrupt – we call these 'runs' and 'twiddles', respectively. Genetic and environmental differences in behavior are associated chiefly with the lengths of runs. Figure 1 illustrates this for the wild type and a nonchemotactic mutant. A number of results of a quantitative run-twiddle analysis are given in table I.

Runs are long in *cheC* 497, short in *unc* 602 and of intermediate length in AW405 (table I). Mutants able to respond to a restricted set of attractants

Table I. Run-twiddle analysis of mutants swimming in a homogeneous, isotropic medium

	Strain and type		
	AW405, wild type	*Unc*602, uncoordinated	*CheC*497, nonchemotactic
Number of bacteria tracked	35[1]	10	14
Total tracking time, min	20	3.0	2.7
Mean speed, μm/sec[2]	14.2±3.4	14.4±3.9	20.0±4.9
Mean twiddle length, sec[3]	0.14±0.19	0.14±0.24	0.10±0.13
Mean run length, sec	0.86±1.18	0.42±0.27	6.3±5.2
Mean change in direction from run to run, degrees	68±36	74±33	33±15
Mean change in direction during runs, degrees	23±23	18±23	35±22
Mean angular speed while twiddling, degrees/word[4]	56±29	54±27	41±32
Mean angular speed while running, degrees/word[4]	14±9	19±9	9±6

1 Experiments done with three different cultures.

2 The values are the means ±1 SD. In the calculation of the mean speed the mean for each bacterium is weighted equally, and the standard deviation is the standard deviation in the mean.

3 In this and in subsequent entries in the table each twiddle or run is weighted equally; the standard deviations are of the same order of magnitude as those found with a single bacterium.

4 The angular speed is the change in the direction of motion from one word (data point) to the next.

Data points (words) were generated at the rate 12.6/sec. The beginning of a run was scored if the angular speed (footnote 4) was less than 35°/word for three successive words. The end of a run was scored if the angular speed was greater than 35°/word for two successive words or if it was greater than 35°/word for one word provided, in the latter case, that the change in the average direction between successive pairs of words was also greater than 35°. The angular speed is sensitive to short-term fluctuations in the data. These depend on the ways in which the bacteria wobble and on the time constants (0.08 sec) of the circuits which precede the analog-to-digital converter. The time constants, the recording rate and the value 35°/word were chosen empirically by comparing results of digital analyses with plots of the kind shown in figure 1.

swim much like the wild type; the motion of the serine-blind mutant AW518 is essentially identical to that of AW405; the aspartate-blind mutant AW539 has somewhat shorter twiddles and somewhat longer runs (0.11 ± 0.18 and 1.3 ± 2.1 sec, respectively). The speed is nearly uniform during runs, but the bacteria slow down or stop on twiddling (fig. 2). The mean change in direction from the end of one run to the beginning of the next is less than 90° (table I). If the bacteria chose a new direction at random, the probability of an angle change between Θ and $\Theta + d\Theta$ would be $1/2 \sin\Theta \, d\Theta$, the mean value of Θ would be 90° and the standard deviation would be 39.2°. The distribution observed, however, is skewed toward small angles (fig. 3). If changes in direction were random, the skew would be toward large angles because the digital analysis ignores the smallest changes (table I, legend). Changes in direction also occur during runs (table I). The drift is about what one would expect from rotational diffusion: the root-mean-square angular deviation of a 2-μm diameter sphere occurring in t sec in a medium of viscosity 2.7 cp at 32°C is 29 $\sqrt{t}$ degrees.

The shortest twiddles and the shortest runs are the most probable (fig. 4). The distribution of twiddle lengths is exponential (fig. 4a). The distribution of run lengths is exponential for *unc* 602 (not shown) but only approximately so for AW405 (fig. 4b). If for AW405 one allows for variations in mean run length for different bacteria, the curvature in the semi-log plot of the aggregate run-length data vanishes (fig. 4c). From calculations of auto-correlation functions of sequences of twiddles and of sequences of runs we conclude that twiddles and runs of different length occur at random. The statistics are Poisson; for a given organism in a given isotropic environment the probability per unit time of the termination of a twiddle or the termination of a run is a constant.

The wild type is known to have chemoreceptors for serine, for aspartate and for a number of sugars [2]. If serine is added to suspensions of AW405 (no gradients), the run-length distributions remain exponential but shift dramatically toward longer runs (fig. 5); the twiddles are suppressed. Cal-culations of the autocorrelation functions indicate that runs of different length still occur at random. The shift does not occur with aspartate (fig. 5), even though the chemotactic responses to aspartate and serine are nearly the same [9, 11]. The shift due to serine involves the serine chemoreceptor; it does not occur in the serine-blind mutant AW518. The shift is not a metabolic effect, since it can be generated by a nonmetabolite sensed by the serine chemoreceptor (experiment done with α-amino isobutyric acid [11] and the aspartate-blind mutant AW539, which also shows the shift with

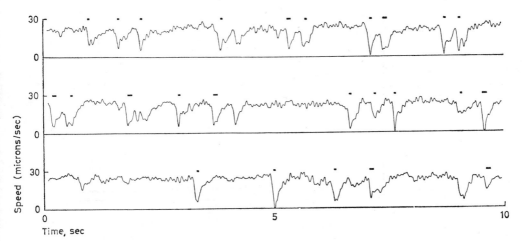

Fig. 2. The speed of the wild-type bacterium of figure 1 displayed by an analog monitor. The recording has been divided into three parts, each 9.8 sec long; the figure should be read from left to right and from the top down. Twiddles occurred during the intervals indicated by the bars. Note the consequent changes in speed. The longest run can be seen at the left end of the bottom trace. It appears in the upper panel of figure 1 angling downwards and slightly to the left, 5 runs from the end of the track. It is 45 words or 3.57 sec long.

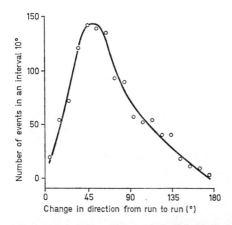

Fig. 3. Distribution of changes in direction from the end of one run to the beginning of the next for the wild-type bacteria of table I. The distribution was constructed from 1,166 events by summing the numbers falling in successive 10° intervals. If the analysis is confined to the shortest twiddles, the distribution is skewed even farther toward small angles (mean ± SD 62±26°).

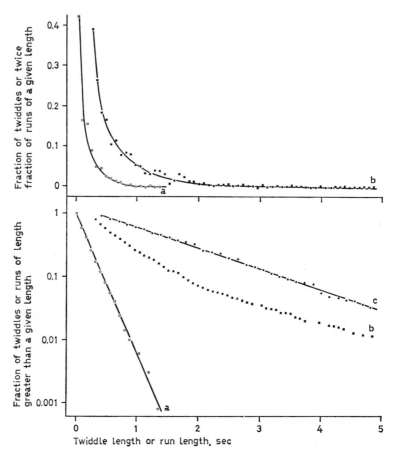

Fig. 4. Top: the fractional number of twiddles (a) or runs (b) of different lengths for the wild-type bacteria of table I. There were 1,201 twiddles and runs. All the twiddles are plotted, but runs longer than 5 sec are not (lengths 5.1, 5.4, 6.0, 6.4, 7.0, 7.1, 7.1, 7.2, 7.9, 7.9, 7.9, 12.9 and 24.2 sec). Bottom: the same data plotted as the logarithm of the fractional number of twiddles (a) or runs (b) of length greater than a given length [6]. Curve c was obtained by scaling the run lengths of each bacterium so that its mean run length was equal to the ensemble mean.

serine). It does not occur in the uncoordinated mutant *unc* 602. ADLER [2] notes that 'serine slightly inhibits chemotaxis toward all other attractants, and this inhibition remains unexplained'. If chemotaxis results from the suppression of twiddles (see below), serine should inhibit chemotaxis generally, provided that the mutants tested have a functional serine receptor.

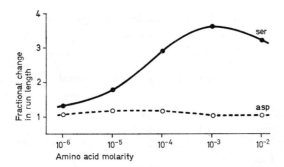

Fig. 5. Changes in mean run length caused by serine (•) or aspartate (o). The bacteria (AW405) were diluted in the tracking medium (fig. 1, legend) to an absorbance of 0.02 (590 nm). Aliquots of this suspension were mixed with equal volumes of tracking medium containing L-serine or L-aspartate (Calbiochem A grade). Controls were made by omitting the amino acids. The ratios of the mean run lengths observed in the presence and in the absence of the amino acid are plotted as a function of concentration.

Serine has other effects on the motion of the wild type which are less dramatic but which have a similar concentration dependence. The mean speed increases by about 40%; the mean twiddle length, the mean change in direction from run to run and the mean angular speed while running all decrease by about 40%. With aspartate there is a slight increase in speed, but the other changes are not significant.

Motion in Gradients

Gradients were generated by diffusion of attractants from capillary tubes of the kind used by ADLER [2], which we inserted through a flat side wall of the tracking chamber. In preliminary experiments we found the response (the number of bacteria entering a capillary in 1 h) to be negligible when bacteria were used at an optical density of 0.1 (590 nm; about 10^8 bacteria/ml); the clouds of bacteria which accumulated near the mouths of the capillaries sank. At optical densities of order 0.01, the response to an attractant was proportional to the optical density, and the dependence on the concentration of the attractant was similar to that described by MESIBOV and ADLER [11] (experiments at 32° with serine, aspartate, α-amino iso-butyric acid and α-methyl-DL-aspartate, all in tracking medium).

Table II. Run-twiddle analysis of the wild type swimming in gradients

Attractant	Serine		Aspartate	
	control	gradient	control	gradient
Number of bacteria tracked	11	34	23	24
Total tracking time, min	7.1	14.8	11.1	11.0
Mean run length, sec	0.83 ± 0.88	1.67 ± 2.56	0.83 ± 0.90	0.90 ± 1.56
Mean concentration of attractant, μM	0	9.5 ± 2.7	10	8.4 ± 2.0
Mean distance from mouth of capillary, μm	–	577 ± 112	–	644 ± 88
Mean value of $(\delta C/\delta r)/C$, mm^{-1}	–	2.5 ± 0.4	–	2.4 ± 0.4

The cells were prepared as described in the legend of figure 1, except that the suspension used for the aspartate control also contained 10^{-5} M aspartate. Capillaries were used in the gradient experiments but not in the controls. For each data point we computed the distance from the mouth of the capillary to the bacterium being tracked (r), the angle between the direction of motion of the bacterium and the gradient (the 'inclination', 0° for motion radially down the gradient), the concentration of the attractant at the bacterium [C, as defined by equation (4) from ref. 8, with $r_c = 0.01$ cm, $C_0 = 2.0 \times 10^{-3}$ M, $D_{serine} = 1.0 \times 10^{-5}$ cm^2/sec and $D_{aspartate} = 0.89 \times 10^{-5}$ cm^2/sec], the steepness of the gradient at the bacterium $(\delta C/\delta r)$, the time rate of change of the concentration at the bacterium (dC/dt), and the logarithmic derivatives $(\delta C/\delta r)/C$ and (dC/dt)/C. Mean speeds, twiddle lengths, changes in direction and angular speeds are not shown; the values are essentially identical to those for AW405 (table I).

Results of the run-twiddle analysis for the gradient experiments are given in tables II and III. Runs are longer in the gradients (table II) than we would expect from the concentration dependence (fig. 5). This is true for runs which move the bacteria up the gradient but not for runs which move them down the gradient (table III). The differences in the up-gradient and down-gradient data are dramatic when the run-length distributions are examined (fig. 6). For serine, the distribution of runs down the gradient (fig. 6b) is similar to the distribution in a 9-μM isotropic solution; for aspartate, it is indistinguishable from the control. From the data of figure 6 and from calculations of autocorrelation functions of sequences of runs (not separated into subsets), we conclude that the statistics are still Poisson. When a

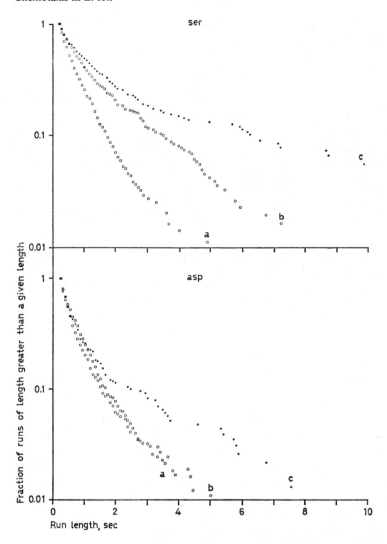

Fig. 6. The data from the serine (top) and the aspartate (bottom) experiments (table II, III) plotted as the logarithm of the fractional number of runs of length greater than a given length. a=Runs in the control experiment; b=runs down the gradient; c=runs up the gradient.

Table III. Analysis of runs which move the bacteria up the gradient or down the gradient

	Attractant			
	serine	serine	aspartate	aspartate
Net displacement of runs	up	down	up	down
Mean concentration, μM	10.0 ± 2.8	9.2 ± 2.6	8.8 ± 1.9	8.1 ± 1.9
Mean run length, sec	2.19 ± 3.43	1.40 ± 1.88	1.07 ± 1.80	0.80 ± 1.38
Mean run length expected from the control run length (table II) and the concentration dependence (fig. 5), sec	1.48	1.45	0.82	0.82

The runs of the gradient experiments (table II) divided into two subsets according to whether the net displacement of a run is toward or away from the mouth of the capillary (up-gradient or down-gradient). The mean speed was only slightly larger for runs up the gradient than for runs down the gradient (2% for serine, 7% for aspartate).

bacterium moves up the gradient the probability per unit time of the termination of a run decreases; when it moves down the gradient the probability reverts to the value appropriate to an isotropic solution of similar concentration. At the concentrations we have studied, the stimulus is sensed and acted upon only when the bacterium swims up the gradient.

Further proof of the assertion that motion away from the capillary is not sensed can be obtained from computation by linear regression [13] of the correlation between the length of a run and the mean value over the run of dC/dt, $(dC/dt)/C$, $-\delta C/\delta r$, or $-(\delta C/\delta r)/C$. These correlations are all positive (correlation coefficients of order 0.12 ± 0.02). If the analysis is confined to runs which move the bacteria up the gradient, the correlation coefficients are larger (of order 0.19 ± 0.03); if it is confined to runs which move the bacteria down the gradient, the coefficients are statistically insignificant (-0.02 ± 0.02). This implies that when the bacterium swims down the gradient there is no functional relationship between the length of a run and the derivatives of the concentration with respect to space or time. This is true both for serine and aspartate.

There is nothing in our data to suggest that the bacteria are able to steer in the direction of the gradient while running or that the motion is topotactic [17]. When they twiddle, the change in direction is still biased toward small angles, as in figure 3, but the angle chosen does not depend on the direction of the gradient; there is no correlation between the inclination at the end of a run (table II, legend) and the change in direction from run to run. Nor is there any correlation between the length of the run and the change in direction.

An accurate calculation of the mean rates at which the bacteria drift up the gradients could be made from the information in tables II and III if we knew the functional dependence of the mean run lengths (for runs up the gradient) on inclination; the data at hand are inadequate. If we assume that the run-length bias is proportional to $\cos \Theta$ ($90° < \Theta < 180°$), the drift rate in serine is about 2.0 μm/sec, and the drift rate in aspartate is about 0.9 μm/sec. The value for serine is in rough agreement with that obtained by DAHLQUIST *et al.* [7] for suspensions of *Salmonella* in exponential gradients of comparable steepness.

Mechanisms

When a bacterium runs, its flagellar filaments work together in a bundle of the kind photographed in *E. coli* by RAMSEY and ADLER [15a] or in *Salmonella* by MITANI and IINO [12]. When it twiddles the bundle probably loosens or comes apart. When the bundle reforms, the cell goes off in a new direction. The direction chosen depends on the change in orientation of the bundle relative to the body of the cell. Smaller changes in direction require smaller changes in orientation and occur in shorter periods of time (fig. 3). The stability of the bundle is improved by interaction with chemoreceptors. The association of an attractant with a receptor increases the stability even more. If the attractant is serine, the stability of the bundle is affected by both the average level of association (fig. 5) and the rate at which it increases (fig. 6). If it is aspartate, only the rate of increase is important (fig. 6).

We do not know what the molecular structure of the twiddle generator is or how it is able to perturb the flagellar bundle. We do know it operates on Poisson statistics (fig. 4) and that its firing rate can be suppressed by chemoreception. It is likely that the generator is built from elements which are missing or defective in generally nonchemotactic mutants. When the generator and the chemoreceptors are uncoupled, the generator runs free and the mutants are uncoordinated.

Summary

A microscope which automatically follows the motion of individual bacteria has been used to study the behavior of *E. coli*. The motion of the wild type appears as an alternating sequence of intervals, called 'runs' and 'twiddles', during which changes in direction are gradual or abrupt. The changes in direction which occur during runs can be explained by rotational diffusion, but those which occur during twiddles cannot. When a bacterium twiddles, it chooses a new direction at random, but there is an internal bias favoring small angles. The lengths of runs and twiddles are governed by Poisson statistics. In isotropic solutions, twiddles are suppressed by serine (if the mutant is chemotactic toward serine) but not by aspartate. In spatial gradients, twiddles are suppressed when the bacteria happen to swim toward higher concentrations of serine or aspartate. The speeds and the magnitudes of the directional changes are not affected.

Acknowledgements

We thank JULIUS ADLER and MARGARET DAHL for mutants and instruction in their handling. PFEFFER assays were done by SUSAN MACFADDEN. This research was supported by grants from the Research Corporation and the National Science Foundation. This paper was reproduced with the kind permission of the editor and publishers of Nature, Lond. *239:* 500 (1972).

Addendum

We describe more fully here the methods used for collecting data, the steps involved in the run-twiddle analysis, including those designed to minimize errors due to Brownian motion, the criteria by which we judge that twiddle generation is a Poisson process, and the statistical methods used to compare runs which carry bacteria up and down gradients. We conclude from a more rigorous analysis than our initial one that the response to negative gradients, although substantially smaller than that to positive gradients, is nonetheless finite.

Data Acquisition

The microscope is described in detail elsewhere [BERG, 1971, henceforth B71]. The signals which provide a measure of the position of a bacterium (the x, y and z *readout* voltages, B71, fig. 1, 5) are amplified by a factor of 10 with circuits having time constants $RC = 0.08$ sec. The amplified signals are displayed by an analog monitor and converted to digital form for storage on computer compatible magnetic tape. The monitor computes the x, y and z velocities and the square root of their sum of squares [the speed, see BERG and BROWN, 1972, henceforth BB72, fig. 2]. The digital system generates data points in the form of 60-bit words (12 bits each for x, y, z and an environmental parameter, 4 for a time digit, and 8 for 8 event markers). The conversion is done with a multiplexer and a single analog-to-digital converter (Burr Brown 4047/43 and ADC 30-12N-BTC); it is complete in 2.4×10^{-4} sec, so the bits are stored in latches until they can be written on magnetic tape. This is done with an incremental recorder (Digi Data 1357–200). As many as 50 words can be written per second, but 12.6 has proved to be adequate for work with *E. coli*. A separate string of words (a record) is written for each bacterium tracked. These are read and analyzed off-line by a Control Data Corporation 6400 computer. Conversions of the x, y and z scales from volts to microns are made with factors determined empirically. The z-axis is a special case, because motion in this direction (the direction of focus) changes the thickness of the medium between the objective and the bacterium being tracked; the displacement of the bacterium is approximately n times larger than the displacement of the chamber, where n is the index of refraction of the medium (1.33). Organisms are maneuvered to the sensitive region of the field by *search* signals [B71, fig. 1, 5] generated by a 3-axis joy stick designed for model airplane control (Royal Electronics, Denver, Colorado, Model KHP-3R). The data aquisition system switches on automatically when the tracking begins [B71, p. 896].

Run-Twiddle Analysis

We compute the x, y and z velocities from the second-order central difference equation [FRÖBERG, 1965]:

$$\frac{dx(J)}{dt} = \frac{2}{3T}[x(J+1)-x(J-1)] - \frac{1}{12T}[x(J+2)-x(J-2)],$$

where x(J) is x for word J, and T is the time interval between words J and J+1. From the velocity components we compute for each word the speed and the direction cosines (the length of the velocity vector and the projections on the x, y and z axes of a unit vector in the direction of the velocity, respectively). From the latter we compute the angle by which the bacterium changes its direction from word J to word J+1; we call this the angular speed [BB72, table I]. The beginning of a run is scored at word J if the angular speed is less than 35°/word for words J, J + 1 and J + 2. The end of a run is scored at word K *either* if the angular speed is greater than 35°/word for words K and K + 1 *or* if it is greater than 35°/word for word K alone, provided in the latter case that the difference between the directions of the vector sums of the velocities for words K-1 and K and words K + 1 and K + 2 is also greater than 35°; we ignore 1-word transients. In our more recent work we have made this condition more stringent by requiring that the change in direction between vectors obtained by summing successive 3-word sets be greater than 35° (see below). The shortest run is three words long, but twiddles can be of any length, including 0.

The changes in direction which occur from run to run or during runs are computed from the vectors obtained by summing the velocity vectors of the first three and the last three words of a run, respectively. Similarly, the values of scaler functions are defined at the beginning and at the end of a run in terms of their mean values for the first three and the last three words, respectively. We include the first and the last runs of a record even though the tracking often begins or ends in the middle of these runs; thus, the analysis tends to underestimate the mean run length. The error is conspicuous only in tracking *cheC* 497, where many organisms are lost before they twiddle. Cells which swim for long distances in the same general direction evade capture more easily and outrun the tracker more often than those which change direction frequently; nevertheless, we have observed single runs as long as 60 sec. Every culture contains a number of cells which hardly move; these we ignore. They also contain some which swim so rapidly they are impossible to track; these we do without. As a general rule we discard records shorter than about 7 sec, and we rarely follow a given organism for more than about 1 min.

The angular speed is sensitive to short-term fluctuations in the data, including those due to Brownian motion. These fluctuations become more severe as the time constants of the circuits which precede the analog-to-digital converter are decreased. The smallest useful time constant (the largest useful recording rate) depends both on the speed and the diffusion constant of the bacterium. Consider a bacterium swimming from left to right at a constant speed, s (fig. 7). After an interval of time, t, it will have swum a distance, st, but it also will have moved by Brownian motion a root-mean-square distance $\sqrt{6Dt}$, where D is its diffusion constant. It is likely that it will have changed direction by an angle θ as large as $\sqrt{6D/s^2t}$. For $D = 0.83 \times 10^{-9}$ cm²/sec (computed from Stoke's law for a 2-μm diameter sphere in a medium

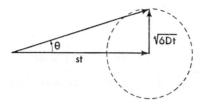

Fig. 7. A bacterium swims from left to right at speed, s, for a period of time, t. Because of Brownian motion, the chance is only 0.68 that it will be found within the dashed sphere, radius $\sqrt{6Dt}$.

of viscosity 2.7 cp at 32 °C) and t = 0.08 sec fluctuations as large as $\Theta = 35°$ will be common if the speed is less than about 3.6 μm/sec. At speeds greater than this, values of the angular speed in excess of 35°/word will occur mainly as 1-word transients, and most of these will be ruled out by demanding that the change in direction between successive 2-word or 3-word vectors be greater than 35°.

These expectations have been confirmed experimentally by tracking nonmotile *E. coli* (flagellated cells paralyzed by exposure to the sulfhydryl reagent *p*-chloromercuriphenyl sulfonate) while superimposing drift velocities electrically (by adding currents to the summing junctions of the amplifiers which precede the analog-to-digital converter). The results are summarized in table IV. At speeds ≥ 6 μm/sec the number of runs is reduced dramatically by the stratagem of ruling out 1-word transients in the angular speed when the change in direction between the corresponding successive 2-word or 3-word vectors is less than 35°. We have recomputed the tracking data for AW405 [BB72, table I] using the 3-word rather than the 2-word scheme: only 50 of the 1,200 previously acceptable twiddles were rejected; the results listed in the table [BB72, table I] changed by at most $\pm 5\%$. We conclude that the analysis for motile bacteria is not significantly biased by Brownian motion.

Table IV includes twiddles which arise from digitizing errors. These can be demonstrated experimentally by recording just the electrically simulated drift velocities. When the x, y and z drift velocities are constant and of equal magnitude, twiddles occur at speeds of about 4–5 μm/sec; they are common at speeds of about 3 μm/sec. The reason for this is apparent if one replaces the sphere shown in figure 7 by a cube 2 bits (0.4 μm) on a side (2 bits because the ends of the vector, st, are each uncertain by 1 bit). Digitizing errors can be reduced by using a 14 or 16 bit analog-to-digital converter, or by digitizing the velocities computed by the analog monitor; the uncertainties

Table IV. Run-twiddle analysis of paralyzed bacteria with drift velocities added electrically

Mean speed μm/sec	Mean twiddle length, sec			Mean run length, sec		
	none[1]	2-word[1]	3-word[1]	none[1]	2-word[1]	3-word[1]
11.0	0	0	0	15	>30	>30
8.9	0.01	0.02	0.02	3.5	7.8	10.5
6.3	0.08	0.09	0.09	0.94	1.5	4.1
4.4	0.24	0.22	0.14	0.35	0.39	0.41

1 Scheme used to reject transients in the angular speed.
Data at 12.6 words/sec, RC=0.08 sec, value of the angular speed used to distinguish a run from a twiddle 35°/word, transients in the angular speed accepted or rejected in accord with the magnitude in the change in direction of successive 2-word or 3-word vectors.

due to Brownian motion are more fundamental. Some *E. coli* run smoothly in the direction of their long axes [as did the bacterium shown in figure 1, BB72] while others tumble end-over-end or shake from side to side. Wobbles of this kind are less apparent at higher viscosities where, surprisingly, the bacteria swim more rapidly [about 30% faster at 2.7 cp than at 0.8 cp; see also SHOESMITH, 1960]. Since the diffusion constants also are smaller at higher viscosities, the experiment can be improved by the addition of methyl cellulose [BB72, fig. 1 legend].

As noted earlier [BB72, table I legend], the criterion for the success of the digital analysis is a comparison of the computer output with plots of the trajectories. This criterion is less subjective than might be expected, because the twiddles are quite distinctive. As discussed before, the changes in direction which occur during runs can be explained adequately by rotational diffusion, whereas those which occur during twiddles cannot. In other words, when the bacteria run, they swim about as straight as they can; when they twiddle, they alter their course actively and abruptly. We have tracked a number of *E. coli* using different time constants and different recording rates, and we have run the digital analysis with different threshold values for the angular speed. The mean run length is the parameter most sensitive to these changes (e.g., doubling in one experiment when the time constant was increased from RC = 0.08 to 0.16 sec); however, distributions of changes in direction from run to run remain skewed toward smaller angles [BB72, fig. 3; the mean shifting in the experiment cited from 66 to 64°], and the distributions of run lengths and twiddle lengths remain Poisson [BB72, fig. 4].

Distribution of Run Lengths

If an event is rare (if the probability of its occurrence during an observation of length, dt, is small) and random (if the probability of its occurrence is the same regardless of when the observation is made), then the intervals between successive events will be distributed in accord with the Poisson interval distribution [PARRATT, 1971]; the frequency distribution of the lengths of the intervals will be exponential; the order of their occurrence will be random. The tracking data meets both these criteria.

The frequency distribution is

$$f_0(t) = \frac{1}{\tau} \exp(-t/\tau), \ t \geq 0,$$

where t is the interval between successive events, and $1/\tau$ is the probability per unit time that the event will occur. The mean interval length, $\bar{t}$, and the root-mean-square deviation from the mean,

$$\sigma = \left[\overline{(t-\bar{t})^2} \right]^{1/2} \tag{1}$$

are both equal to τ. If one only is able to measure time intervals of length greater than t_0, the frequency distribution is given by

$$f(t) = \frac{1}{\tau} \exp[-(t-t_0)/\tau], \ t \geq t_0. \tag{2}$$

The fraction of intervals of length greater than t is

$$g(t) = \int_t^{\infty} f(t')dt' = \exp[-(t-t_0)/\tau], \ t \geq t_0.$$

A semilog plot of g(t) as a function of t is linear with slope $-1/\tau$:

$$\ln g(t) = -t/\tau + t_0/\tau, \ t \geq t_0. \tag{3}$$

An unbiased estimate for τ [KENDALL and STEWART, 1967, p. 5] is

$$\tau = \bar{t} - t_0, \tag{4}$$

where $\bar{t}$ is the experimental mean:

$$\bar{t} = \frac{1}{n} \sum_{i=1}^{n} t_i. \tag{5}$$

Here n is the number of runs for a given bacterium and the t_i are their lengths. In practice, t_0 is set by the time required for an observation (for runs, 0.24 sec or 3 digital words, a limit imposed by the run-twiddle analysis).

Runs observed for individual chemotactic bacteria tracked for comparatively long periods of time fit equation (3) fairly well (fig. 8); however, τ differs from individual to individual (fig. 9). As a result, it is not possible to fit equation (3) simply by pooling the data from different bacteria; this is shown in BB72, figure 4, curve b. Since too few runs were observed for most bacteria for us to test equation (3) very accurately, we devised the following procedure for handling the aggregate data. Consider for each bacterium, j, the frequency distribution, f(x), of the parameter

$$x \equiv (t-t_0)\bar{\tau}/\tau_j,$$

where

$$\bar{\tau} = \frac{1}{J} \sum_{j=1}^{J} \tau_j, \tag{6}$$

with J the total number of bacteria and τ_j the value of τ for the jth bacterium, as defined by equations (4) and (5). Since

$$f(x) = f(t)dt/dx = \frac{1}{\bar{\tau}} \exp(-x/\bar{\tau}), \, x \geq 0,$$

the fraction of intervals of length greater than or equal to x, g(x), is $\exp(-x/\bar{\tau})$, and

$$\ln g(x) = -x/\bar{\tau}, \, x \geq 0;$$

therefore, the semilog plot of g(x) as a function of x should be the same for all the bacteria. If the normalized run lengths, x, are pooled, the plot will be linear provided the distributions of the run lengths, t, are exponential for the bacteria individually, equation (2). The result is shown in BB72, figure 4, curve c (offset to the right a distance t_0).

If the order of occurrence of different intervals is random, the length of a given interval should not depend on the lengths of the intervals which precede it. This idea is expressed mathematically by the requirement that the autocorrelation coefficients of lag k, ϱ_k, be zero for all $k \geq 1$ [KENDALL

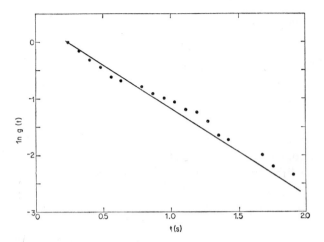

Fig. 8. The natural logarithm of the fraction of runs of length greater than t as a function of t for one bacterium (AW405, 69.0 sec, 73 runs, mean speed 19.9 μm/sec). The plot includes all but the 6 longest runs. The solid line is a graph of equation (3) with $t_0 = 0.24$ sec and $\tau = 0.65$ sec, as computed from equations (4) and (5). The first 29.5 sec of the track of this particular bacterium is shown in BB72 (fig. 1).

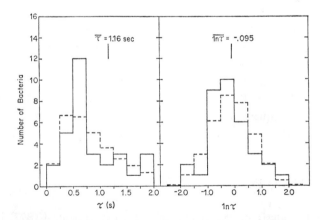

Fig. 9. The distribution of mean run lengths, τ, for the 35 wild-type bacteria of BB72 (table I) plotted on a linear and a logarithmic scale. Solid lines = experiment; dashed lines = the best log-normal fit, equation (8), with $\lambda = \exp(\overline{\ln\tau}) = 0.91$ and $\delta = (\sigma_{\ln\tau})^{-1} = 1.29$, where $\overline{\ln\tau}$ is the experimental mean and $\sigma_{\ln\tau}$ is the experimental SD for the log distribution. Three bacteria (not shown) were observed with $\tau \geq 2.0$ sec; the number expected from the log-normal fit is 4.35.

and STEWART, 1966, p. 404]. The autocorrelation coefficients are defined by

$$\varrho_k = \gamma_k / \sigma^2$$

with

$$\gamma_k = \overline{[(t_i - \bar{t}) (t_{i+k} - \bar{t})]},$$

where t_i is the ith interval, $\bar{t}$ is the mean interval and σ is the root-mean-square deviation from the mean, equation (1). The autocorrelation coefficient of lag k for the jth bacterium, $(\varrho_k)_j$, can be estimated by

$$(\varrho_k)_j = \frac{1}{\sigma_j^2 (n_j - k)} \sum_{i=1}^{n_j - k} (t_{i,j} - \bar{t}_j) (t_{i+k,j} - \bar{t}_j).$$

If the sequence of intervals is random, this estimate is biased [KENDALL and STEWART, 1966, pp. 432–434]; its mean is $-(n_j-1)^{-1}$ and its standard deviation is $(n_j-k)^{-\frac{1}{2}}$. The data from different bacteria can be combined by averaging the $(\varrho_k)_j$ over j, weighting each term inversely as the square of its standard deviation. For random sequences, one càn then show that the kth coefficient has a mean value

$$(N-kJ)^{-1} \left[-J + (k-1) \sum_{j=1}^{J} (n_j-1)^{-1} \right]$$

and a standard deviation $(N-kJ)^{-\frac{1}{2}}$, where N is the total number of runs.

Assuming that runs of different length occur at random, we expect the autocorrelation coefficients for the sequences of runs observed for the wild-type bacteria of BB72, table I, to be of order -0.04 ± 0.03. The computed coefficients ($1 \leq k \leq 10$) are all within about 1 SD of the expected values, so we conclude that the sequences are, indeed, random. Thus, by all the criteria of which we are aware, the generation of twiddles is a Poisson process.

Dependence of Run Lengths on Gradients

Our original analysis of the spatial gradient experiments [BB72] included calculations by linear regression of the correlation between the lengths of runs and the values of the spatial or temporal gradients averaged over those runs. We concluded that a positive spatial or temporal gradient decreased the rate of twiddling and that negative gradients had no observable effect. This analysis now appears to us inadequate, because the equations for linear correlation are strictly interpretable only when the variables are normally distributed [KENDALL and STEWART, 1967, p. 288]. As we have seen, run lengths are exponentially distributed, equation (2). Consider the case in which the decay constant, τ, is related to the variable, x, exactly by $\tau = ax + b$, where a and b are constants. In the limit that the amount of data is large, the correlation coefficient for t and x can be shown to approach the maximum value

$$\left[2 + \left(\frac{\bar{t} - t_0}{a\sigma_x} \right)^2 \right]^{-1/2},$$

where $\bar{t}$ is defined by equation (5) and σ_x is the standard deviation of x. For runs which move the bacteria up the gradient this number is of order 0.3, about 1.5 times larger than the values observed [BB72]. Although the probability that a value as large as 0.2 could arise by chance seems small, we do not know how to estimate this probability.

Fortunately, there exists a more general method by which the responses of each bacterium to different stimuli can be compared, the method of maximum likelihood [KENDALL and STEWART, 1967, pp. 1–46]. In this analysis we look for a dependence of the decay time, $\bar{\tau}$, on the temporal gradient, dC/dt, of the form

$$\bar{\tau} = a(dC/dt) + \tau_0, \tag{7}$$

where τ is defined by equation (6), a and τ_0 are constants, and dC/dt is the time rate of change of the concentration of the attractant as seen by the bacterium [BB72, table II legend]. For each bacterium we plot the lengths of the runs as a function of the values of dC/dt averaged over those runs; the abcissae range from $-(dC/dt)_{max}$ to $+(dC/dt)_{max}$, where $(dC/dt)_{max}$ is the absolute value of the largest gradient encountered by any bacterium. We divide the abcissae into 6 equal segments. From the data points in the ith segment of the jth graph we compute the decay time τ_{ij}, equations (4) and (5), and the mean gradient dC_{ij}/dt; here i ranges from 1 to 6, j from 1 to J, where J is the total number of bacteria. We then assume that for any given value of dC/dt the frequency distribution of the decay times for the different bacteria, $h(\tau)$, is lognormal, i.e., that

$$h(\tau) = \frac{\delta}{\tau\sqrt{2\pi}} \exp[-\tfrac{1}{2} (\delta \ln \tfrac{\tau}{\lambda})^2], \tag{8}$$

where δ and λ are parameters which may depend on dC/dt. The log-normal distribution fits the data reasonably well in every case we have examined,

Table V. Maximum likelihood test of the model $\bar{\tau} = a(dC/dt) + \tau_0$

	Attractant					
	serine	serine	serine	aspartate	aspartate	aspartate
Net displacement of runs	up and down	up	down	up and down	up	down
a, sec²/μM	1.7±0.3	8.3±3.4	1.5±0.4	2.9±0.7	6.3±3.0	3.0±0.7
τ_0, sec	2.8±0.4	1.9±0.3	2.5±0.6	1.2±0.2	0.8±0.2	1.2±0.2
Probability that a=0	0.001	0.02	0.03	0.0004	0.04	0.005
Probability that $a_{up} = a_{down}$			0.06			0.27

The ± values are standard deviations.

e.g. for AW405 in the absence of attractants (fig. 9) and in the presence of different amounts of serine [at each concentration studied in the experiment of BB72, fig. 5]. Since from equation (8) $\bar{\tau} = \lambda \exp(1/2\delta^2)$, equation (7) can be used to express λ in terms of a(dC/dt), τ_0 and δ. The likelihood function is constructed from the product of 6J factors h(τ) evaluated at the points τ_{ij}, dD_{ij}/dt; the best (or ML) values of a, τ_0 and δ are those which make this function maximum.

We have computed the ML values of a and τ_0 (and their standard deviations) for the data of the gradient experiments [BB72], testing two models: (1) a model in which equation (7) is used to fit runs which carry the bacteria up as well as down the gradient, or (2) a model in which equation (7) is used to fit the up data and the down data separately. In addition, we have computed from the likelihood ratio [KENDALL and STEWART, 1967, pp. 224–231] the probability that a = 0 or $a_{up} = a_{down}$. The results are given in table V. We conclude that the bacteria respond more strongly to positive gradients than to negative gradients and that the asymmetry is larger for serine than for aspartate. The conclusion drawn from the earlier more rudimentary analysis [BB72] that when a bacterium swims down a gradient there is no functional relationship between the length of a run and the derivatives of the concentration with respect to space or time is probably incorrect.

Some time ago we began experiments in which the attractant glutamate – glutamate and aspartate use the same receptor [MESIBOV and ADLER, 1972] –

is generated (or destroyed) in an *isotropic* solution by the action of the enzyme alanine amino transferase (EC 2.6.1.2). Responses to positive gradients are substantially larger than those to negative gradients [BROWN and BERG, 1974]. In these experiments the bacteria are never exposed to spatial gradients; therefore, a temporal stimulus is sufficient to induce a response. We hope to learn whether the responses observed in spatial gradients can be explained fully by temporal effects by tracking bacteria in spatial gradients of glutamate under equivalent conditions. Models suggested by the temporal experiments [see also MACNAB and KOSHLAND, 1972] can be tested by integrating the spatial data word by word.

References to Addendum

BERG, H. C.: How to track bacteria. Rev. Sci. Instr. *42:* 868 (1971).

BERG, H. C. and BROWN, D. A.: Chemotaxis in *Escherichia coli* analysed by three-dimensional tracking. Nature, Lond. *239:* 500 (1972) and Antibiotics and Chemotherapy, vol. 19, pp. 55–66 (Karger, Basel 1974).

BROWN, D. A. and BERG, H. C.: Temporal stimulation of chemotaxis in *Escherichia coli*. Proc. nat. Acad. Sci., Wash. *71:* 1388 (1974).

FRÖBERG, C.-E.: Introduction to numerical analysis, p. 168 (Addison-Wesley, Reading 1965).

KENDALL, M. G. and STUART, A.: The advanced theory of statistics, vol. 3 (Griffin, London 1966).

KENDALL, M. G. and STUART, A.: The advanced theory of statistics, 2nd ed., vol. 2 (Hafner, New York 1967).

MACNAB, R. M. and KOSHLAND, D. E.: The gradient-sensing mechanism in bacterial chemotaxis. Proc. nat. Acad. Sci., Wash. *69:* 2509 (1972).

MESIBOV, R. and ADLER, J.: Chemotaxis toward amino acids in *Escherichia coli*. J. Bact. *112:* 315 (1972).

PARRATT, L. G.: Probability and experimental errors in science, p. 228 (Dover, New York 1971).

SHOESMITH, J. G.: The measurement of bacterial motility. J. gen. Microbiol. *22:* 528 (1960).

General References

1   ADLER, J.: Chemotaxis in bacteria. Science *153:* 708 (1966).
2   ADLER, J.: Chemoreceptors in bacteria. Science *166:* 1588 (1969).
3   ARMSTRONG, J. B.: Chemotaxis in *Escherichia coli;* thesis, Wisconsin (1968).
4   ARMSTRONG, J. B.; ADLER, J., and DAHL, M. M.: Nonchemotactic mutants of *Escherichia coli*. J. Bact. *93:* 390 (1967).
5   BERG, H. C.: How to track bacteria. Rev. Sci. Instr. *42:* 868 (1971).

6   BLANC-LAPIERRE, A. and FORTET, R.: Theory of random functions, vol. 1 p. 146 (Gordon & Breach, New York 1965).

7   DAHLQUIST, F. W.; LOVELY, P., and KOSHLAND, D. E., jr.: Quantitative analysis of bacterial migration in chemotaxis. Nature New Biol. *236:* 120 (1972).

8   FUTRELLE, R. P. and BERG, H. C.: Specification of gradients used for studies of chemotaxis. Nature, Lond. *239:* 517 (1972).

9   HAZELBAUER, G. L.; MESIBOV, R. F., and ADLER, J.: *Escherichia coli* mutants defective in chemotaxis toward specific chemicals. Proc. nat. Acad. Sci., Wash. *64:* 1300 (1969).

10 JENNINGS, H. S. and CROSBY, J. H.: Studies on reactions to stimuli in unicellular organisms. VII. The manner in which bacteria react to stimuli, especially to chemical stimuli. Amer. J. Physiol. *6:* 31 (1901).

11 MESIBOV, R. and ADLER, J.: Chemotaxis toward amino acids in *Escherichia coli.* J. Bact. *112:* 315 (1972).

12 MITANI, M. and IINO, T.: Electron microscopy of bunched flagella of the curly mutant of *Salmonella abortivoequina.* J. Bact. *90:* 1096 (1965).

13 PARRATT, L. G.: Probability and experimental errors in science, p. 129 (Wiley, New York 1961; Dover, New York 1971).

14 PFEFFER, W.: Locomotorische Richtungsbewegungen durch chemische Reize. Untersuch. bot. Inst., Tübingen *1:* 363 (1884).

15 PFEFFER, W.: Ueber chemotaktische Bewegungen von Bacterien, Flagellaten und Volvcineen. Untersuch. bot. Inst., Tübingen *2:* 582 (1888).

15a RAMSEY, S. W. and ADLER, J.: Personal communication (1970).

16 ROTHERT, W.: Beobachtungen und Betrachtungen über tactische Reizerscheinungen. Flora *88:* 371 (1901).

17 WEIBULL, C.: Movement; in GUNSALUS and STANIER The bacteria, vol. 1, p. 153 (Academic Press, New York 1960).

Authors' address: Dr. HOWARD C. BERG and DOUGLAS A. BROWN, Department of Molecular, Cellular and Developmental Biology, University of Colorado, *Boulder, CO 80302* (USA)

Antibiotics and Chemotherapy, vol. 19, pp. 79–93
(Karger, Basel 1974)

Mathematical Aspects of Bacterial Chemotaxis

Evelyn Fox Keller[1]

Division of Natural Science, State University of New York, College at Purchase, Purchase, N.Y.

I. Introduction

Although mathematical analysis of biological problems has a long tradition, many biologists remain dubious of the relevance and ultimate value of such an approach. With the extraordinary successes in experimental biology of the past 20 years, the scepticism of biologists has, if anything, increased. Such scepticism is philosophically based on the fact that mathematical analysis is never, in itself, capable of the direct insights into real systems that experiments can achieve – can never tell us what a biological or physical system actually does. Nevertheless, experimentally informed mathematical analysis is frequently capable of syntheses and kinds of understanding not possible with experiment alone, as is well attested by our experience in physics. While such successes in biology may have been few in the past, the belief is growing among some researchers that biology is now reaching the stage in which theoretical efforts are appropriate. I therefore welcome the opportunity of describing some limited successes that have been achieved through mathematical analysis of a particular biological problem – that of bacterial chemotaxis.

The phenomenon of bacterial chemotaxis first came to my attention several years ago through the work of J. Adler and his associates [1–5]. His careful quantitative description of migrating or chemotactic bands in *Escherichia coli* offered a compelling invitation for a theoretical description, if not explanation, of a phenomenon which may be the simplest yet known

---

1   I would like to thank Joseph B. Keller for several helpful discussions.

expression of sensory perception. In response to this challenge, Lee Segel and I attempted to formulate a set of simple equations from which we could deduce the existence and properties of chemotactic bands [11]. My discussion therefore begins with a brief summary of this analysis.

II. Travelling Bands of Chemotactic Bacteria

In Adler's experiments [1–5], motile cells of *E. coli* were placed at one end of a closed capillary tube filled with a medium capable of supporting motility, although not necessarily growth. In time, a sharp band of cells could be seen moving away from the starting point at constant speed. Under certain conditions, the first band was followed by a second band. The explanation of this phenomenon is that bacteria consume some substrate (oxygen in the case of the first band, and some energy source in the case of the second), thus creating a steep gradient in substrate concentration, up which the cells then move. Quantitative measurements were recorded by Adler and his coworkers on such properties of the phenomenon as band shape, concentration gradient and band speed.

To simplify the theoretical analysis we ignore growth (which can be excluded experimentally), and consider only a single band of bacteria, responding to a single substrate, which we will call the critical substrate. In view of the microscopic dimensions of the organisms, it is appropriate to describe bacterial density as a function of position and time, $b(x, t)$. Our considerations are limited to motion in one dimension primarily for simplicity, although this simplification is particularly appropriate for a description of the essentially one-dimensional phenomenon of chemotactic bands. The equation for $b(x, t)$ is premised on the observation that bacterial motion in the absence of a chemical gradient is fundamentally random, and can therefore be described as a diffusion process. This view is supported by the measurements of Adler and Dahl [4]. The flux of bacteria in the absence of a chemical gradient is accordingly taken to be

$$J_{diff} = -\mu \frac{\partial b}{\partial x} \tag{1}$$

where μ is a measure of motility, directly analogous to the diffusion coefficient. We assume that the effect of a chemical gradient is to generate a second, or chemotactic flux, most simply assumed to be proportional to that gradient. The total flux is then

$$J = J_{\text{diff}} + J_{\text{chemo.}} = -\mu(c)\frac{\partial b}{\partial x} + \chi(c)b\frac{\partial c}{\partial x}, \qquad (2)$$

where $c(x, t)$ is the concentration of the critical substrate, and $\chi(c)$ the chemotactic coefficient. Of necessity, the chemotactic flux is proportional to b. Since it is reasonable to expect that, in general, the critical substrate concentration may affect the strength of the motility and chemotactic sensitivity, μ and χ are permitted to depend on c.

In the absence of reproduction, conservation of bacteria requires that

$$\frac{\partial b}{\partial t} = -\frac{\partial J}{\partial x}. \qquad (3)$$

(A simple derivation of this relation can be found in [9].) The final equation for b (x, t) is obtained by combining equations (2) and (3), and is

$$\frac{\partial b}{\partial t} = \frac{\partial}{\partial x}\left[\mu(c)\frac{\partial b}{\partial x} - \chi(c)\frac{\partial c}{\partial x}\right]. \qquad (4)$$

The motivation for equation (4) has been given in relative detail because of its importance. Not only is it an equation to which we will constantly refer in our subsequent analyses of bacterial chemotaxis, but it is also intended to be of more general applicability. For example, SEGEL and I [9] have used the same equation in a theoretical analysis of aggregation in slime mold.

A complete description of a particular system, however, requires an equation for c (x, t). In the case of bacteria, the critical substrate is consumed but not produced by the organism, and the appropriate equation is, therefore,

$$\frac{\partial c}{\partial t} = -k(c)b + D\frac{\partial^2 c}{\partial x^2}, \qquad (5)$$

where $k(c)$ is the consumption rate per bacterium, and D is the diffusion constant of the substrate. When typical values of D are compared to measured values of μ [4], it is found that D is small compared to μ and can, therefore, to first approximation be set equal to 0. Equation (5) then reduces to

$$\frac{\partial c}{\partial t} \approx -k(c)b. \qquad (6)$$

In order to solve equations (4) and (6) it is necessary to specify the functions $\mu(c)$, $\chi(c)$ and $k(c)$. We make the following assumptions:

1. $k(c) = k$, a constant.

The meaning of this assumption is that the rate of depletion of substrate is limited only by the ability of the bacteria to consume it, and not by its availability. It is an experimentally justified assumption for sufficiently high levels of concentration such as those found in front of the chemotactic bands [1], though not for very low levels, such as are found behind the band.

2. $\mu(c) = \mu$, a constant.

This assumption is made, in the absence of any information to the contrary, solely for simplicity.

3. $\chi(c) = \dfrac{\delta}{c}$, where δ is a constant.

The third assumption is analogous to the familiar Weber-Fechner law, and derives particular justification in this example from the mathematical result that equations (4) and (6) yield a travelling band solution only if $\chi(c)$ becomes infinite at $c = 0$ at least as fast as c^{-1}. The reader is referred to [11] for a proof of this assertion.

Equations (4) and (6), together with the appropriate initial and boundary conditions, now constitute a well defined mathematical problem. We look for 'travelling wave' solutions — that is, solutions that are functions only of $z = x - vt$, where v is the constant velocity of the travelling wave or, in this case, migrating band. Equations (4) and (6) then reduce to ordinary differential equations for functions of one variable, z, and can be solved by straightforward integration. Details can be found in our original paper.

The solution can be compared with experiment in a number of ways. Values of the band speed, band width and band shape parameters can be calculated and compared with data, and the agreement found is encouraging. Perhaps the most interesting feature of the results is that it is shown that migrating bands with the proper shape are obtained if and only if $\delta > \mu$. One consequence of this result is the realization that chemotactic bands are not an inevitable consequence of chemotaxis in bacteria, but will result only if the chemotactic response is sufficiently strong.

The major weakness of the analysis described above lies in the assumptions made on the functional form of $\varkappa$, μ and k. The latter two assumptions were made with no direct experimental support. It comes, therefore, as something of a surprise to find that at least one of these assumptions, namely $\chi(c) = \delta c^{-1}$, seems, on the basis of the recent experiments of Dahlquist et al. [7], to have been a very good guess. Dahlquist et al., using a technique they developed for obtaining detailed quantitative data, studied the response of *Salmonella typhimurium* to externally imposed chemical gradients. Using equation (4) to interpret their data, SEGEL and JACKSON [13] have shown that one obtains

$$\chi(c) = \delta(c)\, c^{-1},$$

where $\delta(c)$ is either a constant or a slowly varying function, depending on which data are used. In particular, from the measurement of average velocity in [7], one obtains

$$\delta(c) = \frac{3}{2} + \frac{1}{4}\log_{\cdot 10} c$$

for a range of c from 10^{-6} to 10^{-3} M, where $\delta(c)$ is measured in cm^2/h. Assuming little difference between the chemotactic mechanisms of *E. coli* and *S. typhimurium*, $\chi = \delta c^{-1}$ would seem to be a good approximation. The same experiments do not, unfortunately, yield information on the functional form of $\mu(c)$.

The major goal and accomplishment of the analysis described here is to provide a formal framework which relates a variety of data – a typical goal of biomathematics. The value of such a framework is that it reduces the number of independent parameters and may also provide more precise definitions of key concepts than previously existed. The latter would seem to be particularly relevant to the present problem. The definitions employed here lead to well defined measures of motility and chemotaxis which may well be more meaningful than measures heretofore employed. In particular, chemotactic strength has in the past frequently been measured by the ratio of forward motion to overall motion, a quantity called the chemotactic index. If, however, the motion is fundamentally erratic, or random, then such a measure would be highly dependent on the accuracy of measurement. By contrast, the measure of chemotactic strength introduced here, $\chi(c)$, is independent of the details of the motion, and can be determined solely from the macroscopic response of a population of cells. Nevertheless, this type of analysis suffers from many obvious limitations. Perhaps the most serious of these limitations is that it does not address, directly or indirectly, the kind of question that may be of greatest interest to biologists, namely that of underlying mechanism. In the next section I will describe a very different mathematical approach to biology – one that rests less on purely phenomenological analysis and more on the exploration of particular models.

III. A Model for Chemotaxis without Memory

A major question that the phenomenon of chemotaxis in micro-organisms invites is the following: How can a cell detect a gradient when the concentration difference across its body is necessarily so small? A second paper LEE SEGEL and I wrote offers a possible answer to this question [10].

The gist of our answer is that an individual cell does not need to detect the gradient in order for a chemotactic response to exist in a population. The individual cell may execute a random motion which, when averaged over a large population, gives rise to a macroscopic flux in the direction of the gradient. Physics provides us with a well known analogy in Brownian motion. There a particle moves in response to the instantaneous difference in impact made by bombarding molecules and describes a path that appears quite random. Only its average motion reflects the macroscopic pressure difference. Although this analogy must not be taken literally, it is not difficult to translate it into terms meaningful to a biological organism. What we need to say is that the local response of the cell is dictated by the local concentration of chemical. In general, given that the dimensions of a chemical receptor must be less than the dimensions of the cell, this local concentration would be expected to exhibit significant fluctuations. Our premise is that the fluctuations in the path of an individual cell are a direct reflection of the fluctuations in local estimates of the substrate concentration, while the average path, or the macroscopic flux, is a direct reflection of the macroscopic gradient. A cell with only a single receptor is, without making any measurement of the gradient, as capable as a Brownian particle of responding to a gradient. For generality, however, we consider the possibility of each cell having more than one receptor which, without requiring a direct comparison of estimates at different receptors, provides a means of augmenting the chemotactic response.

As a particular example, we consider a one-dimensional organism undergoing steps of length Δ to the right or left at a frequency determined by the estimate of local concentration made at a receptor. Let us suppose that the organism has two receptors, separated by distance $a\Delta$. An equation for the net cellular flux in the direction of increasing x, J (x), can be obtained in a straightforward way from the average frequency of steps in a particular direction, f (c), where c is the mean concentration of critical substrate. The exact form of the function f(c) will depend on how the receptor measures concentration.

Let b (x) denote the density of cells centered at x. For cells propelled by a pushing motion, such as is the case for bacteria, we have

$$J(x) = \int_{x-\Delta}^{x} f\left[c\left(y - \frac{1}{2}a\Delta\right)\right] b(y)\, dy - \int_{x}^{x+\Delta} f\left[c\left(y + \frac{1}{2}a\Delta\right)\right] b(y)\, dy. \qquad (7)$$

In effect, J (x) is computed by subtracting the number of cells moving to the left per unit time from the number of cells moving to the right. For cells propelled by pulling, such as with pseudopod formation, the same equation is obtained, with the sign of a reversed. If the step size Δ is small compared with macroscopic lengths, the right hand side of (7) can be expanded in powers of Δ, yielding, to lowest order in Δ,

$$J(x) \approx -\Delta^2 \left\{ f[c(x)]\, b'(x) + (a+1)\, f'[c(x)]\, b(x)\, c'(x) \right\}. \qquad (8)$$

We can now rewrite the equation (8) as

$$J(x) = -\mu \frac{\partial b}{\partial x} + \chi b \frac{\partial c}{\partial x} \qquad (9)$$

with

$$\mu = f(c)\, \Delta^2$$

and

$$\chi = -(a+1)\, f'(c)\, \Delta^2.$$

Thus, we have obtained the starting equation of the first section, with the special property

$$\chi(c) = -(a+1)\, \mu'(c). \qquad (10)$$

A similar equation with a similar relation between μ and χ results from consideration of motion in which the step length rather than step frequency is determined by concentration. If both step length and step frequency are permitted to depend on c, equation (9) results, but (10) does not.

The above analysis demonstrates one way in which a micro-organism may exhibit chemotactic properties without any mechanism for detecting a gradient *per se*. Fluctuations in the path of an individual cell may be computed in what is standard fashion for physicists, for which the reader is referred to the original paper [10]. Specific models for chemoreception which generate specific forms for f (c) are also possible to analyze, but remain highly speculative.

IV. A Model for Chemotaxis with Memory

Although it is useful to consider possible theoretical mechanisms for biological function, it is important to bear in mind that organisms do not consult us in evolving their mechanisms, and may well arrive at very different resolutions. Indeed, this seems to be the case with bacteria. In a recent experiment, MACNAB and KOSHLAND [12] have demonstrated in a simple and elegant way that at least some bacteria have a chemical memory, the possession of which vastly simplifies the task of detecting gradients. By directly observing the effects of an abrupt temporal change in attractant concentration on the motion of *S. typhimurium*, they report a marked qualitative, albeit temporary, change in the motion of individual bacteria. While this observation does not exclude the possibility of a Brownian-type response, it removed the necessity of it, at least for *S. typhimurium*. The analysis presented in section III does not include the possibility of a memory system and is consequently not, as it stands, applicable. I will, therefore, now describe an alternative model for bacterial chemotaxis, based directly on the observations of MACNAB and KOSHLAND [12] as well as those of BERG and BROWN [6; see also this volume], which does incorporate memory. For a somewhat different model see the paper by SEGEL [14].

BERG has developed a three dimensional tracking device with which he is able to accurately describe the path of an individual bacterium. The data he and BROWN have obtained on *E. coli* [6] and the observations of MACNAB and KOSHLAND [12] both support a description of bacterial motion originally put forward by DELBRUCK [8]. According to this view, bacteria move along a straight line at uniform speed, stop and reorient themselves (i.e., 'tumble'), and move along a new direction, random with respect to the original direction. Chemotaxis is achieved through the dependence of the frequency of 'tumbling' on concentration and concentration differences, i.e. memory.

A quantitative formulation of this view begins with the bacterial distribution function in position and velocity space. Let f (x, v, t) represent the number of bacteria at a given position x with velocity v, at time t. Conservation of bacteria leads directly to the constitutive equation for f:

$$\frac{\partial f}{\partial t} + v \frac{\partial f}{\partial x} = \int [\sigma (x, v', v, t) \, f (x, v', t) - \sigma (x, v, v', t) \, f (x, v, t)] \, dv', \tag{11}$$

where $\sigma (x, v', v, t)$ is the transition probability, or probability of tumbling, from velocity v' to v per unit time. (We assume no external forces.)

For simplicity, let us assume that, at a given value of x and t, bacteria may be found with only two values of v, namely $\pm v_0$ (x, t). The observed velocity is permitted to depend on x and t in order to allow for the influence of environmental factors such as chemical concentration. This assumption may be expressed mathematically as follows:

$$f(x, v, t) = f^+(x, t) \, \delta \, [v - v_0(x, t)] + f^-(x, t) \, \delta \, [v + v_0(x, t)], \tag{12}$$

where $f^+(x, t)$ and $f^-(x, t)$ represent the spatial density of bacteria moving with velocity $+v_0$ and $-v_0$, respectively, and $\delta(v-v_0)$ is the Dirac function, familiar to physicists, with the property that $\delta(v-v_0) = 0$ except where $v = v_0$. The biological content of the model is contained primarily in the assumptions made about the transition probability σ. The first assumption, which is in rough accordance with the observations of BERG and BROWN [6], is that reorientation is random. That is,

$$\sigma(x, v', v, t) = \sigma(x, v', t). \tag{13}$$

The transition probability depends on the velocity before tumbling, but does not depend on the velocity after tumbling, as if the bacteria known where they are coming from, but not where they are going.

Equations for bacterial density b (x, t) and bacterial flux J (x, t) are obtained from the definitions of b and J by integrating (11) over velocity v. That is,

$$b = \int f(x, v, t) \, dv$$
$$J = \int v f(x, v, t) \, dv. \tag{14}$$

Integration of (1) yields, as the equations for b and J

$$\frac{\partial b}{\partial t} + \frac{\partial J}{\partial x} = 0, \tag{15}$$

$$\frac{\partial}{\partial t}\left(\frac{J}{v_0}\right) + \frac{\partial}{\partial x}(v_0 \, b) = b \, (\sigma^- - \sigma^+) - \frac{J}{v_0}(\sigma^- + \sigma^+), \tag{16}$$

where

$$\sigma^{\pm} = \sigma(x, \pm v_0, t).$$

So far, nothing has been said about chemotaxis. Where does a chemotactic mechanism enter into these equations? Implicitly, we have assumed here that each organism has only one receptor. This assumption is reflected in the fact that bacteria are treated as point particles with no internal parameters. The inclusion of more than one receptor would require an analysis analogous to that of section III. In the present model, organisms may respond to a chemical gradient either through the dependence of velocity v_0 on chemical concentration, $c(x, t)$, or through dependence of the transition probability σ on c. We assume both. A memory mechanism would be expressed through a dependence of σ on current and past chemical concentrations. In particular, we assume

$$\sigma(x, v, t) = \sigma[c(x, -v\tau, t-\tau), c(x, t)] \tag{17}$$

where τ is the memory time of the organism. Chemical influences on velocity are expressed simply by letting $v_0 = v_0[c(x, t)]$.

Not all terms in equation (16) are necessarily of equal magnitude. For internal consistency, τ must be less than the time between 'tumbles' or turns, and for biological efficiency, we suppose it to be of the same of order of magnitude. Hence, $\tau \sim \sigma^{-1}$. On the other hand, τ is presumably much smaller than T, where T is the time scale over which macroscopic changes in b and J occur. In particular, if T is defined so that

$$\frac{\partial J}{\partial t} \sim \frac{J}{T},$$

then $\tau \ll T$. This is equivalent to saying that the particle velocity is large compared to the average bacterial velocity J/b. Accordingly, equation (16) may be approximated by deleting terms of order τ/T. Observing that the term $\dfrac{1}{v_0} \dfrac{\partial q}{\partial t}$ is of order τ/T relative to the other terms, we remain with

$$\frac{\partial}{\partial x}(v_0 b) = b(\sigma^- - \sigma^+) - \frac{J}{v_0}\left(\sigma^- + \sigma^+ + \frac{1}{v_0}\frac{\partial v_0}{\partial t}\right) \tag{18}$$

in place of (16).

At this point it is useful to consider separately the different experimental situations studied. First, let us consider the experiment of MACNAB and KOSHLAND [12] in which the bacteria are subjected to an abrupt temporal change at $t = 0$ in a spatially uniform chemical concentration. That is, $c_x \sim 0$.

In this case

$$\sigma^- - \sigma^+ = 0$$

and

$$\sigma^- + \sigma^+ = 2\sigma[c(t-\tau), c(t)].$$

The latter may be rewritten more conveniently as

$$\sigma^- + \sigma^+ = 2\,\sigma\,[\bar{c}\,(t),\,a\,(t)],$$

where $\bar{c}\,(t)$ is the average concentration, a relatively slowly varying function of time, and $a\,(t)$ is the concentration difference, $[c\,(t)-c\,(t-\tau)]$, a rapidly varying function of time. Here,

$$a\,(t) = \begin{cases} 0 & t<0,\,t>\tau \\ \Delta c & 0<t<\tau. \end{cases}$$

Equation (18) then yields

$$J = -\left[\frac{v_0^2}{2\,\sigma\,(\bar{c},a\,[t]) - \dfrac{1}{v_0}\dfrac{\partial v_0}{\partial t}}\right]\frac{\partial b}{\partial x} \tag{19}$$

which, when coupled with (15) results in the diffusion equation

$$\frac{\partial b}{\partial t} = -\mu\,(t)\frac{\partial^2 b}{\partial x^2}, \tag{20}$$

with the diffusion coefficient, or motility, given by

$$\mu\,(t) = \left[\frac{v_0^2}{2\,\sigma\,(\bar{},a\,[t]) - \dfrac{v_0'\,(c)}{v_0}c_t}\right]. \tag{21}$$

MACNAB and KOSHLAND [12] report that the motion is temporarily more coordinated when $\Delta c>0$, and less coordinated when $\Delta c<0$. This implies that σ decreases with positive a and increases with negative a or, more accurately,

$$\Delta\sigma \equiv \sigma\,(\bar{c},a) - \sigma\,(\bar{c},0) < 0 \text{ when } a>0, >0 \text{ when } a>0.$$

The duration of the period during which $\sigma = 0$, i.e. during which the motion deviates from normal, is τ.

Equation (21) implies in turn that the motility undergoes a temporary increase when $\Delta c>0$ and decrease when $\Delta c<0$, again with the change persisting for period τ. Similarly, if the velocity v_0 increases with c, μ again increases with $\Delta c>0$, and decreases with $\Delta c<0$ the increase now being permanent, with a sharp spurt of the same sign as c_t at $t = 0$. While MACNAB and KOSHLAND [12], using L-serine as attractant, report no dependence of velocity on concentration, the measurements of BERG and BROWN [6] show clearly that, for *E. coli*, the concentration dependence of velocity may well vary with the choice of attractant.

The other, more typical, experimental situation which should be studied is that in which c varies both temporally and spatially, but where the temporal changes in c occur in response to temporal changes in b and J, rather than by external manipulation. This is the situation which prevails in the experiments of ADLER [3], BERG and BROWN [6] and DAHLQUIST *et al.* [7]. Here $\dfrac{1}{c}\dfrac{\partial c}{\partial t} \sim \dfrac{1}{T}$ where, as before, T is the time scale for changes in b and J. Consistency with the approximation made above, in which terms of order τ/T were deleted, now requires the deletion of terms involving c_s. Thus,

$$c\,(x \mp v_0\tau, t - \tau) \approx c\,(x, t) \mp v_0\,\tau\,c_x. \tag{22}$$

Writing

$$\sigma\,(c_1, c_2) \text{ as } \sigma\left(\frac{c_1 + c_2}{2}, c_2 - c_1\right),$$

we have

$$\sigma^+ \mp \sigma^- = \sigma\left(c - \frac{a}{2}, a\right) \mp \sigma\left(c + \frac{a}{2}, -a\right),$$

where $a = v_0\tau c_x$.

If we can assume that σ is a continuous function of a, then, for small a,

$$\sigma^+ - \sigma^- \approx 2\,a\,\sigma, (c, 0) - a\,\sigma_2\,(c, 0)$$

and

$$\sigma^+ + \sigma^- \approx 2\,a\,\sigma\,(c, 0),$$

where the subscripts 1 and 2 denote differentiation with respect to the first and second variables respectively (i.e., the average concentration and the concentration difference). Equation (18) then reduces to

$$J = -\frac{v_0^2}{2\sigma\,(c, 0)}\frac{\partial b}{\partial x} - \left[\frac{v_0'\,(c) + v_0\,\tau\,(2\sigma_2\,[c, 0] - \sigma_1\,[c, 0])}{2\sigma\,(c, 0)}\right]b\,\frac{\partial c}{\partial x}. \tag{23}$$

Combining (23) with (15) we once again retrieve our basic equation (4) of section II, namely

$$\frac{\partial b}{\partial t} = \frac{\partial}{\partial x}\left[\mu\,(c)\,\frac{\partial b}{\partial x} - \chi\,(c)\,b\,\frac{\partial c}{\partial x}\right]$$

with μ (c) and χ (c) now being given by

$$\mu(c) = \frac{v_0^2}{2\,\sigma(c,0)} \tag{24}$$

$$\chi(c) = -\frac{[v_0'(c) + v_0\,\tau\,(2\sigma_2[c,0] - \sigma_1[c,0])]}{2\sigma(c,0)}. \tag{25}$$

Ignoring, for the moment, the contribution of $v'(c)$ to the chemotactic effect, positive chemotaxis results from $\sigma_1 > 0$ and $\sigma_2 < 0$. BERG and BROWN's data [6] indicate that both of these inequalities are characteristic of the motion of E. coli, while MACNAB and KOSHLAND's qualitative observations [12] on S. typhimurium support the first, and do not address the second. On the other hand, a positive $v'(c)$ would tend to reduce the chemotactic response, as might have been anticipated from the analysis of section III.

V. Discussion and Conclusion

There are basically two ways of proceeding from equations (19) and (23). One is to compare the predicted bacterial response to such quantitative observations as were made by DAHLQUIST et al. [7] in order to determine the functional dependence of v and σ on chemical concentration, and from there attempt to deduce something about the possible molecular mechanism that might give rise to such functional forms. The second is to begin by speculating about molecular mechanisms, deduce the quantitative consequences of such mechanisms, and compare them both with observation and the theoretical formulae. In general, one tries both. The very observation of a memory mechanism in bacteria inevitably invites some speculation as to what kinds of molecular mechanisms might be involved. MACNAB and KOSHLAND [12] have pointed out that it suffices to suppose the existence of two chemical responses to the critical substrates with different relaxation times in order to explain memory. They have proposed a particular model as an example. One feature of their experimental observations is perhaps crucial in distinguishing between possible models. There is the suggestion in [12] that the relaxation time for positive Δc is greater than that for negative Δc. Subsequent work apparently confirms this early suggestion [11a]. Furthermore, there is the related observation of BERG and BROWN [6] that a negative spatial gradient exerts less of an effect on the motion of E. coli than a positive gradient. In the context of the analysis presented here, both observations imply that the memory time τ depends on the sign of the concentration difference experienced by the bacterium. If true, such a dependence would constitute a powerful restriction on the class of admissible models. Preliminary quantitative analysis of MACNAB and KOHSLAND's proposed model suggests that it is not, without modifications, capable of yielding the observed inequality in relaxation times. Clearly, further quantitative measurements of bacterial response to known gradients would be helpful in pursuing this question.

The work described in this article might perhaps be described as the random walk of a theoretician through the environs of a concrete biological problem. Each step is triggered by a particular experimental observation. Hopefully, we can detect some net progress in the direction of a sound ultimate understanding of a small piece of biological reality – an understanding clearly not possible without the impetus and guidance of experimental data.

Summary

Mathematical models of bacterial chemotaxis are discussed with reference to such phenomenological observations as migrating bands of *E. coli*, as well as to more fundamental questions of chemotactic mechanisms. In particular, the question of how a microorganism can detect a chemical gradient is considered.

References

1 ADLER, J.: Effect of amino acids and oxygen on chemotaxis in *Escherichia coli*. J. Bact. *92:* 121–129 (1966).
2 ADLER, J.: Chemotaxis in bacteria. Science *153:* 708–716 (1966).
3 ADLER, J.: Chemoreceptors in bacteria. Science *166:* 1588–1597 (1969).
4 ADLER, J. and DAHL, M.: A method for measuring the motility of bacteria and for comparing random and non-random motility. J. gen. Microbiol. *46:* 161–173 (1967).
5 ADLER, J. and TEMPLETON, B.: The effect of environmental conditions on the motility of *Escherichia coli*. J. gen. Microbiol. *46:* 175–184 (1967).
6 BERG, H. C. and BROWN, D.: Chemotaxis in *Escherichia coli*. Analysis by three-dimensional tracking. Nature *239:* 500 (1972).
7 DAHLQUIST, F. W.; LOVELY, P., and KOSHLAND, D. E., jr.: Quantitative analysis of bacterial migration in chemotaxis. Nature New Biol. *236:* 120–123 (1972).
8 DELBRUCK, M.: Personal communication.
9 KELLER, E. and SEGEL, L.: Initiation of slime mold aggregation viewed as an instability. J. theor. Biol. *26:* 399–415 (1970).

10   KELLER, E. and SEGEL, L.: Model for chemotaxis. J. theor. Biol. *30:* 225–234 (1971).
11   KELLER, E. and SEGEL, L.: Traveling bands of chemotactic bacteria. A theoretical analysis. J. theor. Biol. *30:* 235–248 (1971).
11a KOSHLAND, D. E., jr.: Personal communication.
12   MACNAB, R. M. and KOSHLAND, D. E., jr.: The gradient-sensing mechanism in bacterial chemotaxis. Proc. nat. Acad. Sci., Wash. *69:* 2502 (1972).
13   SEGEL, L. A. and JACKSON, J. L.: Theoretical analysis of chemotactic movement in bacteria. J. mechanochem. Cell Motility (in press).
14   SEGEL, L. A.: Theories for memory aided chemotaxis (to be published).

Author's address: Dr. EVELYN FOX KELLER, Division of Natural Science, State University of New York, College at Purchase, *Purchase, NY 10577* (USA)

Chemotaxis in Acrasieae

Antibiotics and Chemotherapy, vol. 19, pp. 96–110
(Karger, Basel 1974)

The Chemotactic Effect of Cyclic AMP and its Analogues in the Acrasieae

T. M. KONIJN

Cell Biology and Morphogenesis Unit, University of Leiden, Leiden

I. Introduction

Vegetative amoebae of the Acrasieae, or cellular slime moulds, feed on bacteria. When the food supply has been exhausted, the amoebae enter an inter-phase period, the duration of which varies in different species. During this period, which lasts several hours in *Dictyostelium discoideum*, morphological, physiological and biochemical changes occur. At the end of the starvation period one or more amoebae, depending on the species, initiate aggregation by attracting neighbouring cells chemotactically. The amoebae pile up in the centre of the aggregate and, depending on the number of cells and the environmental conditions, either migrate all together over the surface or differentiate immediately into a fruiting structure consisting of spores and stalk cells (fig. 1). Migration of the slug-shaped body of amoebae allows a differentiation into stalk and spores on a more favourable spot for propagation. Spores germinate and amoebae, after crawling out of the spore case, start the life-cycle all over again. We will focus on the transition from the unicellular to the multicellular phase.

The suggestion that chemotaxis was responsible for cell aggregation in *Dictyostelium* had been made by the turn of the century [30, 33]. 40 years later RUNYON [36] observed amoebae that were spread on both sides of a dialysing membrane. Cells on the top of the membrane aggregated in a pattern corresponding to the streams and centres beneath the membrane. He provided evidence for a diffusible chemotactic substance. BONNER [4] ruled out other agents that could be responsible for cell aggregation and coined the term 'acrasin' for the chemotactic substances secreted by the amoebae. That acrasin was a free-diffusing agent, was demonstrated by an experiment

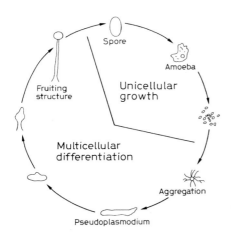

Spore

Amoeba

Unicellular
growth

Fruiting
structure

Multicellular
differentiation

Aggregation

Pseudoplasmodium

Fig. 1. Life-cycle of *Dictyostelium discoideum.*

under flowing water in which downstream amoebae moved to the aggregation centre.

A partial characterisation of the acrasin, produced by amoebae of *D. discoideum*, was undertaken by SHAFFER [39]. The chemotactic agent appeared to be a dialysable, heat-stabile substance which was inactivated in the presence of enzymes secreted by the amoebae themselves. A crude active extract was prepared by methanol extraction [40] or in an acidic solution [42].

Bacteria also secrete chemotactic agents that attract amoebae [2, 19, 37]. Purification of bacterial extracts revealed that adenosine 3′,5′-monophosphate (cyclic AMP) is the attractant for amoebae [25, 26]. The attractant secreted by the amoebae themselves was identified also as cyclic AMP [3, 22, 23]. Though all species seem to secrete cyclic AMP, only the larger species of *Dictyostelium* are attracted by this cyclic nucleotide [21].

Cyclic AMP was isolated from liver about 15 years ago [43]. The intermediating role of cyclic AMP in animals, plants and micro-organisms is now one of the most intensively studied fields in biology. SUTHERLAND and RALL [43] proposed that cyclic AMP functions as a second messenger which mediates the effects of a variety of hormones and other biologically active agents. The hormone is the first messenger which activates membrane-bound adenylate cyclase. This enzyme is supposed to contain two types of sub-units, a regulatory sub-unit at the outside of the cell membrane and a catalytic unit facing the inside of the cell [35]. The regulatory unit interacts with the hormone and varies from tissue to tissue. The catalytic unit catalyses

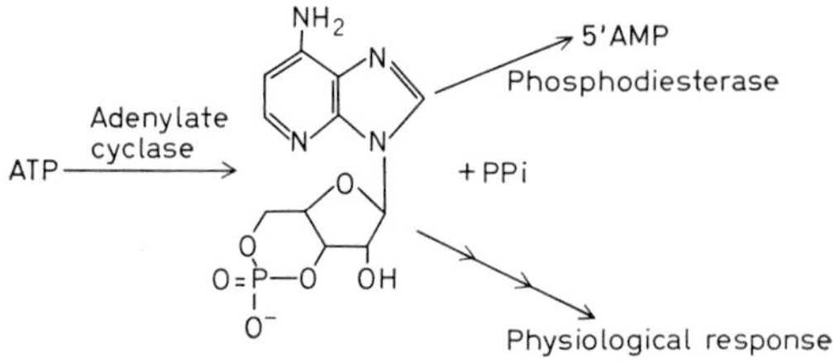

Fig. 2. Components of cyclic AMP mechanism.

the conversion of ATP into cyclic AMP (fig. 2). Cyclic AMP has an affinity to protein kinases which, combined with the cyclic nucleotide, may activate the enzyme phosphorylase. The different effects observed after a rise in the cyclic AMP concentration probably depend on the enzymatic profile of the cell. Possible effects of the increased cyclic AMP level are: enzyme activation, change in membrane permeability or an increased synthesis of RNA and proteins. The intermediator role of cyclic AMP in a large variety of biological systems has been discussed in a monograph [35].

The late discovery of cyclic AMP is partly due to the low concentration in which it is present in the cell. ATP is present at more than 1,000 times higher concentration. An enzyme which keeps the cyclic AMP level low by hydrolysing the $3'$-phosphate bond is adenosine $3', 5'$-monophosphate phosphodiesterase. There is increasing evidence that there is more than one enzyme responsible for phosphodiesterase activity, which may occur in soluble as well as in particulate form [11].

Cyclic AMP, which acts intra-cellularly in other organisms, acts in the cellular slime moulds extra-cellularly. Its action in these primitive organisms is comparable with hormone action in higher organisms.

All the species of the Acrasieae that have been studied show phosphodiesterase activity [6]. In *D. discoideum*, the activity of the cyclic AMP inactivating enzyme varies with the physiological stage of the amoebae [34]. Riedel and Gerisch demonstrated an inhibitor of phosphodiesterase in *D. discoideum* which appears at the end of the growth phase of the amoebae [34]. A mutant, deficient in the inhibitor production, produced more active phosphodiesterase than the wild type and could not aggregate.

In *D. discoideum*, phosphodiesterase occurs in soluble and particulate form [27]. The membrane-bound phosphodiesterase has been assumed to be part of the chemotactic receptor system [27].

Recently, Pan *et al.*, discovered folic acid as a second chemotactic substance in the Acrasieae [31]. Amoebae of *D. discoideum* are not attracted to folic acid when they are close to aggregation, i.e. in their most sensitive stage to cyclic AMP.

For theoretical studies on chemotaxis and cell aggregation the reader is referred to several recent articles in the Journal of Theoretical Biology [9, 10, 17, 18, 38].

Cyclic AMP and guanosine 3',5'-monophosphate (cyclic GMP) have been shown to occur naturally [35]. Other cyclic nucleotides and analogues of cyclic AMP have been synthetised to study the complex effects of cyclic AMP in higher systems, where hormones function as first messengers. In the larger *Dictyostelium* species, cyclic AMP has an analogous function as a hormone and acts also as a first messenger [21].

This review will be limited to a discussion of the chemotactic effect of analogues of cyclic AMP and other cyclic nucleotides, using amoebae of *D. discoideum* as the test organisms.

II. Assay for Attractants of Amoebae

All cyclic nucleotides and their analogues can be tested for their chemotactic activity by using a small population assay. Small drops (0.1 μl) of a suspension of *D. discoideum* amoebae are placed on a non-nutrient hydrophobic agar surface [20]. Only on a water-repellent agar surface do the amoebae stay inside the boundaries of the drop (fig. 4). The area within the margins of the drop becomes immediately hydrophilic. When a second drop of the amoebal suspension is deposited 2 h later, then aggregating cells in the first drop attract amoebae outside the boundary of the drop containing the physiologically younger amoebae (fig. 3). Instead of an aggregate, bacteria or their extracts may act as the attracting drops. By purification of extracts it was shown that cyclic AMP is the chemotactic agent secreted by amoebae and bacteria [3, 22, 23, 25, 26]. The responding amoebae are used in the assay just before their aggregative phase when they are 100 times more sensitive to cyclic AMP than in the vegetative phase [5].

To determine the threshold concentration for a response, small drops (0.1 μl) of a solution of a cyclic nucleotide or an analogue of cyclic AMP were deposited 3 times at 5-min intervals at a distance of about 300 μm from the responding amoebal populations. The response was observed after 5 min and considered positive if more than twice as many amoebae are pressed against the side closest to the attracting drop than on the opposite

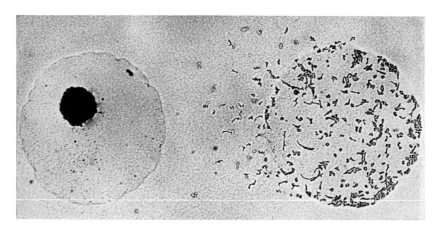

Fig. 3. Chemotaxis in *Dictyostelium discoideum*. The aggregation on the left attracts amoebae outside the boundary of the drop on the right. × 60. From KONIJN [20].

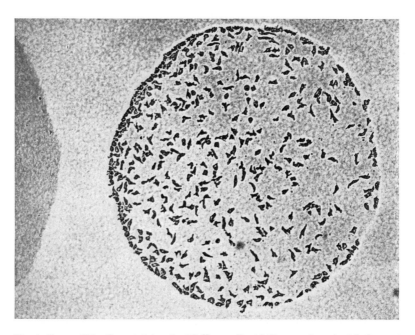

Fig. 4. Drops (0.1 μl) containing 3×10^{-13} g cyclic AMP were deposited 3 times at 5-min intervals to the left of the responding drop. × 95. From KONIJN [20].

side (fig. 4). The attractants were diluted till the threshold activity, which is the activity at which 50% of the responding populations react positively, was reached.

The small population assay for cyclic AMP offers several advantages. The amoebae are tested in their single cell stage and do not need a disaggregation treatment. Close to aggregation the cells are extremely sensitive to cyclic AMP. A few thousand molecules are sufficient for a positive response. A further advantage is that the attractants do not have to permeate through the cell membrane, at least not at appreciable levels, to induce a chemotactic response [28].

The starving amoebae of *D. discoideum* react specifically to cyclic nucleotides and their analogues and not to a gamut of attractants as found in bacteria [1].

III. Chemotactic Activity of Cyclic Nucleotides and Their Analogues

A. Cyclic Nucleotides

Cyclic AMP is the most active chemotactic agent among all cyclic nucleotides. Deposition of only 10^{-12} g of cyclic AMP evokes a positive response (fig. 4). Since the drops containing the cyclic AMP have been placed at a distance of $300\,\mu$m from the amoebal population and will diffuse to the sides and into the agar, only a very small number of molecules will reach an amoeba.

Other cyclic nucleotides are 10^2–10^4 times less active than cyclic AMP (table I). Cyclic nucleotides with a purine base are not always chemotactically more active than those containing a pyrimidine base [21]. An example of the latter, uridine 3′,5′-monophosphate (cyclic UMP), is, after cyclic AMP, the most active cyclic nucleotide and inosine 3′,5′-monophosphate (cyclic IMP), containing a purine base, belongs with deoxythymidine 3′,5′-monophosphate (cyclic TMP) and xanthosine 3′,5′-monophosphate (cyclic XMP) to the least active cyclic nucleotides (table I). From cyclic TMP only the deoxy compound is available. Since deoxy-cyclic AMP is less active than cyclic AMP (table IV), deoxy-cyclic TMP may be less active than cyclic TMP. In other systems deoxy-cyclic nucleotides have been shown to be inactive while the cyclic nucleotides themselves were active [45].

The only other naturally occurring cyclic nucleotide, cyclic GMP, is about as active as cytidine 3′,5′-monophosphate (cyclic CMP) and is in the

Table I. Chemotactic activities of cyclic nucleotides

Cyclic nucleotide	50% of populations react positively between (M)
Cyclic AMP	$10^{-8} - 10^{-9}$
Cyclic UMP	$10^{-6} - 10^{-7}$
Cyclic CMP	$10^{-5} - 10^{-6}$
Cyclic GMP	$10^{-5} - 10^{-6}$
Cyclic IMP	$10^{-4} - 10^{-5}$
Cyclic TMP	$10^{-4} - 10^{-5}$
Cyclic XMP	$10^{-4} - 10^{-5}$

See text for abbreviations.

order of 10^3 times less active than cyclic AMP (table I). Also, in most other biological systems cyclic AMP is more active than cyclic GMP [16].

The various cyclic nucleotides can be hydrolysed by adenosine 3',5'-monophosphate phosphodiesterase from amoebae [8]. The Vm and Km values are similar for the hydrolysis of cyclic AMP, cyclic CMP, cyclic TMP and cyclic UMP, while cyclic GMP and cyclic IMP were hydrolysed at a slower rate [8]. If compared with table I there is no apparent correlation between the chemotactic activity of the various cyclic nucleotides and their hydrolysis by phosphodiesterase.

B. Analogues of Cyclic AMP

1. Substitutions in the Phosphate Moiety

The phosphate ring seems to be an indispensable part of the cyclic AMP molecule for chemotaxis. Within the ring the oxygen atom between the 5'-carbon and the phosphorus (fig. 2) has been replaced by a methylene group with only a slight loss in activity [26]. The threshold activity of the new compound, 5'-methylene adenosine cyclic phosphonate (5'-CH$_2$-cAMP), is in the same range as that of cyclic AMP (table II). The similar activity of the phosphonate and cyclic AMP itself may be due to the slight effect that the replacement of the oxygen by a methylene group has on the conformational characteristics of the parental compound [41]. However, the ability of 5'-CH$_2$-cAMP to activate phosphorylase b kinase is very weak compared

Table II. Chemotactic activities of analogues of cyclic AMP with substitutions in the phosphate moiety

Cyclic nucleotide	50% of populations react positively between (M)
5'-CH$_2$-cyclic AMP	$10^{-8} - 10^{-9}$
5'-NH-cyclic AMP	$10^{-8} - 10^{-9}$
5'-NH-cyclic AMPS I	$10^{-6} - 10^{-7}$
5'-NH-cyclic AMPS II	$10^{-5} - 10^{-6}$
5'-NCH$_3$-cyclic AMP	$10^{-4} - 10^{-5}$
5'-NCH$_2$C$_6$H$_5$-cyclic AMP	$10^{-4} - 10^{-5}$
5'-NC$_4$H$_9$-cyclic AMP	$10^{-4} - 10^{-5}$
5'-N(CH$_2$)$_3$N(CH$_3$)$_2$-cyclic AMP	$10^{-4} - 10^{-5}$
5'-NC$_8$H$_{17}$-cyclic AMP	$10^{-3} - 10^{-4}$
5'-NC$_8$H$_{17}$-OC$_6$H$_4$-NO$_2$-cyclic AMP	$10^{-3} - 10^{-4}$
3'-CH$_2$-cyclic AMP	$10^{-2} - 10^{-3}$

to the activation by cyclic AMP [12]. The importance of the position where an oxygen atom has been replaced becomes clear when the oxygen between the 3'-carbon and the phosphorus (fig. 2) is substituted by a methylene group. The threshold activity of 3'-methylene adenosine cyclic phosphonate drops to a concentration of 10^{-2}–10^{-3}M (table II), which means a reduction in activity with a factor of 10^6.

Replacement of the oxygen at the 5'-carbon position by an unsubstituted amido group results in a chemotactic activity similar to cyclic AMP [24]. In mammalian systems, 5'-amino-adenosine cyclic 3',5'-phosphate (5'-NH-cAMP) is 10–100 times less active than cyclic AMP [14]. Small stereochemical changes as in 5'-N-methylaminoadenosine cyclic 3',5'-phosphate (5'-NCH$_3$-cAMP) in which a methyl group has been added to the amido group reduce the chemotactic activity by a factor of 10^4 [24]. 5'-N-benzyl-aminoadenosine cyclic 3',5'-phosphate (5'-NCH$_2$C$_6$H$_5$-cAMP), 5'-N-*n*-butyl-aminoadenosine cyclic 3',5'-phosphate (5'-NC$_4$H$_9$-cAMP) and 5'-N-γ-dimethylaminopropylaminoadenosine cyclic 3',5'-phosphate [5'-N(CH$_2$)$_3$-N(CH$_3$)$_2$-cAMP] have a similar threshold activity to 5'-NCH$_3$-cAMP (table II). The large variation in the groups that are attached to the amido group does not affect the response of the amoebae [24]. However, when the addition becomes as long as an octyl group, as in 5'-N-*n*-octylaminoadenosine cyclic 3',5'-phosphate (5'-NC$_8$H$_{17}$-cAMP), the chemotactic activity is reduced. The weak attraction of the various derivatives of 5'-NH-cAMP is not due to traces of cyclic AMP or 5'-NH-cAMP [15, 29].

Although the activity of 5'-N-n-octylaminoadenosine cyclic 3',5'-0-p-nitrophenyl phosphate (5'-NC$_8$H$_{17}$-OC$_6$H$_4$NO$_2$-cAMP) is low and its slope of activity at various concentrations differs from that of other 5'-amido analogues [24], it indicates that a negative charge in the phosphate moiety is not required for chemotaxis of amoebae.

The chemotactic activity is 100- to 1,000-fold less than that of 5'-NH-cAMP when the double-bonded oxygen has been replaced by a sulphur atom. The two diastereomeres of 5'-aminoadenosine cyclic 3',5'-thiophosphates (5'-NH-cAMPS$_{II}^{I}$) induce a different chemotactic response and also their slope of activity at various concentrations varies [24]. Since the absolute configuration at the phosphorus atom has not been determined yet, it is not possible to decide which configuration has the lower threshold activity. Their different activities indicate the importance of the stereo-chemistry of the atoms bound to the phosphorus.

2. Substitutions in the Base Moiety

The response of amoebae to 8-piperidino-cyclic AMP is slightly less than to cyclic AMP (table III). Not only a substitution at the 8-position but also at the 6-position of the base may result in a nearly optimal response as can be observed when N^6-benzoyl-cyclic AMP is the attractant. The threshold activity of N^6-benzyl-cyclic AMP is 10^{-6}–10^{-7}M. If one butyryl group has been attached to the 6-amido group, as in N^6-monobutyryl cyclic AMP, the threshold activity is also 10^{-6}–10^{-7}M (table III).

A substitution of hydrogen by bromium at the 8-position gives a loss in activity of a factor of 10^3. An opposite effect was found by FREE et al. [13], who tested 13 8-substituted derivatives of cyclic AMP for their ability to activate steroidogenesis and lipolysis in the rat. In their systems, 8-substitution nearly always increased the activity if compared with the parental compound.

A similar observation was made by POSTERNAK and CEHOVIC [32] who found that in several different hormonal tests the activity of 8-bromocyclic AMP and of N^6-butyryl-cyclic AMP was higher than that of cyclic AMP.

A replacement of the nitrogen atom at the 7-position by a carbon, as in tubercidin-cyclic MP, also weakens the threshold response of the amoebae by a factor of 10^3 (table III). This analogue was also more active than cyclic AMP in other systems such as the stimulation of phosphorylase b kinase [43].

Table III. Chemotactic activities of analogues of cyclic AMP with substitutions in the base moiety

Cyclic nucleotide	50% of populations react positively between (M)
8-Piperidino-cyclic AMP	$10^{-7} - 10^{-8}$
N⁶-benzoyl-cyclic AMP	$10^{-7} - 10^{-8}$
N⁶-benzyl-cyclic AMP	$10^{-6} - 10^{-7}$
N⁶-monobutyryl cyclic AMP	$10^{-6} - 10^{-7}$
8-Bromo-cyclic AMP	$10^{-5} - 10^{-6}$
Tubercidin-cyclic MP	$10^{-5} - 10^{-6}$

Table IV. Chemotactic activities of analogues of cyclic AMP with substitutions in the ribose moiety

Cyclic nucleotide	50% of populations react positively between (M)
Deoxy-cyclic AMP	$10^{-7} - 10^{-8}$
N⁶-0-2-dibutyryl-cyclic AMP	$10^{-4} - 10^{-6}$
Iso-cyclic AMP	$10^{-3} - 10^{-4}$

3. Substitutions in the Ribose Moiety

One of the most effective analogues in biological systems is N^6,0-2-dibutyryl cyclic AMP. In higher organisms it often is more active than cyclic AMP itself. Supposedly the more lipophilic dibutyryl analogue penetrates faster into the cell than cyclic AMP. Besides, dibutyryl cyclic AMP is more resistant to the inactivating enzyme phosphodiesterase. If cyclic AMP does not have to penetrate into the amoebae to induce a chemotactic response, its action will be independent of its permeability through the cell membrane. Therefore, it is not surprising that dibutyryl cyclic AMP is less effective in the cellular slime moulds than in higher organisms (table IV). Also the more lipophilic 5'-amido analogues show a weak chemotactic response (see section III B, 1). The chemotactic activity of dibutyryl cyclic AMP varies depending on the commercial source. Traces of the more active N^6-monobutyryl analogue (see section III B, 2), present in the sample, may cause this variation, especially since dibutyryl cyclic AMP may loose its

Table V. Chemotactic activities of analogues of other cyclic nucleotides

Cyclic nucleotide	50% of populations react positively between (M)
AICAR-cyclic MP	$10^{-6} - 10^{-7}$
8-Bromo-cyclic GMP	$10^{-4} - 10^{-5}$
8-Benzylamino-cyclic GMP	$10^{-3} - 10^{-5}$
8-Benzylamino-cyclic IMP	$10^{-3} - 10^{-5}$
N^6-0-2-dibutyryl cyclic GMP	$10^{-3} - 10^{-4}$

See text for abbreviations.

2'-0-butyryl-residue rapidly [44] and contain up to 5% N^6-monobutyryl cyclic AMP.

The threshold activity of deoxy-cyclic AMP is 10^{-7}–10^{-8} M (table IV). A reduction in activity of the order of 10^5 has been observed in iso-cyclic AMP in which the ribose moiety is attached to the 3-position instead of the 9-position of the purine base. The opposite effect occurs in certain hormonal systems [7], in which iso-cyclic AMP may be as active or even more active than cyclic AMP.

C. Analogues of Other Cyclic Nucleotides

Replacement of hydrogen at the 8-position of cyclic GMP by bromium results in a weaker chemotactic response of the amoebae (table V). The loss in activity is small if compared with the 10^3-fold difference in activity between 8-bromo-cyclic AMP and cyclic AMP; however, cyclic AMP itself is a much stronger attractant than cyclic GMP. Attachment of a benzyl amino group at the 8-position, as in 8-benzylamino-cyclic GMP, also lowers the chemotactic activity (table V). Attachment of the benzyl amino group at the 8-position of cyclic IMP, as in 8-benzylamino-cyclic IMP, has a similar effect.

Addition of butyryl groups at the N^6- and 2'-position of the molecule, as in N^6-0-2-dibutyryl cyclic GMP, reduced the chemotactic activiy drastically (table V).

Surprising was the high activity of 5-amino-1-ribofuranosylamidazole-4-carboxamide 5'-phosphate (AICAR) cyclic MP (table V), the base of which is quite different from the base of cyclic AMP. AICAR-cyclic MP is more active than several other cyclic nucleotides.

IV. Conclusions

By using a small population assay in a study of the relation between the chemotactic activity and the molecular structure of cyclic nucleotides and their analogues, the following conclusions were reached.

1. The phosphate moiety seems to be essential to chemotaxis of amoebae of the larger *Dictyostelium* species. Only cyclic nucleotides and their analogues have been shown to attract aggregative amoebae of *D. discoideum*.

2. Several cyclic nucleotides and analogues are more active than cyclic AMP itself in other biological systems. No other cyclic nucleotides or analogues of cyclic AMP evoke in the Acrasieae a greater chemotactic response than cyclic AMP.

3. Changes in the phosphate, base or ribose moiety affect to various degrees the chemotaxis of amoebae.

4. Substitution of oxygen in the phosphate moiety at the 5'-ribose position by a methylene or an amido group has no, or only a slight, effect on the chemotactic response of the amoebae. A similar substitution of oxygen by a methylene group at the 3'-ribose position reduces the chemotactic activity by a factor of 10^6.

5. The chemotactic activity of an analogue with a substitution at the 5'-position is reduced drastically by adding a protruding group to the substitute at the 5'-position. The type of group that has been added does not seem to have a major effect on chemotaxis.

6. Substitution of the double-bound oxygen atom in the phosphate moiety by a sulphur atom results in a much weaker response of the amoebae.

7. The more lipophilic analogues are not the stronger attractants; probably because of the extra-cellular effect of these compounds better penetration of lipophilic analogues does not lead to a stronger response.

8. The molecular receptor mechanisms are sensitive to stereo-chemical alterations in the base moiety.

9. Substitutions within the base moiety always result in a weaker response of the amoebae and slight changes may result in a large decrease of chemotactic activity.

10. The importance of changes in the base moiety show that the components of this moiety represent some of the structural features that react in some way with the receptor mechanism of the amoebae.

11. The mechanisms responsible for changes in cyclic AMP activity by various functional groups are not known yet.

References

1 ADLER, J.: Chemoreceptors in bacteria. Science *166:* 1588–1597 (1969).

2 ARNDT, A.: Rhizopodenstudien. III. Untersuchungen über *Dictyostelium mucoroides* Brefeld. Roux' Arch. EntwMech. *136:* 681–744 (1937).

3 BARKLEY, D. S.: Adenosine-3′,5′-phosphate. Identification as acrasin in a species of cellular slime mold. Science *165:* 1133–1134 (1969).

4 BONNER, J. T.: Evidence for the formation of cell aggregates by chemotaxis in the development of the slime mold *Dictyostelium discoideum.* J. exp. Zool. *106:* 1–26 (1947).

5 BONNER, J. T.; BARKLEY, D. S.; HALL, E. M.; KONIJN, T. M.; MASON, J. W.; O'KEEFE, G., and WOLFF, P. B.: Acrasin, acrasinase, and the sensitivity to acrasin in *Dictyostelium discoideum.* Develop. Biol. *20:* 72–87 (1969).

6 BONNER, J. T.; HALL, E. M.; NOLLER, S.; OLESON, F. B., and ROBERTS, A. B.: Synthesis of cyclic AMP and phosphodiesterase in various species of cellular slime molds and its bearing on chemotaxis and differentiation. Develop. Biol. *29:* 402–409 (1972).

7 CEHOVIC, G.; MARCUS, I.; VENGADABADY, S. et POSTERNAK, T.: Sur la préparation de l'acide iso-adénosine-3′,5′-phosphorique (iso-AMP cyclique) et sur certaines de ses propriétés biologiques. C. R. Soc. phys. Hist. nat., Geneva *3:* 135 (1968).

8 CHASSY, B. M.; LOVE, L. L., and KRICHEVSKY, M. I.: Purification, properties and biological role of an extracellular nucleoside-3′,5′-cyclic phosphate phosphodiesterase from *Dictyostelium discoideum.* Fed. Proc. *28:* 842 (1969).

9 COHEN, M. H. and ROBERTSON, A.: Wave propagation in the early stages of aggregation of cellular slime molds. J. theor. Biol. *31:* 101–118 (1971).

10 COHEN, M. H. and ROBERTSON, A.: Chemotaxis and the early stages of aggregation in cellular slime molds. J. theor. Biol. *31:* 119–130 (1971).

11 ROBERTIS, E. DE; RODRIGUEZ DE LORES ARNAIZ, G.; ALBERICI, M.; BUTCHER, R. W., and SUTHERLAND, E. W.: Subcellular distribution of adenyl cyclase and cyclic phosphodiesterase in rat brain cortex. J. biol. Chem. *242:* 3487–3493 (1967).

12 DRUMMOND, G. I. and POWELL, C. A.: Analogues of cyclic 3′,5′-AMP as activators of phosphorylase b kinase and as substrates for cyclic 3′,5′-nucleotide phospho-diesterase. Molec. Pharmacol. *6:* 24–30 (1970).

13 FREE, C. A.; CHASIN, M.; PAIK, V. S., and HESS, S. M.: Steroidogenic and lipolytic activities of 8-substituted derivatives of cyclic 3′,5′-adenosine monophosphate. Bio-chemistry *10:* 3785–3789 (1971).

14 JASTORFF, B. and BÄR, H. P.: Personal communication (1972).

15 JASTORFF, B. und KREBS, T.: Analoge des Adenosin-(3′,5′)-cyclophosphats mit Stickstoff und Schwefelatomen im Phosphatring. Chem. Ber. *105:* 3192–3202 (1972).

16 JOST, J.-P. and RICKENBERG, H. V.: Cyclic AMP. Annu. Rev. Biochem. *40:* 741–774 (1971).

17 KELLER, E. F. and SEGEL, L. A.: Initiation of slime mold aggregation viewed as an instability. J. theor. Biol. *26:* 399–416 (1970).

18 KELLER, E. F. and SEGEL, L. A.: Model for chemotaxis. J. theor. Biol. *30:* 225–234 (1971).

19 KONIJN, T. M.: Effect of bacteria on chemotaxis in the cellular slime molds. J. Bact. *99:* 503–509 (1969).

20 KONIJN, T. M.: Microbiological assay of cyclic 3′,5′-AMP. Experientia *26:* 367–369 (1971).

21 KONIJN, T. M.: Cyclic AMP as a first messenger; in GREENGARD, ROBISON and PAOLETTI Advances in cyclic nucleotide research, vol. 1, pp. 17–31 (Raven Press, New York 1972).

22 KONIJN, T. M.; BARKLEY, D. S.; CHANG, Y. Y., and BONNER, J. T.: Cyclic AMP. A naturally occurring acrasin in the cellular slime molds. Amer. Naturalist *102:* 225–234 (1968).

23 KONIJN, T. M.; CHANG, Y. Y., and BONNER, J. T.: Synthesis of cyclic AMP in *Dictyostelium discoideum* and *Polysphondylium pallidum*. Nature, Lond. *224:* 1211–1212 (1969).

24 KONIJN, T. M. and JASTORFF, B.: The chemotactic effect of 5′-amido analogues of adenosine 3′,5′-monophosphate in the cellular slime molds. Biochim. biophys. Acta *304:* 774–780 (1973).

25 KONIJN, T. M.; MEENE, J. G. C. VAN DE; BONNER, J. T., and BARKLEY, D. S.: The acrasin activity of adenosine-3′,5′-cyclic phosphate. Proc. nat. Acad. Sci., Wash. *58:* 1152–1154 (1967).

26 KONIJN, T. M.; MEENE, J. G. C. VAN DE; CHANG, Y. Y.; BARKLEY, D. S., and BONNER, J. T.: Identification of adenosine-3′,5′-monophosphate as the bacterial attractant for myxamoebae of *Dictyostelium discoideum*. J. Bact. *99:* 510–512 (1969).

27 MALCHOW, D.; NÄGELE, B.; SCHWARZ, H., and GERISCH, G.: Membrane-bound cyclic AMP phosphodiesterase in chemotactically responding cells of *Dictyostelium discoideum*. Europ. J. Biochem. *28:* 136–142 (1972).

28 MOENS, P. and KONIJN, T.M.: Cyclic AMP as a cell surface activating agent in *Dictyostelium discoideum*. FEBS – letters (in press).

29 MURAYAMA, A.; JASTORFF, B.; CRAMER, F., and HETTLER, H.: 5′-amido analogs of adenosine 3′,5′-cyclic monophosphate. J. org. Chem. *36:* 3029–3033 (1971).

30 OLIVE, E. W.: Monograph of the Acrasieae. Proc. Boston Soc. nat. Hist. *30:* 451–513 (1902).

31 PAN, P.; HALL, E. M., and BONNER, J. T.: Folic acid as second chemotactic substance in the cellular slime molds. Nature New Biol. *237:* 181–182 (1972).

32 POSTERNAK, T. and CEHOVIC, G.: Derivatives and analogues of cyclic nucleotides. Ann. N.Y. Acad. Sci. *185:* 42–49 (1971).

33 POTTS, G.: Zur Physiologie des *Dictyostelium mucoroides*. Flora *91:* 281–347 (1902).

34 RIEDEL, V. and GERISCH, G.: Regulation of extracellular cyclic-AMP-phosphodiesterase activity during development of *Dictyostelium discoideum*. Biochem. biophys. Res. Commun. *42:* 119–123 (1971).

35 ROBISON, G. A.; BUTCHER, R. W., and SUTHERLAND, E. W.: Cyclic AMP (Academic Press, New York 1971).

36 RUNYON, E. H.: Aggregation of separate cells of *Dictyostelium* to form a multicellular body. Collect. Net *17:* 88 (1942).

37 SAMUEL, E. W.: Orientation and rate of locomotion of individual amoebae in the life cycle of the cellular slime mold *Dictyostelium mucoroides*. Develop. Biol. *3:* 317–336 (1961).

38   SEGEL, L. A. and STOECKLY, B.: Instability of a layer of chemotactic cells, attractant and degrading enzyme. J. theor. Biol. *37:* 561–585 (1972).
39   SHAFFER, B. M.: Acrasin. The chemotactic agent in cellular slime molds. J. exp. Biol. *33:* 645–657 (1956).
40   SHAFFER, B. M.: Properties of acrasin. Science *123:* 1172–1173 (1956).
41   SUNDARALINGAM, M. and ABOLA, J.: Molecular conformation of adenosine 3′,5′-monophosphonate monohydrate. Nature New Biol. *235:* 244–245 (1972).
42   SUSSMAN, M.; LEE, F., and KERR, N. S.: Fractionation of acrasin. A specific chemotactic agent for slime mold aggregation. Science *123:* 1171–1172 (1956).
43   SUTHERLAND, E. W. and RALL, T. W.: Fractionation and characterization of a cyclic adenine ribonucleotide formed by tissue particles. J. biol. Chem. *232:* 1077–1091 (1958).
44   SWISLOCKI, N. I.: Decomposition of dibutyryl cyclic AMP in aqueous buffers. Analyt. Biochem. *38:* 260–269 (1970).
45   WALAAS, O.; WALAAS, E., and OSAKI, S.: The effect of nucleoside 2′,3′-cyclophosphates and nucleoside 3′,5′-cyclophosphates on UDP glucose. α-1,4-glucan α-4-glucosyltransferase and its kinase; in WHELAN Control of glycogen metabolism, pp. 139–152 (Academic Press, New York 1968).

Author's address: Dr. THEO M. KONIJN, Cell Biology and Morphogenesis Unit, University of Leiden, Kaiserstraat 63, *Leiden* (The Netherlands)

Chemotaxis in Leucocytes

Antibiotics and Chemotherapy, vol. 19, pp. 112–125
(Karger, Basel 1974)

The *in vitro* Assessment of Leucocyte Chemotaxis[1]

H. U. KELLER, M. W. HESS and H. COTTIER

Department of Pathology, University of Bern, Bern

I. What Are the Essential Requirements for Measuring Chemotaxis?

Chemotaxis has been defined as 'a reaction by which the direction of locomotion of cells or organisms is determined by substances in their environment. If the direction of movement is not definitely toward or away from the substance in question, chemotaxis is indifferent or absent' [27]. Thus only directional migration can be used as a measure for chemotaxis. A review of the literature shows that terms such as 'chemotaxis' and 'random migration' have often been used in a confusing and misleading way. Consequently, inadequate techniques have sometimes been applied to measure these leucocyte activities. Some techniques were found to be unsatisfactory for other reasons discussed below and it is therefore appropriate to consider the basic requirements for assessing leucocyte chemotaxis *in vitro*.

A. Distinction between Random and Directional Migration

Directional migration is a specific feature of the chemotactic response, while random migration can be observed without chemotactic stimulation or following chemotactic stimulation in absence of a gradient [19]. The technique used must, therefore, distinguish between random and directional

---

1   Part of the work presented in this review has been supported by the Swiss National Foundation for Scientific Research.

migration and the assessment of chemotaxis must be based on the directional response. This implies that a concentration gradient can be established in the test system and that the cells have to migrate an appreciable distance in order to express their directional response. It is worthwhile to remember here that migration rates up to 9.4 μm/min have been noted *in vivo* [6] and up to 32 μm/min *in vitro* [27]. Directional as well as random migration can be affected by the quality of the culture medium. It is, therefore, important to culture cells in a medium providing optimal conditions for the expression of their migratory activity in order to exclude any unspecific increase of cell movement in and towards a more favourable environment.

B. Identification of the Responding Cell Type(s)

It has been demonstrated that the chemotactic response can be cell-specific [12, 20, 23, 37], though some cytotaxins seem to act on more than one cell type [38]. The identification of the reactive cell(s) is, therefore, an essential part of the assessment and must be achieved by any method measuring chemotaxis. Though such information can be collected by comparing responses of different purified cell populations, it is often more convenient and instructive to test mixtures of different cell types under comparable conditions and to determine the responding cell type(s) by direct microscopical observation.

C. The Responding Cells Should Be Representative for the Behaviour of the Cell Population Studied

With some methods, measurement of chemotaxis is based on measuring the distance traveled by the most advanced cells which may be representative for only a small fraction of the total cell population. It is reasonable to assume that any cell population consists of individuals of different age, life history and functional capacity, and their chemotactic responsiveness may vary accordingly. For example, neutrophil granulocytes show a decreased chemotactic response following phagocytosis [15, 22]. If the measurement of chemotaxis is based on the performance of a small fraction of the total cell population, e.g. those which have not engulfed particles and therefore respond normally to chemotactic stimulation, the result could be misleading.

D. Reliable Quantitation and Reproducibility

Such a postulate seems trivial but it deserves special attention with respect to chemotaxis. For a long time, progress in the field was hampered because no versatile technique for quantitating leucocyte chemotaxis was available. The problem appeared to be solved when BOYDEN [5] developed a new technique measuring the passage of cells through a filter membrane, but it turned out that cells become detached in variable numbers from the lower side of these filters and the method had to be modified in order to be sufficiently reliable [16, 17].

The chemotactic responsiveness of leucocytes from patients with recurrent infections should be evaluated together with their capacity to engulf and kill micro-organisms intracellularly in order to localise defects in the defence of the host. This requires that reproducible values expressing the chemotactic responsiveness must be available to distinguish between normal and pathological conditions.

E. The *in vitro* System Should Be Relevant to *in vivo* Conditions

The mechanisms regulating the chemotactic response have been successfully analysed under *in vitro* conditions. While many of the cytotaxins have thereby been identified, little is known about possible inhibitors and/or inactivators and the precise role of this large number of factors in the complex process of leucocyte accumulation in various types of inflammatory lesions. Results obtained by *in vitro* experiments have, nevertheless, tentatively been extrapolated to leucocyte accumulation *in vivo*. This applies in particular to clinical studies relating impaired chemotaxis *in vitro* to increased susceptibility to infections. It is, therefore, appropriate to consider the differences which exist between the *in vitro* system and *in vivo* conditions.

Such differences occur in the assessment of cytotaxin formation in fresh plasma or serum. It has been amply demonstrated that interaction of various materials such as antigen-antibody complexes, bacteria or endotoxins results in the formation of heat-stable cytotaxins [3, 5, 18]. It is, in fact, impossible to avoid heating (30 min at 56 °C) the plasma or serum before placing it into the chamber because unspecific activation will otherwise occur to an uncontrollable extent. However, this precludes the detection of heat-labile cytotaxins and/or inhibitors which may be present.

Many workers find it convenient to use diluted serum – usually at a concentration of 10% – in the test solution, though it is by no means certain that this experimental design approximates *in vivo* conditions sufficiently to permit any extrapolations. Also, species differences in the chemotactic response may limit the *in vivo* significance of the experiments [12, 17, 23].

II. In vitro *Techniques for Measuring Chemotaxis*

A. The Buffy Coat System

Various modifications of this method such as the tissue culture technique of KIÄR [25] and MEIER and SCHÄR [28] or the slide-cell technique [24] have been used for assessing chemotaxis and/or random migration [29]. MEIER and SCHÄR in an excellent study showed, however, that chemotactic agents fail to influence the direction of migrating leucocytes in the buffy coat system or comparable techniques. They concluded that physicochemical factors determine the rate and direction of locomotion in such populations of packed cells. BERTHRONG and CLUFF [1] noted that leucocytes obtained from endo-toxintreated rabbits migrate normally from uncentrifuged lung or spleen fragments and migrate normally in the blood on the warm stage prior to centrifugation. Their migration is, however, inhibited in comparison to centrifuged cells from untreated rabbits. This confirms the notion that packed cells behave differently from single cells. Furthermore, the buffy coat system relies on measuring the front of the emigrating cells and the responding cells are in general not identified microscopically. Thus the buffy coat system is not suitable for measuring directional and/or random migration.

B. The Capillary Tube Method

Two entirely different systems have been used. One of them measures the emigration of cells packed into a capillary tube [26] and is, therefore, subject to similar limitations as the buffy coat systems.

In a second type of capillary tube test, the tubes containing the attractant are placed into a chamber containing the responding cells and chemotaxis is assessed by counting the cells which have accumulated in the tube. PFEFFER [31] has successfully used this system for quantitating the chemotactic behavior of plant sperms. Later, this technique was used for *in vivo* demonstration of leucocyte chemotaxis but it was abandoned after PFOEHL

[32] showed that particles such as erythrocytes, sand or starch accumulated in these tubes, possibly as a result of convection currents.

C. The Slide Coverslip Technique

This technique has been introduced by COMMANDON [7] who recorded the movement of leucocytes in a blood film spread between a slide and a coverslip by time-lapse cinematography. The technique was extensively used to quantitate the directional migration of leucocytes by McCUTCHEON [27] and by HARRIS [10]. The latter modified this system by incorporating the test material into a plasma film which was allowed to clot. Directional and random migration can be demonstrated by this method. Quantitation, however, involves observation of the migrational pathways of individual leucocytes, is tedious and relies on a small percentage of the total cell population; furthermore, it is difficult to achieve a well defined gradient in this system.

D. The Boyden Technique

1. General Considerations

BOYDEN [5] designed a chamber consisting of two compartments separated by a cell-permeable membrane. It is easy to establish a reasonably well-defined gradient between these two compartments and to study its effect on leucocyte migration. Random as well as directional migration can be demonstrated [8, 14, 19] and the responsive cells can be identified by microscopical examination of the stained filters. The measurement can be based on a representative proportion of the total population as a large majority of the chemotactically reactive cells respond to an appropriate stimulus by migrating to the other side of the filter where they can be counted. Thus, the first three of the essential requirements stated in the introduction to this paper need not be discussed any further.

Many difficulties have been encountered as far as reliable and reproducible quantitation is concerned. Boyden's one-filter technique is based on the assumption that all cells which have migrated through the filter remain at its lower surface where they can be counted. We have noted, however, that a variable proportion of the cells on the bottom side of the membrane becomes detached, rendering this method unreliable. Modified techniques

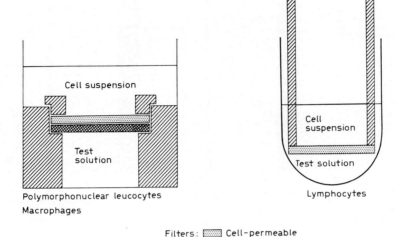

Polymorphonuclear leucocytes
Macrophages

Lymphocytes

Filters: ▦ Cell-permeable
▨ Cell-impermeable

Fig. 1. Two different modifications of Boyden's technique *in vitro* designed to evaluate the total number of cells which have passed through the cell-permeable filter. *Left:* two-filter system in a conventional Boyden chamber. The upper filter is cell-permeable, the lower one cell-impermeable. Chemotaxis is evaluated by adding the cells per high power field found on the lower side of the cell-permeable and the upper side of the cell-impermeable filter. *Right:* a cell-permeable millipore filter is glued to one end of a plastic tube cut from a syringe. This tube receives the cell suspension and it is placed into another tube containing the test solution in such a way that equal levels of medium are obtained in both compartments. Chemotaxis is evaluated by adding the total number of cells counted on the lower surface of the filter and in the test solution. This system has been used for assessing lymphoid cell migration, but with some modifications it may also be applied with other cell types.

(fig. 1) which count all cells migrating through the membrane offer a more accurate assessment of leucocyte migration [16, 17]. One of these procedures eliminates loss by the use of two filters, the lower one being impermeable for leucocytes (fig. 2). Chemotaxis is then evaluated by adding the number of cells per high power field found on the lower surface of the cell-permeable membrane and those settled on top of the impermeable filter. The other modified technique involves counting the cells which have dropped into the test solution as well as those which have remained attached to the cell permeable filter [17]. Since the rate of detachment depends also on the cell type studied, this problem will accordingly be discussed in more detail in the following paragraphs.

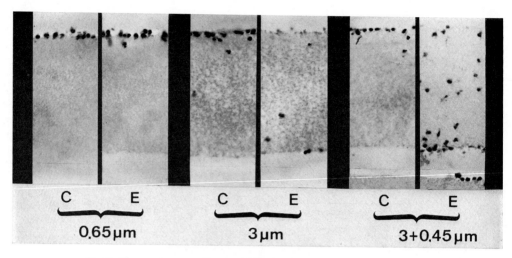

Fig. 2. The assessment of neutrophil chemotaxis using different filters. Sections through filters from experiments performed with either a single membrane of 0.65 or 3 μm pore size or a two-filter system (3 and 0.45 μm pore size). The test compartment contained Gey's solution in the controls (C) and 10% rabbit serum treated with immune complexes in the experimental chambers (E). While few neutrophils migrate into filters of 0.65 μm pore size, a large proportion migrates into and through filters of 3 μm pore size in response to chemotactic stimulation. Many of the neutrophils which have reached the lower surface of the 3 μm filter are lost unless the two-filter system is used.

Some workers have been using a one-filter system measuring the migration of cells into rather than through the filter, thereby avoiding loss of cells from the lower side. WARD *et al.* [36] have been using filters of small pore size (0.65 μm), and since leucocytes do not manage to migrate through the entire thickness of the filter, they have been counting the number of cells found within the filter. Quantitation is difficult because the cells have to be counted on different optical levels within the filter. Furthermore, only a small proportion of the neutrophil granulocytes which are actually capable of responding to chemotactic stimulation as evidenced by experiments with filters of larger pore size managed to penetrate into filters of 0.65 μm pore size (fig. 2) [16].

ZIGMOND and HIRSCH [39] proposed a one-filter system employing filters of larger pore size (3 μm) and a short incubation time which precludes passage of cells through the entire thickness of the filter. Chemotaxis is quantitated by measuring the distance from the top of the filter to the furthest two cells at the same focal plane. The evaluation is thus based on the performance of a small proportion of the total cell population; it remains to be evaluated whether results obtained by this technique are always representative for the entire population.

2. Neutrophil Chemotaxis

Optimal test conditions are a prerequisite for an efficient chemotactic response. Studies with rabbit leucocytes have shown that the pore size of the cell-permeable filter should be at least 3 μm in order to allow free passage of neutrophils [21]. Furthermore, cells suspended in 2% human serum albumin (HSA) in Gey's solution respond better to chemotactic stimulation than neutrophils cultured in homologous serum, 6% dextran, physiogel or Gey's solution alone [13].

It has been mentioned that in Boyden's one-filter system a variable proportion of the cells which have passed through the permeable filter becomes detached. Loss of neutrophil granulocytes may be quite significant. Using 10% activated serum as attractant, an average (10 experiments) of 42% of the rabbit neutrophils which had reached the lower side of the filter within the first 3 h had become detached and as many as 62% had been lost from this site at 6 h incubation time. In 3 of these 10 experiments, 90% or more of the neutrophils had become detached after 6 h. Loss of cells may vary with the pore size, the thickness and the material of the filter membrane, the incubation time, the test solution and even between replicate trials of a given experiment. Loss of cells can be observed at any time, once the neutrophils have reached the lower surface of the filter. It can, therefore, not be prevented by shortening the incubation time. While under certain conditions, which can not be defined in general terms because of the complexity of the system, the one-filter system can provide a quantitative assessment of chemotaxis, under other conditions, this procedure cannot assure even a semi-quantitative assessment and may give results in which a strongly active agent actually yields lower cell counts than a weaker agent [16]. Furthermore, a reliable distinction between inhibition of chemotaxis and cell loss is impossible unless all cells which have passed through the filter can be counted.

GALLIN *et al.* [9; see also this volume] described a two-filter system employing two cell-permeable filters of equal pore size (3 μm) and measuring the passage of ^{51}Cr-labeled leucocytes on and into the second filter by assessing its radio-activity. This procedure does not permit identification of the responding cell type(s). Moreover, we have noted that a variable proportion (up to 38%) of the neutrophils which have passed through the upper membrane stick to its lower side instead of passing on to the second cell-permeable filter [KELLER, unpublished results]. Consequently, the assessment of chemotaxis is based on a variable fraction rather than the total number of responding cells.

With the one-filter technique, the chemotactic response of rabbit peritoneal exudate cells varies from one experiment to the next [13]. The variability is reduced, but not eliminated, by the use of the two-filter system. The variable responsiveness of exudate neutrophils can be tentatively attributed to phagocytosis [15, 22], to inhibitors present in the inflammatory exudate and/or cytotaxin-induced 'deactivation' [35]. But peripheral blood granulocytes from normal human blood donors which are presumably not exposed to these factors to any appreciable extent, nevertheless show wide variations in their chemotactic responsiveness. Consequently, chemotactic values defining normal and pathological responses are not available as yet for clinical studies.

Various differences between the *in vitro* system and *in vivo* conditions need to be considered for the assessment of neutrophil chemotaxis. Cytotaxin formation in serum or plasma is usually evaluated with test solutions which have been heated prior to filling the Boyden chambers in order to prevent subsequent uncontrollable spontaneous activation. It should be noted, however, that evidence for the existence of heat-labile inhibitors of neutrophil chemotaxis in plasma of patients suffering from nephritis has been presented [30]. Heat-labile neutrophil cytotaxins have not been demonstrated convincingly, though their existence cannot be excluded.

Diluted serum or plasma can be used efficiently to demonstrate cytotaxin formation but it may be questioned if these test conditions are reasonably close enough to *in vivo* conditions to be used in clinical studies. We have occasionally encountered rabbit sera activated by immune complexes which, for unknown reasons, showed an unusual dose-response curve. Chemotactic values increased with serum concentrations up to 10% and decreased at higher concentrations even though the serum did not exert a toxic effect on rabbit neutrophils as judged by the trypan blue exclusion test. Thus the inhibitory effect is only detected with undiluted serum and may pass unnoticed if the serum is tested at a concentration of 10%, possibly because dilution of the serum changes the balance between cytotaxins and inhibitors [KELLER, HESS and COTTIER, unpublished results].

We have also found factors interfering with neutrophil migration in all sera from normal human donors which we have tested. The inhibitory action of serum can be demonstrated in two ways: (1) human neutrophils become chemotactically unresponsive following exposure to 50% heated normal

Table I. Inhibition of human neutrophil migration by the addition of normal human serum to the test solution

Test solution	Neutrophils per field, mean ± SD
Gey's solution	24 ± 2
1.25% activated serum	136 ± 39
2.5% activated serum	198 ± 12
5.0% activated serum	250 ± 54
10% activated serum	284 ± 59
55% activated serum	469 ± 42
50% normal serum	50 ± 34
50% normal serum plus 5% activated serum	74 ± 20

Activated serum: normal human serum incubated (10 min at 37 °C) with immune complexes (ovalbumin-rabbit anti-ovalbumin).

human serum, having no chemotactic activity of its own, and (2) the chemotactic response to 5% serum treated with immune complexes can be inhibited by the addition of 50% heated normal serum but not by the same concentration of serum treated with immune complexes (table I). It remains to be evaluated whether activation of the serum by immune complexes is not only associated with cytotaxin formation but also with removal of inhibitors and/or inactivators. Cells treated with heated normal serum show impaired migration but no signs of increased cell death, as judged by total cell counts and the trypan blue exclusion test; other functions such as phagocytosis and intracellular reduction of nitroblue tetrazolium are not affected. Since neutrophils isolated from heparinized blood are chemotactically responsive unless they are exposed to normal serum, we can conclude that plasma rather than serum should be used for assessing the chemotactic responsiveness in patients.

Finally, one has to take into account that species differences in the chemotactic response of neutrophils have been reported [12, 23]; therefore chemotactic data cannot necessarily be extrapolated from one species to others.

3. Eosinophil Migration

The experimental conditions for quantitating eosinophil granulocyte migration have not yet been studied satisfactorily. Filters with a pore size of 8 μm have been recommended [12, 23]. Loss of eosinophils from the lower side of the filter has not yet been evaluated. Therefore, we do not know if the values obtained with the one-filter technique are sufficiently reliable.

4. Monocyte and Macrophage Migration

Chemotaxis of mononuclear phagocytes has been studied using human blood monocytes [33] or peritoneal exudate cells from experimental animals [20, 21, 34, 37]. It has not been evaluated, however, whether the chemotactic responsiveness of different types of mononuclear phagocytes is comparable. Evidence suggesting functional differences between mineral oil induced and starch-induced peritoneal exudate macrophages has been presented [34], but these findings have not been confirmed [4, 20].

The test conditions required for the demonstration of macrophage chemotaxis are somewhat different from those used for neutrophils. Macrophages appear to move more slowly and, therefore, require a longer incubation time in the chambers unless the distance of migration is reduced by the use of a thinner (10 μm) filter [11, 33]. Filters of relatively large (8 μm) pore size are recommended [20]. Only certain batches of the filters made of mixed esters of cellulose proved suitable for demonstrating macrophage migration [2, 21]. In our experience, loss of cells from the lower side of the cell-permeable filter is less important with macrophages as compared to other cell types. Using 10 mg/ml of casein as attractant and an incubation period of 6 h, we have noted that 24% (average of 5 experiments) were lost from the lower side of the filter. It is, however, premature to generalise from these observations, since changes in the filter material and/or the test solution may increase cell loss.

Differences between serum and plasma are also relevant in assessing macrophage chemotaxis. Oil-induced rabbit exudate macrophages respond chemotactically to immune complex treated plasma but not to serum treated in a similar way [4]. It remains to be evaluated whether this is due to differences in the mechanisms of cytotaxin formation and/or to inhibitors.

5. Lymphocyte Migration

Attempts to assess lymphocyte migration with the Boyden method have failed for a long time. This failure may, in part, be explained by the fact that lymphocytes are significantly less adherent to the filter membranes than other types of leucocytes. Usually more than 90% of the lymphocytes which have reached the lower side of the cell-permeable filter become detached and the small number of remaining cells is by no means proportional to the total number of lymphocytes which have passed through the filter. Since the adherence of lymphocytes to a second (cell-impermeable) filter was also found to be poor, it proved to be more accurate to assess lymphocyte passage by counting the cells found in the test solution in addition to those remaining at the lower surface of the filter. Filters of 5 and 8 μm pore size were found to be suitable. Lymphocytes require different culture conditions than other cell types. 10% homologous serum proved to be a better cell-suspending medium than 2% HSA, Gey's solution has been replaced by medium RPMI 1640 and the chambers had to be incubated in an atmosphere containing 5% CO_2. Though active passage of lymphocytes can be demonstrated with this system, so far no evidence for directional migration has been obtained [17].

Summary

The essential requirements for assessing leucocyte chemotaxis *in vitro* are: (1) distinction between random and directional migration; (2) identification of the responding cell type(s); (3) the responding cells should be representative for the behaviour of the cell population studied; (4) reliable quantitation and reproducibility, and (5) the *in vitro* system should be relevant to *in vivo* conditions. The efficiency of various *in vitro* techniques for measuring chemotaxis and/or random migration is analysed on the basis of these criteria. The Boyden technique, which has been found to be suitable for assessing random as well as directional migration, is analysed with respect to various technical factors such as the source of cells, the cell suspending medium, the filters, the composition of the test solution and the procedure for assessing the response. The data show that test conditions critically influence the experiments and provide a basis for a reliable standardised test system. The results are also analysed in relation to *in vivo* conditions.

References

1 BERTHRONG, M. and CLUFF, L. E.: Studies of the effect of bacterial endotoxins on rabbit leucocytes. I. Effect of intravenous injection of the substances with and without induction of the local Shwartzman reaction. J. exp. Med. *98:* 331–349 (1953).

2 BOREL, J. F.: Studies on chemotaxis. Effect of subcellular leucocyte fractions on neutrophils and macrophages. Int. Arch. Allergy *39:* 247–271 (1970).

3 BOREL, J. F.; KELLER, H. U., and SORKIN, E.: Studies on chemotaxis. XI. Effect on neutrophils of lysosomal and other subcellular fractions from leucocytes. Int. Arch. Allergy *35:* 194–205 (1969).

4 BOREL, J. F. and SORKIN, E.: Differences between plasma and serum mediated chemotaxis of leucocytes. Experientia *25:* 1333–1335 (1969).

5 BOYDEN, S.: The chemotactic effect of mixtures of antibody and antigen on poly-morphonuclear leucocytes. J. exp. Med. *115:* 454–466 (1972).

6 CLIFF, W. J.: The acute inflammatory reaction in the rabbit ear chamber with particular reference to the phenomenon of leukocytic migration. J. exp. Med. *124:* 543–556 (1966).

7 COMMANDON, J.: Tactisme par l'amidon sur les leucocytes. Enrobement du charbon. C. R. Soc. Biol. *82:* 1171–1174 (1919).

8 CORNELY, H. P.: Reversal of chemotaxis *in vitro* and chemotactic activity of leuco-cyte fractions. Proc. Soc. exp. Biol. Med. *122:* 831–835 (1966).

9 GALLIN, J. I.; CLARK, R. A., and KIMBALL, H. R.: Granulocyte chemotaxis. An improved *in vitro* assay employing ^{51}Cr-labeled granulocytes. J. Immunol. *110:* 233–240 (1973).

10 HARRIS, H.: Chemotaxis of granulocytes. J. Path. Bact. *66:* 135–146 (1953).

11 HORWITZ, D. A. and GARRETT, M. A.: Use of leucocyte chemotaxis *in vitro* to assay mediators generated by immune reactions. I. Quantitation of mononuclear and polymorphonuclear leucocyte chemotaxis by polycarbonate (nucleopore) filters. J. Immunol. *106:* 649–655 (1971).

12 KAY, A. B.: Studies on eosinophil leucocyte migration. II. Factors specifically chemotactic for eosinophils and neutrophils generated from guinea-pig serum by antigen-antibody complexes. Clin. exp. Immunol. *7:* 723–737 (1970).

13 KELLER, H. U.: Studies on chemotaxis. III. Modification of Boyden's technique for the evaluation of chemotactic agents. Immunology, Lond. *10:* 225–230 (1966).

14 KELLER, H. U.: Chemotaxis and its significance for leucocyte accumulation. Agents Actions *2:* 161–169 (1972).

15 KELLER, H. U. and BOREL, J. F.: Chemotaxis of phagocytes; in DI LUZIO The reti-culoendothelial system and immune phenomena, pp. 53–58 (Plenum Publishing, New York 1971).

16 KELLER, H. U.; BOREL, J. F.; WILKINSON, P. C.; HESS, M. W., and COTTIER, H.: Re-assessment of Boyden's technique for measuring chemotaxis. J. Immun. Methods *1:* 165–168 (1972).

17 KELLER, H. U.; HESS, M. W., and COTTIER, H.: Migration of leucocytes including lymphocytes; in BRAUN and UNGAR 'Non-specific' factors influencing host re-sistance, pp. 190–195 (Karger, Basel 1973).

18 KELLER, H. U. and SORKIN, E.: Studies on chemotaxis. II. The significance of normal sera for chemotaxis induced by various agents. Immunology, Lond. *9:* 441–447 (1965).

19 KELLER, H. U. and SORKIN, E.: Studies on chemotaxis. IV. The influence of serum factors on granulocyte locomotion. Immunology, Lond. *10:* 409–416 (1966).

20 KELLER, H. U. and SORKIN, E.: Studies on chemotaxis. VI. Specific chemotaxis in rabbit polymorphonuclear leucocytes and mononuclear cells. Int. Arch. Allergy *31:* 575–586 (1967).

21 KELLER, H. U. and SORKIN, E.: Studies on chemotaxis. IX. Migration of rabbit leucocytes through filter membranes. Proc. Soc. exp. Biol. Med. *126:* 677–680 (1967).

22 KELLER, H. U. and SORKIN, E.: Studies on chemotaxis. X. Inhibition of chemotaxis of rabbit polymorphonuclear leucocytes. Int. Arch. Allergy *34:* 513–520 (1968).

23 KELLER, H. U. and SORKIN, E.: Studies on chemotaxis. XIII. Differences in the

chemotactic response of neutrophil and eosinophil polymorphonuclear leucocytes. Int. Arch. Allergy *35:* 279–287 (1969).

24 KETCHEL, M. M.; FAVOUR, C. B., and STURGIS, S. H.: The *in vitro* action of hydrocortisone on leucocyte migration. J. exp. Med. *107:* 211–218 (1958).

25 KIÄR, S.: A biological method of measuring using leucocytes as indicators. Arch. exp. Zellforsch. *1:* 289–354 (1925).

26 McCALL, C. E.; CAVES, J.; COOPER, R., and DeCHATELET, L.: Functional characteristics of human toxic neutrophils. J. infect. Dis. *124:* 68–75 (1971).

27 McCUTCHEON, M.: Chemotaxis in leucocytes. Physiol. Rev. *26:* 319–336 (1946).

28 MEIER, R. und SCHÄR, B.: Ursächliche Bedingungen der durch Polysaccharide ausgelösten Leukozytenemigrationsförderung und chemotaktischen Reaktion in Gewebekulturen. Arch. exp. Path. Pharmakol. *234:* 102–119 (1958).

29 MILLER, M. E.; OSKI, F. A., and HARRIS, M. B.: Lazy leucocyte syndrome. A new disorder of neutrophil function. Lancet *i:* 665–669 (1971).

30 PAGE, A. R.; GEWURZ, H.; PICKERING, R. J., and GOOD, R. A.: The role of complement in the acute inflammatory response; in MIESCHER and GRABAR 5th Int. Symp. Immunopathology, pp. 221–230 (Schwabe, Basel 1968).

31 PFEFFER, W.: Locomotorische Richtungsbewegung durch chemische Reize. Untersuch. bot. Inst. Tübingen *1:* 363–482 (1884).

32 PFOEHL, J.: Chemotaxis der Leukozyten *in vitro*. Zbl. Bakt. Parasitkde. *24:* 343–345 (1898).

33 SNYDERMAN, R.; ALTMAN, L. C.; HAUSMAN, M. S., and MERGENHAGEN, S. E.: Human mononuclear leukocyte chemotaxis. A quantitative assay for humoral and cellular chemotactic factors. J. Immunol. *108:* 857–860 (1972).

34 WARD, P. A.: Chemotaxis of mononuclear cells. J. exp. Med. *128:* 1201–1221 (1968).

35 WARD, P. A. and BECKER, E. L.: The deactivation of rabbit neutrophils by chemotactic factor and the nature of the activatable esterase. J. exp. Med. *127:* 693–709 (1968).

36 WARD, P. A.; COCHRANE, C. G., and MÜLLER-EBERHARD, H. J.: The role of complement in chemotaxis of leucocytes *in vitro*. J. exp. Med. *122:* 327–346 (1965).

37 WILKINSON, P. C.; BOREL, J. F.; STECHER-LEVIN, V., and SORKIN, E.: Macrophage and neutrophil specific chemotactic factors in serum. Nature, Lond. *222:* 244–247 (1969).

38 WISSLER, J. H.; STECHER, V. J., and SORKIN, E.: Biochemistry and biology of a leucotactic binary serum peptide system related to anaphylatoxin. Int. Arch. Allergy *42:* 722–747 (1972).

39 ZIGMOND, S. H. and HIRSCH, J. G.: Effects of cytochalasin B on polymorphonuclear leucocyte locomotion, phagocytosis and glycolysis. Exp. Cell Res. *73:* 383–393 (1972).

Authors' address: Dr. H. U. KELLER, Dr. M. W. HESS and Dr. H. COTTIER, Department of Pathology, University of Bern, *CH-3008 Bern* (Switzerland)

Antibiotics and Chemotherapy, vol. 19, pp. 126–145
(Karger, Basel 1974)

A Modified Millipore Filter Method for Assaying Polymorphonuclear Leukocyte Locomotion and Chemotaxis

SALLY H. ZIGMOND

Rockefeller University, New York City, N.Y., and Strangeways Research Laboratory, Cambridge
With a mathematical appendix by BRUCE KNIGHT.
(Rockefeller University, New York City, N.Y.)

Introduction

Polymorphonuclear leukocytes (PMN) accumulate at inflammatory sites. Presumably substances which act on PMN are liberated at these sites and diffuse out into the nearby region. The gradients thus established could contribute to leukocyte accumulation by at least two mechanisms. Substances diffusing out might stimulate random locomotion when present at low concentrations; these same or different substances when present in high concentrations at the focus of the site might inhibit locomotion and thus trap any leukocytes which had migrated there. Alternatively, cells could sense the direction of the chemical gradient and orient their locomotion along it. Such an orientation of locomotion, chemotaxis, is known to exist under certain *in vitro* conditions [3]. To what extent one or both of these different mechanisms play a role *in vivo* is uncertain. Information obtained about the cellular mechanisms of leukocyte accumulation in inflammation may be relevant to other situations of cell orientation and accumulation for instance in embryological development.

To study these phenomena quantitative measures of random locomotion and of directed locomotion are required. In addition, one would like to be able to make these measurements on cells exposed to various chemical gradients and concentrations. With microscopic or cine-microscopic observations on individual cells, one can analyse the temporal and spatial patterns of the movements as well as the concomitant morphology [3, 10, 12, 14].

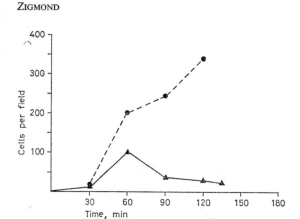

Fig. 1. Cells per field at lower surface of filter vs. time. Ordinate: number of cells per 40 × field on the bottom of the millipore filter (▲), or the total number of cells per 40 × field on the bottom of the filter plus those on the coverslip at the bottom of the chamber (•). Abscissa: time of incubation of millipore filter chambers (minutes). Horse cells in Gey's medium were placed on top of the filter. 30% horse serum in Gey's medium was present below the filter. The diameter of the coverslip at the bottom of the chamber (where cells which had traversed and fallen off the filter landed) was the same as the diameter of the millipore filter (i.e., 25 mm) [from ZIGMOND and HIRSCH, ref. No. 14].

Although rich in information content, these methods are sometimes technically difficult and often tedious and time-consuming when quantitative data is required. Quantitative information on certain aspects of locomotion can often be more readily obtained by analysis of changes in the distribution of a population of cells over a given time-period. It seemed that the Boyden millipore technique might be modified into an efficient measure of rate of locomotion of a cell population as well as a measure of the degree of orientation of the locomotion in a chemical gradient. In the Boyden method, PMN are allowed to crawl through the 3-μm pores of a millipore filter [2]. These pores are large enough to allow the cells to crawl through but too small for the cells to fall in passively [8]. The filter separates two compartments and by varying the media in these it is possible to establish a gradient across the filter. BOYDEN [2] scored the cell movement by counting the number of cells resting on the bottom of the filter at the end of an incubation time. Unfortunately, such counts are not very reproducible in part because some cells fall off the filter into the lower compartment. Since cells fall off, the number of cells counted does not increase continuously with time but reaches a peak and then declines (fig. 1). Thus, it is always possible that at the time of assay a large rapidly moving peak is past and the cell count underestimates the motility of the cells. Furthermore, any substance which made cells more adhesive to the filter preventing them from falling off would increase the cell count even if it had no effect on cell locomotion. These problems can be remedied by recovering cells which have fallen off into the lower chamber. When these cells are counted and added to the counts of cells on the bottom of the filter the total cell number increases continuously with time as shown in the upper curve of figure 1. Such an assay procedure has been worked out by KELLER et al. [6].

Even with an accurate cell count it is necessary to determine whether the system can be used to distinguish quantitative effects on the rate of random locomotion and/or the directionality of the locomotion. The existence of chemotaxis is not established by an increase in the number of cells crossing the filter toward a test substance. Accelerated migration of cells through the filter could reflect true chemotaxis (i.e., directed rather than random locomotion), an increased rate of random locomotion, or both. Reduced migration under conditions in which the test substance was placed in the upper chamber with the cells and control media beneath the filter or where the test substance was placed in both chambers would strengthen the suggestion that the substance was chemotactic [11]. However, these controls are not adequate since the results could also be explained if the substance in question stimulated random locomotion in low concentrations, but inhibited it at high concentrations [1].

Assay of Locomotion

A modified millipore procedure that enables one to distinguish between influence of a substance on rate of locomotion and direction of movement has been devised. The first step in developing this procedure was to examine the movement of cells into the filter when the same material was present homogeneously on both sides of the filter. Locomotion was assayed by measuring the distance into the filter that the front of the cell population moved during a given incubation. The experimental details have been described previously [13, 14]. A standard number of leukocytes in 1 ml of medium were placed on a 3-μm pore size millipore filter (SSWP 025 00) in a Sykes-Moore chamber, the lower compartment of which was filled with 0.9 ml of the same medium. The chambers were incubated at 37 °C for relatively short times (usually 30 min) during which no cells traversed the filter entirely. The filters were scored by measuring with the micrometer on the fine focus

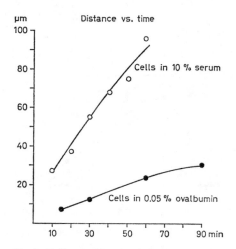

Fig. 2. Ordinate: distance (micrometres) into the filter of the cell front (two PMN in one focal plane). Abscissa: time of incubation of the millipore filter chambers (minutes). Horse cells were incubated in the millipore chambers for varying lengths of time. In each instance, the media used to suspend the cells (either 10% horse serum in Gey's medium or 0.05% ovalbumin in Gey's medium) was also present below the millipore filter [from ZIGMOND and HIRSCH, ref. No. 14].

knob the distance from the top of the filter to the furthest plane of focus which contained at least two cells in one focal plane. The measure was made for 5 fields across each of duplicate filters. Examination of 100 such duplicates showed that the mean distance score of an individual filter differed from the mean score of both duplicates by an average of 8%. In contrast, counts of the cells on the bottom of the filter after longer incubations revealed that the individual filters differed from the duplicate mean by 23%.

A time-course of the movement of the cell front using the distance measure is shown in figure 2. There was a continual increase in the distance moved. The time-course was nearly linear between 10 and 60 min but showed a slight decrease in progression with increasing times. This pattern was similar for highly motile and relatively immotile cells.

Both the reproducibility and the continuously increasing time-course made the distance measure a more satisfactory measure of cell movement than the cell number on the bottom of the filter. In addition, the distance measure was found to be less sensitive to differences in the cell concentration placed on the filter. A 10-fold increase in cell concentration (1×10^6 to 1×10^7 cells/filter) resulted in less than a 2-fold increase in the distance measure

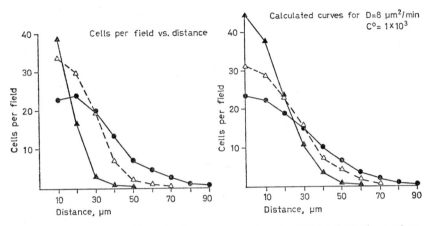

Fig. 3. Ordinate: number of cells per one-quarter of a 40 × field. Abscissa: micrometres from the top of the filter to the level being counted. Left figure: 6 × 10⁶ horse white blood cells in 4% normal horse serum were placed on top of a millipore filter; 4% serum was also present below the filter. Chambers were incubated for 20, 40 or 60 min. The number of cells in focus per one-quarter of a 40 × field were then counted at successive planes, each 10 μm from the top of the filter. Each point represents the mean of 10 fields counted, 5 on each duplicate filter. Right figure: the number of particles (c) at various distances from the origin calculated for a population of 1 × 10³ particles per one-quarte) field (Co) and a diffusion coefficient (D) of 8 μm²/min at times (t) of 20 (▲–▲), 40 (△---△r and 60 (•—•) min. The calculations were made from the equation below for particles moving in one direction out of a thin layer [4] [from ZIGMOND and HIRSCH, ref. No. 14].

$$c = \frac{C_0 dx}{\sqrt{\pi D t}} \; e^{\frac{-x^2}{4Dt}}.$$

during a 30-min assay. In contrast, the number of cells on the bottom of the filter was nearly proportional to the cell number placed on the filter.

It was critical to know if the distance measure was a true indication of the movement of the whole population or whether the cells being scored represented a rapidly moving sub-group of the population. To examine this question the distribution of the cell population throughout the filter was studied after various times of incubation. As seen in figure 3, the distribution of cells in the filter and its change with time, paralleled those of a uniform population of particles moving randomly. This is perhaps most clearly seen in figure 4; there is a linear relationship between the log. of the number of cells at a given depth in the filter and that distance squared, a characteristic of random movement. In addition, the shape of the time-course and the effect of varying the cell concentration on the distance measure are approxi-

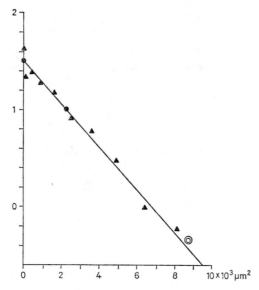

Fig. 4. Log.$_{10}$ of cells per field vs. (distance)^2. Ordinate: log.$_{10}$ of the number of cells per one-quarter of a $40 \times$ field after a 60-min incubation. Abscissa: square of the distance from the top of the filter to the field counted (square micrometres). The data are taken from the experiment described in figure 3. Each ▲ is the mean of 5 determinations of the number of cells in focus per one-quarter of a $40 \times$ field at successive 10-μm distances into the filter. The double circle is the distance to the front two cells/$40 \times$ field for this filter. The D (=8.6 μm^2/min) was calculated from the slope of the line on this graph. The line was drawn by the method of least squares (•) [from ZIGMOND and HIRSCH, ref. No. 14].

mately what one would expect for the front of a population undergoing a random walk. As seen in figure 4, the experimentally determined distance to the front two cells fell near the line determined by the least squares method on data from the total population. Thus, the distance measure was representative of the whole population. From the slope of the line in figure 4, the diffusion coefficient, a measure of cell locomotion in the millipore, was determined to be 8.6 μm.

Assay of Chemotaxis

Since the distance measure seemed to be a reasonable assay of cell locomotion in the millipore, we examined the feasibility of using it to study directed locomotion or chemotaxis. However, first some of the characteristics of a gradient set up across a filter were determined. A calculation showed that the time required for a gradient of a molecule the size of albumin to be established (to 99% completion) across the 130 μm of the filter was about 45 sec, insignificant in a 30-min assay [9]. The rate of decay of the gradient is slow. When a solution of Trypan blue was placed beneath the filter and water was placed above the filter the difference in the optical density between the top and the bottom solutions had only decreased by 10% at the end of a 30-min incubation. In critical studies, a very stable gradient could probably be maintained by perfusing the top and bottom compartments with their respective media.

Evidence of directed locomotion or chemotaxis of cells in the millipore using the distance measure would be convincing only when interpretations of the results based on alterations in the rates of random locomotion had been ruled out. By placing the same media above and below the filter the effects of different concentrations of a given material on random locomotion can be determined. Then cells can be placed in a gradient of this material and their response measured. If the test substance had no effect on random locomotion, or if its effect was essentially concentration independent, an enhanced migration of the cell front in a positive gradient and retarded migration in a negative gradient would be evidence of chemotaxis [7]. If the material has a concentration-dependent effect on random locomotion a quantitation of this dependence will allow one to predict how far, on the basis of random locomotion alone, the cells should move in any given gradient. This predicted score could then serve as the control and be compared with an experimental score for the cells moving in that gradient.

To illustrate this procedure, studies of locomotion were performed using normal autologous horse serum. The serum, obtained from clotted whole blood, was heated at 56°C for 30 min and then stored frozen. As shown in figure 5, the serum stimulated the random locomotion of horse blood PMN. Between 2 and 8% serum in Gey's medium there was a striking dose-dependent stimulation of locomotion. Increasing the percentage of serum above 16 decreased the levels of locomotion slightly. The effects of the serum on locomotion were rapid and at concentrations below 50% were reversible; incubation of cells in 100% serum for 30 min slightly retarded the subsequent movement of the cells in 8% serum. Different batches of serum differed to some extent in the degree to which they stimulated the locomotion but the dose response relationship was always similar with the peak stimulation of locomotion at about 10% serum.

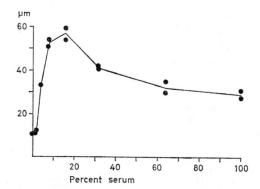

Fig. 5. Distance moved vs. serum concentration. Ordinate: micrometres moved into filter by the cell front during 30-min incubation. Abscissa: concentration of serum present both above and below the millipore filter.

The serum was examined for a chemotactic effect on the cells by allowing cells to migrate in various serum gradients. (The results of these tests are shown in table I.) Test of cells moving in a homogeneous serum concentration were also repeated to obtain information on the levels of random locomotion simultaneously with the gradient tests. These levels of random locomotion were used to calculate the distance the cells would be expected to move in a given gradient due to the effects of serum on random locomotion and are shown in the lower portion of table I. The method of calculating the expected distances is given in the appendix; a simpler but somewhatless precise method was given previously [14].

The calculation of the distance randomly locomoting cells would move assumes that the gradient across the filter is linear, stable and extends just across the 130 μm of the filter. In addition, it assumes that the effects on locomotion of all concentrations in the gradient are known, i.e. that no sharp peak of stimulatory or inhibitory activity has been missed and that the levels of locomotion at untested concentrations can be estimated by the line joining the known level of locomotion. Finally, the effects on locomotion are assumed to be rapid and reversible. Since in these experiments the chambers were not perfused, it is unlikely that a stable gradient was limited to the 130 μm of the millipore filter. The resulting error in predicted score could not be responsible for the differences between experimental and predicted values since, in some instances, the movement of cells in a positive gradient was greater than the movement in any homogeneous concentration tested.

Table I. Evaluation of chemotaxis

Percent serum above filter	Percent serum below the filter							
	1	2	4	8	16	32	64	100
A. Mean micrometre penetration into the filter, observed								
1	11			32	46	71	77	
2		11						
4			33					
8		31		53		49	49	
16					55			
32						42		
64							33	
100								30
B. Calculated micrometre penetration to be expected on the basis of effects on random locomotion alone								
1	11*			11	11	11	17	43
2		11*						
4			33*					
8		46		53*		54	51	
16					55*			
32						42*		
64							33*	
100								30*

A. Mean distance to the front two polymorphonuclear leukocytes (PMN) was determined on duplicate filters after a 30-min incubation with various serum concentrations above and below the millipore filter.

B. Distances to the front two PMN was calculated for a randomly moving population whose rate of movement was a function of the serum concentrations. The calculations, described in the appendix, used rates of random locomotion that were determined experimentally by cells moving in the absence of a gradient (scores marked with *). For each, the distance moved was considered in 5-μm segments except the final 5 μm which was subdivided into 1-μm segments for each calculation. A value of $S = 4.52$ was determined in an additional experiment.

As can be seen in table I, the direction of all differences greater than 10 μm between experimental and calculated values suggested a chemotactic response. Additional experiments confirmed this observation and also indicated that chemotaxis was most readily observed when cells were in low concentrations of serum (less than 10% serum). A quantitation of the degree of chemotaxis in the various gradients would be difficult to interpret in

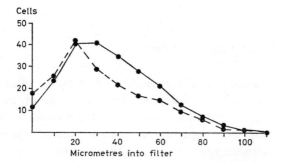

Fig. 6. Cell number in focus at various distances. Ordinate: mean number of cells in focus in one-quarter of a 40 × field. Duplicate filters are plotted. Abscissa: micrometres into the filter to the field counted. Cells in Gey's medium were placed on top of the filter and a 25-percent serum was placed below the filter. Chambers were incubated for 30 min.

these experiments since serum is a complex material probably containing more than one chemotactic factor and perhaps some inhibitory factors. However, a method for quantitating the chemotactic response is included in the appendix.

Additional evidence for the existence of a chemotactic influence of a given material can be obtained by analysing the cell distribution throughout the filter, as was done in figure 3. In the case of random locomotion the peak of cells remains at the top of the filter regardless of the rate of locomotion. In the presence of a chemotactic flux, the peak of cells would be expected to move into the filter. In figure 6 with no serum above the filter and 25% serum below the filter it can be seen that the cells do in fact move en masse toward the source of the gradient.

With the modifications described the millipore method is a reproducable assay of PMN locomotion and can be used to detect a chemotactic flux even in the midst of varying effects on cell locomotion. With knowledge of the random-walk type pattern in which cells move in the filter, one can analyse the filters to determine whether a given material has an effect on the number of cell locomoting, the direction of locomotion or the rate of locomotion.

Acknowledgement

I would like to acknowledge the guidance and help of Dr. JAMES HIRSCH in whose laboratory this work was done. In addition I would like to thank Dr. ALAN WOODIN for criticising the manuscript. The author was a fellow of the Leukaemia Society of America during preparation of this manuscript.

Appendix

Mathematical Methods by Bruce Knight

The experimental measurement of the *toe* of the distribution of cells, which are dispersing under their own independent random motions, leads not only to a laboratory procedure which is convenient and accurate, but also leads naturally to methods of data analysis which are tractable in the hands of an ordinary researcher without recourse to a computer. We now discuss three kinds of computations which are of natural interest, and show how measurement in the toe leads to simplified computational procedures. The first computation is for the diffusion coefficient, given spatially uniform diffusion conditions in the experiment. The second computation is for the location of the toe of the distribution, under the hypothesis that there is no chemotaxis but assuming that the diffusion coefficient is position-dependent in a known and specified way; if chemotaxis is important this computation will yield a result which is in disagreement with what is observed experimentally. The third computation is to deduce a velocity of chemotactic motion which, in conjunction with the specified diffusion, will give a distribution toe which agrees with the experimental measurement.

The second and third procedures are derived from considering 'integration over paths'. For each procedure I will give a sketch of the derivation, followed by detailing how the computation is to be carried out.

Under spatially uniform conditions, a population of cells, of which each executes an independent random walk starting at position $x = 0$ when $t = 0$, will disperse in a distribution $f(x,t)$ which evolves in accordance with the partial differential equation

$$\frac{\delta f}{\delta t} = \frac{1}{4} D \frac{\delta^2 f}{\delta x^2},\tag{1}$$

where D is the diffusion coefficient. The solution of equation (1) is the famous Gaussian probability distribution

$$f(x,t) = e^{-\frac{x^2}{Dt}} \cdot \left(\frac{n}{\sqrt{\pi D t}}\right),\tag{2}$$

where n is the total number of cells. The interpretation of f is that $f(x,t)dx$ is the number of cells expected to fall between x and $x + dx$ at time t. The unusual convention of inserting the factor 4 in equation (1) keeps it out of the rest of the equations, and in particular eliminates nuisance multiplications by 4 in the hand arithmetic which follows.

For one experimental run (which will be used as a standard) let f be measured experimentally at time t_1. If $y = \log_e f$ is plotted against $z = x^2$ a straight line will result, whose y intercept at $z = 0$ will be $\dfrac{n}{\sqrt{\pi D t_1}}$ and whose slope dy/dz will be $\dfrac{1}{\sqrt{D t_1}}$, according to equation (2). Thus, since t_1 is known, the numerical values of n and D may be derived from the numerical values of intercept and slope. The measurement of f vs. x is a substantial project.

The diffusion, D, gives a measure of the random motile activity of the cells. Changes of D with changes in chemical environment are of particular interest. Thus, it is desirable to devise an experiment which evaluates D with less labour than the procedure just discussed.

Let us call the 'toe' of the distribution that region for which the exponent in equation (2) is bigger than about 4, so that the exponential factor is about 1/50 or less. It is easy to check by numerical example that so long as we are in the toe of the distribution, the first term in equation (2) (the exponential term) depends far more sensitively upon D than does the second term (the square-root term). Following this lead, it is easy to show theoretically that for properly designed experiments we can ignore the variation in the second term when comparing two experiments with different diffusion coefficients. The properly designed experiment is an experiment in which we find, in the toe of the distribution, a distance x at which exactly a fixed criterion number of cells appear in the plane of focus of the microscope. Let x_s be this distance in the standard experiment discussed above, and let x_1 establish the criterion count in another experiment. Since the criterion count is equivalent to a particular value of f, we may equate the right-hand side of equation (2) for both experiments, obtaining

$$\exp\left(-\frac{x_1^2}{D_1 t_1}\right) = \exp\left(-\frac{x_s^2}{D_s t_1}\right), \tag{3}$$

where we have ignored the dependence of (2) upon the second term. It immediately follows from (3) that

$$D_1 = \left(\frac{x_1}{x_s}\right)^2 D_s, \tag{4}$$

which is the simple formula to use with the simple experiment. Ignoring the second-term dependence of equation (2) upon D is equivalent to mis-measuring x_1, and in fact is equivalent to choosing an x_1 at which f deviates from its criterion value by the ratio $\sqrt{D_S/D_1}$. Since f drops very abruptly in the toe of the distribution, this distance error typically will be small compared to experimental limitations in measuring x_1.

Of particular interest is the question of whether the cells show chemotaxis – whether they move preferentially along a chemical concentration gradient. However, experiments of the type discussed above show that the *diffusion coefficient* may change by a factor of several with changes in chemical concentration, and this effect will complicate the interpretation of experiments designed to reveal chemotaxis. The typical experiment involves establishing a known chemical gradient and, thus, establishing a diffusion coefficient which changes with position in a known way. It is natural to ask how a distribution of cells will evolve with time in such an environment *without* the complicating influence of chemotaxis.

If we call the position-dependent diffusion coefficient D(x), and if the cells show no local directional preference in their random walk behaviour, it is not hard to show that the distribution evolves according to the equation

$$\frac{\delta f}{\delta t} = \frac{1}{4} \frac{\delta^2}{\delta x^2} D(x) f, \tag{5}$$

of which equation (1) is a special case. Although a simple solution like equation (2) is unavailable, nonetheless for the toe of the evolving distribution there exists a generalisation of (2) which allows the same kind of simplifying strategy.

If at time t we start a cell at position x, and then if we choose a Δt and Δx so small that changes in D(x) are unimportant, equation (2) tells us the relative likelihood that the cell will move a distance Δx in a time Δt:

$$P = e^{-\frac{\Delta x^2}{D(x)\,\Delta t}}. \tag{6}$$

The word 'relative' was used to indicate that we have not included the normalising factor. For later notational convenience let $\Delta x/\Delta t = \dot{x}$ and let

$$L(x,\dot{x}) = \frac{\dot{x}^2}{D(x)}. \tag{7}$$

Then equation (6) becomes

$$P = e^{-L\,\Delta t}. \tag{8}$$

A cell in random motion *might* move from position $x = 0$ at time $t = 0$ to position $x = x_1$ and $t = t_1$ by taking a specified path (position vs. time) which we may call $x(t)$. As the relative likelihood of taking that path is the product of the probabilities of traversing all its pieces, we have from equation (8)

$$\Pi P = e^{-\int_o^{t_1} L \, dt} . \tag{9}$$

The total probability of a cell arriving at position x_1 at time t_1 is the sum of the probabilities of arriving by all possible paths, whence

$$f(x_1, t_1) = \int e^{-\int_o^{t_1} L[x(t)] \, dt} D x(t), \tag{10}$$

where $Dx(t)$ stands for an appropriate differential measure over all paths such that $x(t_1) = x_1$. Though the foregoing sketched argument is far less than a proof of equation (10), nonetheless (10) is indeed a solution to equation (5). The method of integration in function spaces, as applied to diffusion-type equations, is discussed, for example, by KAC [5][2]. Equation (10), in fact, is very general. It follows from the fact that probabilities of independent occurrences multiply, while independent probabilities of achieving a particular end-result in different ways add. A given choice of the likelihood-determining function L corresponds to choosing a given diffusion-type equation as, for example, the choice of L given by equation (7) corresponds to the diffusion equation (5).

Give the integral in the exponent of equation (10) the name

$$S = \int_o^{t_1} L[x(t)] \, dt . \tag{11}$$

Evidently the numerical value of S will depend on which path $x(t)$ we choose. The minimum value S_m that S may have will correspond to the most likely path $x_m(t)$. This most likely path will play a critical role below. Adding and subtracting $L[x_m(t)]$ to $L[x(t)]$ in equation (10) gives the very useful expression

$$f(x_1, t_1) = e^{-S_m} \int e^{-\int_o^{t_1} \left\{ L[x(t)] - L[x_m(t)] \right\}} D x(t) = e^{-S_m} N. \tag{12}$$

2   MARK KAC, Probability and related topics in physical sciences, chapt. IV; see also Appendix II, particularly the discussion involving equations (15) through (19).

Now we observe that for a cell in the toe of the distribution the relative probability of achieving that position by any path which departs much from the most probable path is greatly reduced from the relative probability of the most probable path. For the special case of equation (1) this is easily shown by direct computation. To restate this property: if x_1 is in the toe of the distribution, then only a narrow cluster of paths lying close to the most probable path will contribute to the integral over paths in equation (10). In a manner of speaking, the probability of reaching the point x_1 is the probability of reaching it by the most probable path multiplied by the effective measure of the cluster. That is the content of the second line of equation (12). If t_1 is held fixed, the probability of the best path, e^{-S_m} is a sensitive function of the end-point x_1, while the measure of the cluster N shows an insensitive dependence. In fact, equation (12) may be used to solve equation (1) exactly; in this case these sensitive and insensitive factors are exactly the sensitive and insensitive factors which already appeared in equation (2). The moral is that, in the analysis of a criterion-count experiment, neglect of variation in N should again be equivalent to making an error in the criterion distance x_1; a positional error which is small because the distribution falls off so rapidly in its toe.

Once we agree that N in equation (12) may be adequately approximated by its value determined in the standard experiment discussed earlier, the only remaining job is to evaluate e^{-S_m}, which in turn involves finding the most probable path $x_m(t)$. To minimise S in equation (11) by choosing the optimal $x(t)$ is a standard problem in the calculus of variations, and amounts to solving the Euler-Lagrange differential equation

$$\frac{d}{dt} \frac{\delta L}{\delta \dot{x}} - \frac{\delta L}{\delta x} = 0 \tag{13}$$

for $x = x_m(t)$. (We drop the subscript m henceforth.) Because L does not depend explicitly on t, the second-order equation (13) is equivalent to the first-order equation

$$\dot{x} \frac{\delta L}{\delta \dot{x}} - L = E, \tag{14}$$

from which equation (13) may be retrieved by one differentiation with respect to time. In equation (14) the value of the integration constant E is related to the end-point of the optimal path. If we choose L as in equation (7) then

(14) becomes

$$\frac{\dot{x}^2}{D(x)} = E, \tag{15}$$

(and, thus, by a happy chance $L = E$ is constant along the optimal path). Equation (15) may be arranged in the differential form

$$dt = \frac{1}{\sqrt{E}} \frac{dx}{\sqrt{D(x)}}, \tag{16}$$

which integrates to the optimal x vs. t relation

$$t - t_0 = \frac{1}{\sqrt{E}} \int_{x_0}^{x} \frac{dx}{\sqrt{D(x)}}. \tag{17}$$

If the starting point is $t_0 = 0$, $x_0 = 0$ and the end-point is $t = t_1$, $x = x_1$ then equation (17) determines E by

$$\sqrt{E} = \frac{1}{t_1} \int_{x_0}^{x_1} \frac{dx}{\sqrt{D(x)}}. \tag{18}$$

Because $L = E$, equation (11) gives

$$S = E\, t_1 = \frac{1}{t_1} \left(\int_{0}^{x_1} \frac{dx}{\sqrt{D(x)}} \right)^2 \tag{19}$$

where we have used equation (18) in the last step. *This is the expression to be used for the hand calculation of exp(−Sm) in equation (12).* As we have mentioned, the other factor, N, is to be approximated by its value obtained in the standard experiment. We note in equation (19) that if the diffusion coefficient is constant, then

$$S = \frac{x_1^2}{D\,t_1} \tag{20}$$

in agreement with equation (2).

What we seek in equation (19) is a value of x_1 which will determine a value of S such that e^{-S} agrees with its criterion-count value in the standard experiment. If this calculated value of x_1 departs significantly from what is observed, then we must reconsider the hypothesis that there was no chemotaxis, which led to equation (5). The evaluation of x_1 is to be done by trying a sequence of values of x_1 in equation (19), until the criterion value of e^{-s} is achieved. The only non-obvious part of this procedure is the numerical evaluation of the integral.

The integral is of the form

$$I(x_\varkappa) = \int_0^{x_\varkappa} F(x)\,dx, \tag{21}$$

where in our case $F(x) = \frac{1}{\sqrt{D(x)}}$ A method of evaluation recommended for both its simplicity and its precision is the following: choose a sequence of values $x_0 = 0$, x_1, x_3, x_2... Then $I(x_0) = 0$, and for any other index, $\varkappa$,

$$I(x_\varkappa) = I(x_{\varkappa-1}) + \frac{F(x_\varkappa) + F(x_{\varkappa-1})}{2} \cdot (x_\varkappa - x_{\varkappa-1}). \tag{22}$$

To choose the evaluation points $x_\varkappa$ sensibly remember that the procedure approximates the area under the smooth curve $F(x)$ by the area under the set of trapezoids formed by drawing straight lines between the $F(x_\varkappa)$. One's natural tendency is to pick too many points and waste labour. This possibility may be checked by repeating the calculation, using only every second point.

The third and final calculation is to deduce a velocity of chemotaxis which is consistent with the data, in case the null hypothesis fails. If we superimpose a preferred 'drift' velocity on the random motions of the cells, their distribution will evolve according to the equation

$$\frac{\delta f}{\delta t} = \frac{1}{4} \frac{\delta^2}{\delta x^2} D(x) f - v \frac{\delta f}{\delta x}, \tag{23}$$

which reduces to equation (5) if the drift velocity is set to zero. (We assume that V is a constant for the sake of simplicity, and on the conjecture that the data do not merit a more detailed assumption.) The likelihood determining function L which corresponds to the diffusion equation (23) is

$$L(x, \dot{x}) = \frac{(\dot{x} - v)^2}{D(x)}, \tag{24}$$

whence equation (14) gives in this case

$$\frac{\dot{x}^2 - v^2}{D(x)} = E. \tag{25}$$

Re-arrangement yields the differential form

$$dt = \frac{dx}{\sqrt{E\,D(x) + v^2}}. \tag{26}$$

The unknowns will untangle a bit if we let

$$V = a\sqrt{E}.\tag{27}$$

Then

$$dt = \frac{1}{\sqrt{E}}\frac{dx}{\sqrt{D(x) + a^2}}.\tag{28}$$

which leads, as in equation (18), to

$$\sqrt{E} = \frac{1}{t_1}\int_0^{x_1}\frac{dx}{\sqrt{D(x) + a^2}}.\tag{29}$$

In this problem L is *not* constant along the optimal path, and the simple evaluation of S from equation (11) is a more subtle matter than before. We proceed as follows: we solve equation (25) for $\dot{x}^2$ as a function of x and substitute in equation (24) to obtain L in terms of x. Then using equation (26) to express dt in terms of x we obtain

$$S = \int_0^{t_1} L\,dt = \int_0^{x_1}\frac{dx}{D(x)}\frac{[\sqrt{E\,D(x) + v^2} - v]^2}{\sqrt{E\,D(x) + v^2}}$$

$$= \sqrt{E}\int_0^{x_1}\frac{dx}{D(x)}\frac{[\sqrt{D(x) + a^2} - a]^2}{\sqrt{(Dx) + a^2}}\tag{30}$$

$$= \frac{1}{t_1}\left(\int_0^{x_1}\frac{dx}{\sqrt{D(x) + a^2}}\right)\left(\int_0^{x_1}\frac{dx}{D(x)}\frac{[\sqrt{D(x) + a^2} - a]^2}{\sqrt{D(x) + a^2}}\right),$$

where equation (29) was used in the final step. Notice that if $a = 0$ (whence $v = 0$ by equation [27]) then this result reduces to equation (19). However, in equation (30) the value of x_1 is the experimentally determined distance at which the criterion count of cells was achieved. *In equation (30) we must choose a value of* a *such that S achieves its criterion value. In the process of evaluating equation (30), (29) is evaluated for* $\sqrt{E}$*, and the velocity of chemotaxis* V *follows from (27).*

The evaluation of equations (29) and (30) is by the numerical integration procedure of equation (22) and is not particularly laborious. We conclude by showing how a good value of a may be found in a few guesses.

We know that the advance of the toe arises jointly from diffusion and drift velocity. Thus, if we exaggerate the effect of diffusion by setting the diffusion coefficient everywhere equal to its maximum value, we should underestimate drift velocity. Conversely, setting the diffusion everywhere to its minimum should overestimate drift velocity. But for constant D equations (29) and (30) take particularly simple forms. Equation (29) becomes

$$\sqrt{E} = \frac{x_1}{t_1} \frac{1}{\sqrt{D + a^2}}, \tag{31}$$

while equation (30) becomes

$$S = \frac{x_1^2}{D t_1} \left(1 - \frac{1}{\sqrt{\frac{D}{a^2} - 1}} \right)^2, \tag{32}$$

which can be solved directly for a. If we let $S_0 = x_1^2/Dt_1$ (the value S would have if there were no drift) then

$$a = \sqrt{\frac{D}{\left(\frac{1}{1 - \sqrt{S/S_0}} \right)^2 - 1}}. \tag{33}$$

This may be evaluated with the maximum and minimum values of $D(x)$ and then equations (31) and (27) may be used to obtain bounds on V. These easily evaluated bounds on V may be information enough. If they are not we may proceed as follows: a trial value of a may be chosen which lies half way between the two values which followed from equation (33), and that trial value may be inserted in the full equation (30). Now equation (30) should share a feature more obviously possessed by equation (32) – that an increase in a will lead to a *decrease* in S. If equation (30) gave us (say) too large a value of S for our first trial value of a, then we may increase our next trial value of a halfway up to the larger estimate that followed from equation (33). By the obvious extension of this procedure we may reduce the unknown gap in the value of a by a factor of $1/2^n$ in n trials. More sophisticated procedures involving interpolation will do even better.

Summary

Methods have been devised for measuring the random and directed locomotion (chemotaxis) of PMN in a modified Boyden millipore filter system. It has been shown that in the absence of a gradient, cells move into the filter in a random-walk like manner. Under such conditions, the distance from the top of the filter to the furthest plane of focus containing at least two cells is a reproducible measure of cell locomotion. Evaluation of chemotaxis by cells moving in a gradient of a given substance was possible after the effects of the substance on random locomotion had been determined. Equations have been worked out to aid in the detection and quantitation of directed locomotion.

References

1   BECKER, E. L.: The relationship of the chemotactic behavior of the complement-derived factors, C3a, C5a and C$\overline{567}$, and a bacterial chemotactic factor to their ability to activate the pro-esterase 1 of rabbit polymorphonuclear leukocytes. J. exp. Med. *135:* 376–387 (1972).
2   BOYDEN, S.: Chemotactic effect of mixtures of antibody and antigen on polymorphonuclear leukocytes. J. exp. Med. *115:* 453–466 (1962).
3   HARRIS, H.: Role of chemotaxis in inflammation. Physiol. Rev. *34:* 529–562 (1954).
4   JACOBS, M. H.: One-dimensional diffusion processes in infinite and semi-infinite systems; in: Diffusion processes, p. 90 (Springer, New York 1967).
5   KAC, M.: Probability and related topics in physical sciences, chapt. IV, appendix II (Interscience Publishers, New York 1959).
6   KELLER, H. U.; BOREL, J. F.; WILKINSON, P. C.; HESS, M. W., and COTTIER, H.: Reassessment of Boyden's technique for measuring chemotaxis. J. immunol. Meth. *1:* 165–168 (1972).
7   KELLER, H. U. and SORKIN, E.: Studies on chemotaxis. IV. The influence of serum factors on granulocyte locomotion. Immunology, Lond. *10:* 409–416 (1966).
8   KELLER, H. U. and SORKIN, E.: Studies on chemotaxis. IX. Migration of rabbit leucocytes through filter membranes. Proc. Soc. exp. Biol. Med. *126:* 677 (1967).
9   LONGSWORTH, L. G.: Temperature dependence of diffusion in aqueous solutions. J. physiol. Chem. *58:* 770–773 (1954).
10   McCUTCHEON, M.: Chemotaxis in leukocytes. Physiol. Rev. *26:* 319–336 (1946).
11   PHELPS, P.: Polymorphonuclear leukocyte motility *in vitro*. III. Possible release of a chemotactic substance after phagocytosis of urate crystals by polymorphonuclear leukocytes. Arthrit. Rheumat. *12:* 197–204 (1969).
12   RAMSEY, W. S.: Analysis of individual leukocyte behaviour during chemotaxis. Exp. Cell Res. *70:* 129–139 (1972).
13   ZIGMOND, S. H. and HIRSCH, J. G.: Effects of cytochalasin B on polymorphonuclear leucocyte locomotion, phagocytosis and glycolysis. Exp. Cell Res. *73:* 383–393 (1972).
14   ZIGMOND, S. H. and HIRSCH, J. G.: Leucocyte locomotion and chemotaxis. New methods for evaluation, and demonstration of a cell-derived chemotactic factor. J. exp. Med. *137:* 387–410 (1973).

Author's address: Dr. SALLY H. ZIGMOND, Section of Cell Biology, Yale Medical School, 333 Cedar Street, *New Haven, CT 06510* (USA)

Antibiotics and Chemotherapy, vol. 19, pp. 146–160
(Karger, Basel 1974)

Radioassay of Granulocyte Chemotaxis

Studies of Human Granulocytes and Chemotactic Factors

J. I. Gallin[1]

Laboratory of Clinical Investigation, National Institute of Allergy and Infectious
Diseases, National Institutes of Health, Bethesda, Md.

I. Introduction

There is increasing evidence that granulocyte chemotaxis is an important
host defense mechanism. Abnormalities of granulocyte chemotaxis have been
associated with a number of clinical diseases which have an increased
incidence of bacterial infections as one of their characteristic features [1–7].
Defective generation of chemotactic stimulants from serum has also been
demonstrated in clinical situations in which infection is a major complication
of the underlying illness [8–10]. Despite these facts, there have been relatively
few clinical studies of granulocyte chemotaxis. This may, in part, be related
to difficulty in the quantitative assessment of this aspect of granulocyte
function. Most laboratories evaluate *in vitro* leukocyte chemotaxis utilizing
modifications of Boyden's original chamber [11] in which the number of
granulocytes migrating through a single micropore filter has been correlated
with granulocyte chemotactic function. This method is associated with
considerable variability even during the best controlled studies [12] and time-
consuming filter staining and tedious microscopic counting of micropore
filters greatly limits the number of studies that can routinely be performed.

During the past year my colleagues, Robert A. Clark, Harry R. Kimball and I have developed a radioassay of human granulocyte chemotaxis
employing ^{51}Cr-labeled leukocytes and a double micropore filter system [12].
This method offers considerable advantages over the conventional Boyden
chamber technique including (1) elimination of filter staining and microsco-

1   The author wishes to thank Dr. David Alling for valuable statistical advice and
Dr. Allen Kaplan, Dr. Michael Frank and Dr. Sheldon Wolff for critical review
of the manuscript.

pic counting of micropore filters thereby enabling rapid assessment of many samples; (2) a means for standardization of neutrophil chemotaxis among different laboratories; (3) elimination of variability among observers and subjective individual bias; (4) reliability over a wider range of cell responses, and (5) ability to monitor cellular events in each compartment of the Boyden chamber.

The purpose of this chapter is to briefly describe the new radioassay of chemotactic activity and its application to studies on patients with either granulocytes having abnormal chemotactic behavior or serum deficient in the ability to generate chemotactic activity upon appropriate stimulation.

II. Method

The method has been described in detail [12] and will be only briefly reviewed. Granulocytes are obtained by either dextran sedimentation [2] or Hypaque-Ficoll separation [13] of whole blood. After two cycles of hypotonic-saline lysis to remove residual erythrocytes and one wash in cold modified Hanks' solution [14] the granulocytes are labeled with ^{51}Cr by incubating the cells at 37 °C with agitation for 1 h with 1 μCi of $Na_2Cr^{51}O_2$ in 0.9% saline/10^6 granulocytes. After labeling, the cells are washed three times with cold modified Hanks' solution (MHS) and then adjusted to 2.3 × 10^6 cells/ml in Gey's balanced salt solution [2]. Any conventional Boyden chamber is suitable for the assay with the only modification being that two 5-μm micropore filters are used instead of one to separate the upper and lower chamber compartments. This double micropore filter system provides a means for selectively counting all the cells that migrate into the lower of the two filters. The ^{51}Cr-labeled cells are added to the upper compartment and the chemotactic stimulus to the lower. After 3 h of incubation at 37°C in 100% humidity with 5% carbon dioxide [2], the upper and lower filters (LF) are then rinsed in saline and assayed for ^{51}Cr activity in a γ-counter. The number of cells migrating into the LF is proportional to the total ^{51}Cr in the LF. Variability of isotope incorporation into the leukocytes (related to the specific activity of ^{51}Cr) is adjusted to an arbitrary value of 10,000 counts/min/10^6 granulocytes; chemotactic activity is then expressed as corrected counts per minute LF (cor cpm LF)[2]. In this way results can be standardized and compared between individual experiments.

2    $$\text{cor cpm LF} = \frac{\text{observed cpm LF} \times 10,000}{\text{cpm}/10^6 \text{ granulocytes}}.$$

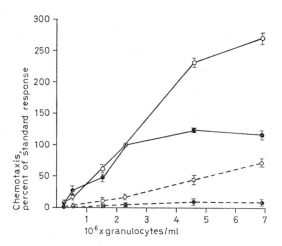

Fig. 1. Comparison of the chemotactic responses by the ^{51}Cr radiolabel (o) and conventional Boyden chamber (●) techniques to variable granulocyte concentrations added to the chambers (two experiments). The standard response is that obtained at 2.3×10^6 granulocytes/ml. The bars denote the standard error of the mean at each granulocyte concentration as denoted by the open or closed circles. ——= Endotoxin-activated serum; ––= control. [From GALLIN *et al.*, 12, reproduced by permission of the publishers.]

Compared with the conventional Boyden method the radioassay is equally sensitive to varying potencies of chemotactic stimuli and, as shown in figure 1, is more responsive to large cell loads. Elimination of filter staining and microscopic counting makes it possible for one observer to assay up to 200 chambers/day. Furthermore, as shown in figure 2, the events in each compartment of the chemotactic chamber are readily monitored.

As of December, 1972, we have used this assay in over 100 different experiments. In the following section, our definition of normal granulocyte chemotaxis and the chemotactic activity of serum is presented.

III. Normal Chemotaxis

To define normal chemotaxis, it is necessary to describe and distinguish granulocyte function from the chemotactic activity of serum and to be familiar with the inherent day-to-day variability of the assay used. The assessment of granulocyte chemotaxis or the generation of serum chemotactic factors has been routinely performed utilizing endotoxin (*Escherichia*

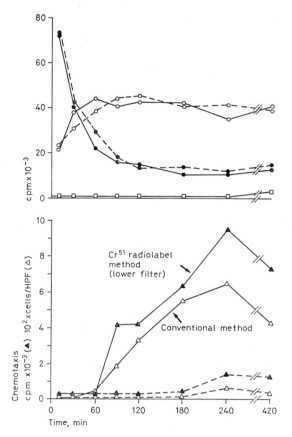

Fig. 2. Time-course of granulocyte chemotaxis in the ^{51}Cr radiolabel and conventional Boyden chamber techniques. The chemotactic factor was endotoxin-acitvated normal human serum. HPF=high power field; cpm=counts per minute; ——=chemotactic factor present; ---=controls (no factor); • =upper compartment fluid; o =upper filter; □ =lower compartment fluid. [From GALLIN *et al.*, 12, reproduced by permission of the publishers.]

coli) activated normal human serum [15, 16]. The mean normal granulocyte responses in 100 experiments among 25 normal subjects of both sexes (age 21–64 years) was 2,411 ± 127 SE cor cpm LF (range 1,174–4,432) when incubated for 3 h. Analysis of these data revealed that in experiments in which the same serum was used to attract cells from several individuals (mean chemotaxis 2,666 cor cpm LF, range 1,528–3812) there was significantly more variability (p<0.05, Siegel-Turkey test for relative spread [17])

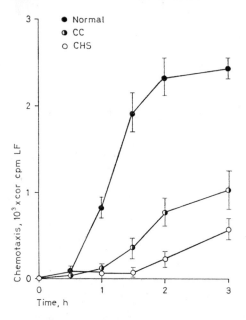

Fig. 3. Kinetics of granulocyte migration into the lower filter of the radioassay chamber of a normal subject, a patient with recurrent infections and chronic mucocutaneous candidiasis (CC) and a patient with Chediak-Higashi syndrome (CHS). The bars denote the standard error of the mean at each time for quadruplicate chambers; cor cpm LF = corrected counts per minute lower filter.

compared with experiments in which different sera were used to attract cells from one individual (mean chemotaxis 2,608 cor cpm LF, range 2,265–2,979). Therefore, variability among cells is greater than the variability among sera. On the other hand, variation was about the same in experiments in which one serum was used to attract cells from different individuals (regardless of whether these were done on the same day or on different days). Accordingly, day-to-day variability is believed to be minor compared to cell-to-cell variability.

IV. Abnormal Granulocyte Chemotaxis

Cellular defects in leukocyte chemotaxis have been described in diabetes mellitus [1], Chediak-Higashi syndrome [2], rheumatoid arthritis [3], agammaglobulinemia [4], acute infections with 'toxic neutrophils' [5], the lazy leukocyte syndrome [6] and in a patient with mucocutaneous candidiasis and recurrent pyogenic infections in the absence of other evidence of impaired neutrophil function [7]. We have had the opportunity to use the ^{51}Cr chemotactic assay to evaluate two patients with Chediak-Higashi syndrome [18, 19] and one patient with recurrent infections and chronic mucocutaneous candidiasis [7]. The mean value for granulocyte chemotaxis in 5 experiments on two patients with Chediak-Higashi syndrome was 348 cor cpm LF (range 233–575) and in 5 experiments in a patient with recurrent bacterial infections was 1,239 cor cpm LF (range 719–1,800), representing 14 and 51% of normal, respectively. As shown in figure 3, the kinetics of the defective granulocyte migration into the lower filter of the chemotactic chamber emphasizes the magnitude of the abnormality. Studies are currently in progress to evaluate granulocyte chemotaxis in a number of clinical conditions associated with impaired host defenses. To date, granulocyte chemotaxis has been normal in two patients with cyclic neutropenia, three patients on immunosuppressive therapy, one patient with chronic granulomatous diease of childhood, and three patients with chronic mucocutaneous candidiasis.

V. Generation of Chemotactic Factors from Human Serum

Chemotactic factors can be generated from normal human serum by activation of the complement pathway [15, 16, 20, 21] or the kinin-generating system [22, 23]. The inter-relations between the kinin-generating and complement pathways and the current knowledge of their respective contributions to the inflammatory process has been the subject of a recent review [24]. In the following sections, recent studies of these two pathways from our laboratory using the ^{51}Cr radioassay are presented.

A. Complement Pathway

Two major pathways of complement fixation have been delineated. The 'classical' pathway is activated by antigen-antibody complexes [20, 25, 26] and requires both calcium and magnesium [27]; in contrast, the alternate pathway is apparently activated in the absence of antibody [28–34] and requires magnesium, but not calcium [27]. In the former pathway complement components are activated by the enzyme C-3 convertase which is assembled by the catalytic action of activated C1 (C1) on C4 and C2 [25]. In the latter pathway terminal complement components are activated by conversion of C3 proacticator to C3 activator without a requirement for antibody or C1, C4 or C2 [28]. Immune complexes and endotoxins have been shown to activate complement in the sera of mice [20], rabbits [11, 20], guinea pigs [20], mink [35] and man [36], and result in the formation of factors chemotactic for granulocytes. The major chemotactic factor resulting from such complement activation is a heat-stable small molecular weight (15,500) cleavage product of C5 designated C5a [37, 38]. Small amounts of a heat-labile cleavage product of C3 (C3a) having a molecular weight of 6,000 and a trimolecular complex of the fifth, sixth and seventh components of complement (C567) with a molecular weight 160,000 have also been reported [20, 39].

The availability of a strain of guinea pigs with a total deficiency of C4 [40, 41] has made it possible to distinguish generation of chemotactic factors by the two major pathways of complement activation. Bacterial endotoxins have been shown to activate the alternate pathway in the serums of un-immunized normal and C4-deficient animals [30]. Immune complexes, on the other hand, activate the classical and alternate pathways in the serum from normal animals but only the alternate pathway in the serum of C4-deficient animals [30]. No detectable difference in the chemotactic activity of serum from normal and C4 deficient guinea pigs was demonstrated after 1 h of activation with either *E. coli* endotoxin or antigen antibody complexes, suggesting that the alternate pathway adequately supported normal chemotactic factor generation. However, kinetic analysis of the generation of C5a in the guinea pig provided a simple method for distinguishing the classical and alternate complement pathways [42]. The kinetics of the former were characterized by rapid (within 5 min) formation of chemotactic factor where as the latter had a latency of 15–20 min before chemotactic activity was detectable.

Dr. MICHAEL FRANK and I have extended these observations to studies of human serum in which the results are similar. As shown in figure 4, formation of chemotactic factors generated from normal human serum activated with immune complexes was rapid with significant chemotactic activity present 5–10 min after activation. The plateau seen at 10–20 min was present in every experiment [also reported in guinea pigs, 42] and may represent inhibition of early complement components by products of the alternate complement pathway [43]. Generation of chemotactic factors from human serum with *Salmonella typhosa* endotoxin was characterized by a

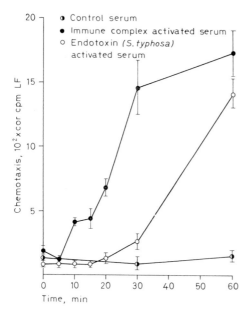

Fig. 4. Generation of chemotactic factors from normal human serum by immune complexes or *S. typhosa* endotoxin activation. Control serum contained no immune complexes or endotoxin and activated samples contained either 300 µg endotoxin or 500 µg immune complexes/ml serum. Incubation was at 37 °C for the indicated times. All samples were then heated at 56 °C for 30 min. The test cells were human granulocytes. The differences between immune complex activated and endotoxin-activated serum were significant at 10–30 min (p<0.01) but were not significant at 60 min (p>0.05).

15- to 20-min latency before the appearance of chemotactic activity (fig. 4). The kinetics of *E. coli* endotoxin activation were different from that of *S. typhosa* endotoxin and, as shown in figure 5, were similar to the kinetic pattern observed with immune complex activation. Antibody to *E. coli* but not *S. typhosa* was detected in the serum used in the experiment shown in figure 5 (bentonite floculation test) and the difference in the kinetics of chemotactic factor formation with the two endotoxins may be related to the larger amount of *E. coli* antibody present which activates the classical complement pathway in normal serum.

To further evaluate the association of a lag period preceding the generation of chemotactic activity with activation of the alternate pathway a kinetic analysis of chemotactic factor formation was studied in human serum deficient in C2 (<0.05% normal C2 titer). As shown in figure 6, when C2-deficient serum was activated with *E. coli* endotoxin there was a 20-min

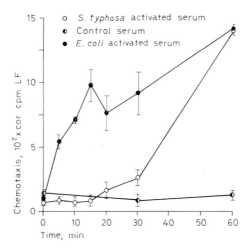

Fig. 5. Generation of chemotactic factors from normal human serum by *E. coli* or *S. typhosa* endotoxin activation. Control serum contained no endotoxin and activated samples contained 300 μg endotoxin/ml serum. Incubation was at 37 °C for the indicated times. All samples were then heated at 56 °C for 30 min. The test cells were human granulocytes. The differences between *S. typhosa* and *E. coli* endotoxin activated sera were significant at 5–30 min (p<0.01) but were not significant at 60 min (p>0.05).

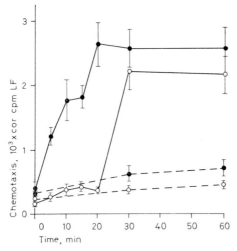

Fig. 6. Generation of chemotactic factors in normal and C-2 deficient human serum by activation with *E. coli* endotoxin (——). Control serum (---) contained no endotoxin and activated samples contained 300 μg endotoxin/ml serum. Incubation was at 37 °C for the indicated times. All samples were then heated at 56 °C for 30 min. The test cells were human granulocytes. The differences between normal (•) and C-2 deficient (○) activated serum were significant at 5–20 min (p<0.01) but not significant at 30 and 60 min (p>0.05).

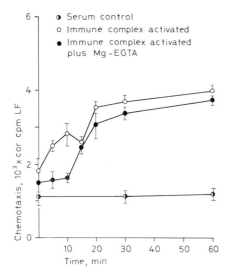

Fig. 7. Generation of chemotactic factor from normal human serum with immune complexes in the presence and absence of Mg^{++}-EGTA. Control serum contained neither immune complexes or Mg^{++}-EGTA. Activated samples contained 500 μg immune complexes/ml serum with or without Mg^{++} (5×10^{-3} M)-EGTA (10^{-2} M) as denoted by the open and closed circles, respectively. Incubation was at 37°C for the indicated times. All samples were then heated at 56°C for 30 min. The test cells were human granulocytes. The differences between the immune complex activated samples with or without Mg-EGTA were significant at 5 and 10 min (p<0.05) but not significant after 15 min of activation (p>0.05).

latency before formation of chemotactic factor; similar results occurred with the C2-deficient serum activated with immune complexes. Addition of purified C2 to the C2-deficient serum restored normal kinetics. In a related series of experiments the kinetics of generation of chemotactic activity from normal serum by immune complexes was examined in the presence of magnesium (5×10^{-3} M)-ethylene-(*bis*)-glycol-tetraacetic acid (10^{-2} M) (EGTA) in order to chelate calcium and preferentially block the classical but not alternate complement pathway. The time-course of generation of chemotactic activity was converted to the pattern characteristic of *S. typhosa* endotoxin activation (fig. 7).

Using molecular sieve chromatography (Sephadex G-75) we have analyzed the chemotactic factors generated by endotoxin or immune complex activation. In both instances, the chemotactic activity was associated with elution volumes corresponding to a molecular weight of about 12,000–

Table I. Generation of chemotactic factor from human plasma by activation of the kinin-generating system

Plasma	Chemotaxis[1] (cor cpm LF)
Normal	
Nonactivated	$1,501 \pm 107$
Kaolin-activated	$3,682 \pm 249$
Prekallikrein-deficient	
Nonactivated	$1,586 \pm 101$
Kaolin-activated	$2,079 \pm 55$
Kaolin-activated plus prekallikrein	$2,603 \pm 18$

1 Mean $\pm$ SE; cor cpm LF = corrected counts per minute lower filter.

16,000. This material was heat-stable at 56 °C and was inactivated with goat antihuman C5 but not anti-C3 and therefore represents the C5 cleavage product, C5a. There were no detectable quantitative differences in chemotactic material obtained by either endotoxin or immune complex activated sera.

B. Kinin-Generating System

Kallikrein, the serum enzyme which digests kininogen to yield bradykinin, recently has been shown to possess chemotactic activity for human neutrophils [22]. This enzyme is generated by activation of prekallikrein by activated Hageman factor or fragments derived from activated Hageman factor [44, 45]. Dr. ALLEN KAPLAN and I have recently studied the generation of chemotactic factors from normal and prekallikrein deficient human plasma utilizing kaolin to activate the Hageman factor [46, 47]. As shown in table I, the total chemotactic activity generated by kaolin-activated plasma is comparable to the activity generated by endotoxin or immune complexes. However, generation of chemotactic factors with kaolin is clearly distinguished from the complement pathways. Ethylene-diamine-tetraacetic acid (EDTA), which chelates both calcium and magnesium [48] and thereby inhibits both pathways of complement-related chemotactic factor formation [49], has no effect on the ability of kaolin to generate chemotactic activity.

Furthermore, as shown in table I, human plasma deficient in prekallikrein is unable to generate normal chemotactic activity after kaolin activation unless the absent substrate is added back to the plasma. Normal chemotactic activity is generated when prekallikrein deficient serum is activated with endotoxin or immune complexes demonstrating at least a partially intact complement system. The *in vivo* physiological role of the chemotactic properties of kallikrein to host-defense mechanisms remains to be established.

Summary

The above studies demonstrate that the ^{51}Cr radiolabel chemotactic assay is a relatively simple and objective means for studying leukocyte chemotaxis in both normal and pathological conditions. Application of this method to studies of normal human chemotaxis revealed a relatively narrow range of normal and little day-to-day variability. Analysis of this variability revealed that there is more variability among the response of different granulocytes to a constant chemotactic stimulus than among the chemotactic activity of different sera to a single cell source. Utilizing the ^{51}Cr radioassay, the abnormal granulocyte chemotactic behavior reported in Chediak-Higashi syndrome [2] and a patient with recurrent pyogenic infections and mucocutaneous candidiasis [7] has been confirmed.

The ^{51}Cr chemotactic assay has also been used to assess the generation of chemotactic activity from human serum and plasma. The *in vitro* generation of two distinct chemotactic factors were examined; the complement product (C5a) and kallikrein, an enzyme of the kinin-generating pathway. Kinetic analysis of complement-related chemotactic factor formation, utilizing immune complexes or endotoxin to activate normal sera in the presence or absence of EGTA as well as kinetic analysis of activation of C2-deficient human serum, provided an easy means of distinguishing the classical (antibody-mediated) complement pathway from the alternate pathway. Such kinetic analysis is necessary to detect clinically important abnormalities since, after 60 min of generation time, normal chemotactic activity may be present despite complete absence or inhibition of one complement pathway. The chemotactic factor generated by either pathway of complement activation appears to be predominately attributable to C5a.

References

1   MOWAT, A. J. and BAUM, J.: Chemotaxis of polymorphonuclear leukocytes from patients with diabetes mellitus. New Engl. J. Med. *284:* 621–627 (1971).
2   CLARK, R. A. and KIMBALL, H. R.: Defective granulocyte chemotaxis in the Chediak-Higashi syndrome. J. clin. Invest. *50:* 2645–2652 (1971).
3   MOWAT, A. G. and BAUM, J.: Chemotaxis of polymorphonuclear leukocytes from patients with rheumatoid arthritis. J. clin. Invest. *50:* 2541–2549 (1971).
4   STEERMAN, R. L.; SNYDERMAN, R.; LEIKIN, S. L., and COLTEN, H. R.: Intrinsic

defect of the polymorphonuclear leukocyte resulting in impaired chemotaxis and phagocytosis. Clin. exp. Immunol. 9: 839–852 (1971).

5 McCall, C. E.; Caves, J.; Cooper, R., and DeChatelet, L.: Functional characteristics of human toxic neutrophils. J. infect. Dis. 124: 68–75 (1971).

6 Miller, M. E.; Oski, F. A., and Harris, M. B.: Lazy leukocyte syndrome. Lancet i: 665–669 (1971).

7 Clark, R. A.; Root, R. K.; Kimball, H. R., and Kirkpatrick, C. H.: Defective neutrophil chemotaxis and cellular immunity in a child with recurrent infections. Ann. intern. Med. 78: 515–519 (1973).

8 Page, A. R.; Gewurz, H.; Pickering, R. J., and Good, R. A.: The role of complement in the acute inflammatory response. Immunopathology 5: 221–2J8 (1967).

9 Keller, H. U. and Sorkin, E.: Chemotaxis of leukocytes. Experientia 24: 641–652 (1968).

10 Clark, R. A.; Kimball, H. R., and Decker, J. L.: Neutrophil chemotaxis in systemic lupus erythematosis, Ann. rheum. Dis. (in press).

11 Boyden, S.: The chemotactic effect of mixtures of antibody and antigen on polymorphonuclear leukocytes. J. exp. Med. 115: 453–466 (1962).

12 Gallin, J. I.; Clark, R. A., and Kimball, H. R.: Granulocyte chemotaxis. An improved in vitro assay employing ⁵¹Cr labeled granulocytes. J. Immunol. 110: 233–240 (1973).

13 Boyum, A.: Isolation of mononuclear cells and granulocytes from human blood. Scand. J. clin. Lab. Invest. 97: suppl. 21, pp. 77–89 (1968).

14 Mickenberg, I. D.; Root, R. K., and Wolff, S. M.: Leukocyte function in hypogammaglobulinemia. J. clin. Invest. 49: 1528–1538 (1970).

15 Snyderman, R.; Gewurz, H., and Mergenhagen, S. E.: Interactions of the complement system with endotoxin lipopolysaccharide. J. exp. Med. 128: 259–275 (1968).

16 Gewurz, H.; Shin, H. S., and Mergenhagen, S. E.: Interactions of the complement system with endotoxic lipopolysaccharide. Consumption of each of the six terminal complement components. J. exp. Med. 128: 1049–1057 (1968).

17 Siegel, S. and Tukey, J. W.: A nonparametric sum of ranks procedure for relative spread in unpaired samples. J. amer. Statist. Ass. 55: 429–445 (1960).

18 Blume, R. A. and Wolff, S. M.: The Chediak-Higashi syndrome. Studies in four patients and a review of the literature. Medicine, Balt. 51: 247–280 (1972).

19 Wolff, S. M.; Kimball, H. R., and Clark, R. A.: The Chediak-Higashi syndrome. Birth Defects Orig. Art. Ser. 8: 69–73 (1972).

20 Ward, P. A.; Cochrane, C. G., and Müller-Eberhard, H. J.: The role of serum complement in chemotaxis of leukocytes in vitro. J. exp. Med. 122: 327–346 (1965).

21 Stecher, V. J. and Sorkin, E.: Studies on chemotaxis. XII. Generation of chemotactic activity for polymorphonuclear leukocytes in sera with complement deficiencies. Immunology, Lond. 16: 231–239 (1969).

22 Kaplan, A. P.; Kay, A. B., and Austen, K. F.: A prealbumin activator of prekallikrein. III. Appearance of chemotactic activity for human neutrophils by the conversion of human prekallikrein to kallikrein. J. exp. Med. 135: 81–97 (1972).

23 Weiss, A. P.; Gallin, J. I., and Kaplan, A. P.: Fletcher factor deficiency: A diminished rate of Hageman factor activation caused by absence of prekallikrein with abnormalities of coagulation, fibrinolysis, chemotactic activity and kinin generation. J. clin. Invest. 53: 622–633 (1974).

24 RUDDY, S.; GIGLI, I., and AUSTEN, K. F.: The complement system of man. New Engl. J. Med. *287:* 489–495, 545–549, 592–596, 642–646 (1972).

25 MÜLLER-EBERHARD, H. J.; HADDING, U., and CALCOTT, C. A.: Current problems in complement research. Immunopathology *5:* 179–188 (1967).

26 BOREL, J. F. and SORKIN, E.: Differences between plasma and serum mediated chemotaxis of leukocytes. Experientia *25:* 1333–1335 (1969).

27 FINE, D. P.; MARNEY, S. R.; CALLEY, D. G.; SERGENT, J. S., and DESPREZ, R. M.: C3 shunt activation in human serum chelated with EGTA. J. Immunol. *109:* 807–809 (1972).

28 GÖTZE, O. and MÜLLER-EBERHARD, H. J.: The C3-activator system. An alternate pathway of complement activation. J. exp. Med. *134:* 90s–108s (1971).

29 PHILLIPS, J. K.; SNYDERMAN, R., and MERGENHAGEN, S. E.: Activation for γ2 globulin, C1, C4 and C2 in the consumption of terminal complement components by endotoxin-coated erythrocytes. J. Immunol. *109:* 334–341 (1972).

30 FRANK, M. M.; MAY, J.; GAITHER, T., and ELLMAN, L.: *In vitro* studies of complement function in sera of C4 deficient guinea pigs. J. exp. Med. *134:* 176–187 (1971).

31 MAY, J. E. and FRANK, M. M.: Complement-mediated tissue damage. Contribution of the classical and alternate complement pathways in the Forssman reaction. J. Immunol. *108:* 1517–1525 (1972).

32 SANDBERG, A. L.; SNYDERMAN, R.; FRANK, M. M., and OSLER, A. G.: Production of chemotactic activity by guinea pig immunoglobulins following activation of the C3 complement shunt pathway. J. Immunol. *108:* 1227–1231 (1972).

33 MAY, J. E.; KANE, M. A., and FRANK, M. M.: Host defense against bacterial endotoxin-contribution of the early and late components of complement to detoxification. J. Immunol. *109:* 893–895 (1972).

34 MAY, J. E.; KANE, M. A., and FRANK, M. M.: Immune adherence by the alternate complement pathway. Proc. Soc. exp. Biol. Med. *141:* 287–290 (1972).

35 CLARK, R. A.; KIMBALL, H. R., and PADGETT, G. A.: Granulocyte chemotaxis in the Chediak-Higashi syndrome of mink. Blood *39:* 644–649 (1972).

36 SNYDERMAN, R. and MERGENHAGEN, S. E.: Characterization of polymorphonuclear leukocyte chemotactic activity in serum activated by various inflammatory agents; in INGRAM Proc. 5th Int. Symp. Canad. Soc. Immunology, p. 117 (Karger, Basel 1972).

37 SNYDERMAN, R.; SHIN, H. S.; PHILLIPS, J. K.; GEWURZ, H., and MERGENHAGEN, S. E.: A neutrophil chemotactic factor derived from C5 upon interaction of guinea pig serum with endotoxin. J. Immunol. *103:* 413–422 (1969).

38 SNYDERMAN, R.; PHILLIPS, J., and MERGENHAGEN, S. E.: A polymorphonuclear leukocyte chemotactic activity in rabbit serum and guinea pig serum treated with immune complexes evidence for C5a as the major chemotactic factor. Infect. Immun. *1:* 521–525 (1970).

39 WARD, P. A.: Chemotaxis of polymorphonuclear leukocytes. Biochem. Pharmacol. *17:* suppl., pp. 99–105 (1968).

40 ELLMAN, L. I.; GREEN, I., and FRANK, M. M.: Genetically controlled total deficiency of the fourth component of complement in the guinea pig. Science *170:* 74–75 (1970).

41 FRANK, M. M.; ELLMAN, L., and GREEN, I.: A new genetically-controlled complement abnormality. C4 deficiency in the guinea pig. In: Biological activities of complement, pp. 256–259 (Karger, Basel 1972).

42 CLARK, R. A.; KIMBALL, H. R., and FRANK, M. M.: Generation of chemotactic factors in normal and C4 deficient guinea pig serum by activation with endotoxin and immune complexes. Immunopath. clin. Immunol. *1:* 414–425 (1973).

43 KOETHE, S. I.; GIGLI, I., and AUSTEN, F. F.: Blockage of a 142 intermediate by products of the alternate pathway. Fed. Proc. *31:* 787 (1972).

44 KAPLAN, A. P. and AUSTEN, K. F.: A prealbumin activator of prekallikrein. J. Immunol. *105:* 802–811 (1970).

45 KAPLAN, A. P. and AUSTEN, K. F.: A prealbumin activator of prekallikrein. II. Deprivation of activators of prekallikrein from active Hageman factor by digestion with plasmin. J. exp. Med. *133:* 696–712 (1971).

46 RATNOFF, O. D. and DAVIE, E. W.: The purification of activated Hageman factor (activated factor XII). Biochemistry *1:* 967–975 (1962).

47 ZIMMERMAN, T. S. and ARROYAVE, C. M.: Participation of complement in initiation of blood coagulation and in the normal coagulation process; in AUSTEN and BECKER Biochemistry of the acute allergic reactions. 2nd Int. Symp., pp. 321–322 (Blackwell, Oxford 1971).

48 DANIEL, E. E. and IRWIN, J.: On the mechanism whereby EDTA, EGTA, DPTA, oxalate, desferrioxamine, and 1,10-phenantroline affect contractility of rat uterus. Canad. J. Physiol. Pharmacol. *43:* 111–136 (1965).

49 PATTEN, E.; GALLIN, J. I.; CLARK, R. A., and KIMBALL, H. R.: Effects of cell concentration and various coagulants on neutrophil migration. Blood *41:* 711–719 (1973).

Author's address: Dr. JOHN I. GALLIN, MD, Laboratory of Clinical Investigation, National Institute of Allergy and Infectious Diseases, National Institutes of Health, *Bethesda, MD 20014* (USA)

Antibiotics and Chemotherapy, vol. 19, pp. 161–178
(Karger, Basel 1974)

Localised Leukocyte Mobilisation in the Rabbit Ear

An in vivo *Cell Migration Technique Using Plastic Collection Chambers*

CAMILLE FEURER and J. F. BOREL

Biological and Medical Research Division, Sandoz Ltd., Basel

Introduction

Much progress has been achieved in the qualitative and quantitative measurement of cell migration towards chemotactic agents *in vitro* since the development of the Boyden technique [BOYDEN, 1962]. Successful demonstration of *in vivo* leukocyte chemotaxis from blood capillaries into a focal heat-injured extravascular region came from BUCKLEY [1963]. The cine-microscopic observation, however, seems too tedious for general use in testing chemotactic and inhibitory substances *in vivo*. Several workers have, therefore, investigated other possibilities for a versatile and reliable method for measuring chemotaxis *in vivo*, particularly in experimental animals. Most of the techniques developed until now are, however, not very satisfactory and often produce conflicting results. To study the emigration of granulocytes from skin lesions in man, REBUCK and CROWLEY [1955] developed the skin window method. GOWLAND [1964] has successfully used a skin window-box, a technique which was subsequently modified by others [SOUTHAM and LEVINE, 1966; SENN *et al.*, 1969]. The present paper deals with a modified plastic skin collection chamber applied to the rabbit ear. Results obtained by this method, which allows a quantitative assessment of leukocyte emigration *in vivo*, are reported and related to those obtained *in vitro* with the Boyden chamber technique [BOYDEN, 1962].

Material and Method

Animals. Female rabbits of the conventional Swiss Hare breed weighing 2½–3 kg were used throughout these experiments.

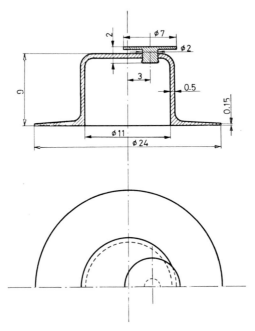

Fig. 1. Design of the plastic collection chamber for the rabbit ear. Polyacetylene cap with enlarged edge. Aperture on top closed by a plastic press-button (dimensions in millimetres).

Fresh normal rabbit serum (NRS) and Gey's solution. Rabbits were exsanguinated and the blood allowed to clot. Individual sera were centrifuged free of cells, pooled, filtered through a 0.45-μm millipore filter and stored at $-20\,°C$. If used as a positive control, the NRS was thawed at $37\,°C$ with no further treatment. For use in negative controls or as diluting medium Gey's solution was modified by a 5-fold increase in dextrose concentration and the addition of antibiotics (200 IU/ml penicillin from Novo Industri A/S, Kopenhagen, Denmark, and 50 μg/ml dihydro-streptomycin sulfate from Calbiochem, Los Angeles, Calif., USA) [KELLER and SORKIN, 1967].

Design of the plastic collection chamber. Polyacetylene caps of 12 mm diameter and 0.8 ml capacity were modified by enlarging the edge to a width of 5–7 mm with a thin plastic ring sealed to it. This was done to ensure tight adhesion of the chamber to the ear surface. A small aperture, which could be closed by a plastic press-button, was drilled into the top to allow filling and emptying of the chamber (fig. 1).

Skin lesion. The whole ear was shaved and cleaned with 70% ethanol. The area to be abrased was demarcated by fixing on the ear a piece of tape with a window of 1 cm² cut out. The lesion itself was produced by repeatedly sticking firmly onto the skin and tearing off a piece of 'Poroplast' plaster (Int. Verbandstoff-Fabrik, Schaffhausen, Switzerland) until the epidermis was removed and the lesion appeared exudating and reddish.

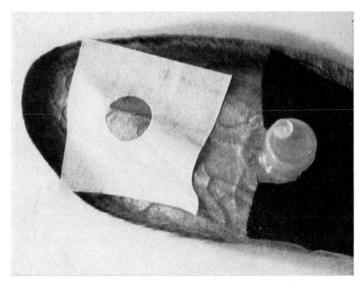

Fig. 2. Illustration of the fixation of a plastic collection chamber on the rabbit ear. Left: demarcation with tape of area to be abrased. Right: chamber in position. White semi-circle is the enlarged basis of the chamber glued to the ear. Only one strip of black tape is shown.

By this procedure the basal cell layer was more or less removed, but the underlying corium and capillary network remained intact. The lesions never showed any signs of bleeding.

Fixation of the chamber. When the lesion was prepared and the tape removed, a sterile chamber was placed over it, glued to the ear with 'Cementit' paste (Merz & Benteli AG, Berne, Switzerland) and held in place until the glue started to dry. 'Cementit' in no way influenced the leukocyte emigration. In order to maintain the chamber very firmly in position and to avoid leakage, the fixation was strengthened by sticking over the edge of the chamber and on the outside of the ear two wide strips of black 'Tesa' tape (Tesa-Band 541, Beiersdorf AG, Hamburg, FRG) (fig. 2). Up to 4 chambers per rabbit or 2 per ear can be positioned. To prevent the animals from scratching off the chambers, they were kept in narrow wooden boxes during the entire experimental period.

Filling of the chamber and collection of leukocytes. Once the chambers had been tightly fixed, they were filled with test fluid by introducing through the opening a fine-gauge needle connected to a tuberculin syringe. Approximately 0.8 ml of fluid could be injected into the chamber until all air had escaped. The aperture was then closed with the press-button and the rabbit left alone in a quiet room until the time for cell collection, usually 6 h later. To harvest the emigrated cells, the entire fluid in the chamber was first flushed twice directly over the lesion using a needle with a blunt, thickened end. Immediately afterwards, the fluid was aspirated into a syringe and the leukocytes were counted in a haemocytometer. After collection, the chamber could be either removed or refilled with the old or a new sample. Rabbits were rested for 3 weeks before being used for another experiment, to ensure complete healing of the wound.

Results

Effects of NRS and Gey's Solution

GOWLAND [1964] and SENN et al. [1969; SENN, 1972] reported the high leukotactic activity of fresh autologous serum in man. The effect of fresh autologous and homologous NRS on localised leukocyte mobilisation was therefore investigated in the rabbit ear using plastic collection chambers. Fresh NRS samples were used without previous inactivation as routinely carried out for in vitro experiments. Pooled NRS was as effective as autologous serum and subsequently only pooled samples were used.

The concentration of the NRS was of importance for the leukocyte mobilisation as shown in table I. Ten per cent NRS showed very little activity, whereas 50% NRS gave positive results. The best results, however, were obtained with undiluted fresh NRS (100%) which was used in all experiments as the standard positive control.

When undiluted fresh NRS was inactivated for 30 min at 56 °C before being filled into a chamber, the leukocyte count decreased to approximately 15% of the corresponding value for the same, but untreated, fresh NRS (average of 4 trials). Furthermore, a comparison between untreated fresh NRS and NRS treated with antigen-antibody complexes showed that both sera gave approximately similar cell counts. Activation by immune complexes did not noticeably modify the chemotactic activity of fresh NRS in vivo, in contrast to previous in vitro findings.

Gey's solution, formerly used in vitro as the negative control, produced values varying from 0 to 1,000 leukocytes/mm^3. As these counts were always considerably below the corresponding ones for NRS (cf. also fig. 3), it was maintained as the standard negative control.

Individual Variations

The variation between values obtained both for different rabbits and within a single individual were compared. Table II gives a representative

Table I. Effect of concentration of fresh normal rabbit serum (NRS) on the localised leukocyte mobilisation with collection chambers on the rabbit ear

Concentration of fresh NRS (pool) %	Leukocytes/mm³ in single animals			
	RN 6,702	RN 6,712	RN 8,748	RN 8,652
10	4,370	630	250	940
50	8,650	11,000	9,200	7,800
100	10,400	11,400	17,800	22,800

RN = rabbit No.

Table II. Individual variations between and within rabbits in the localised leukocyte mobilisation induced by fresh NRS

Rabbits No.	Leukocytes/mm³				Mean value + SD
	exp. 1	exp. 2	exp. 3	exp. 4	
8,534	2,140	3,490	1,490	5,580	3,175 ± 1,807
8,545	2,410	3,410	4,180	–	3,333 ± 888
8,606	3,380	3,140	3,830	–	3,450 ± 350
8,536	5,020	7,270	8,600	5,600	6,623 ± 1,627
7,721	8,760	7,800	14,800	12,600	10,990 ± 3,279
8,652	2,070	12,800	8,600	22,800	11,568 ± 8,693
8,531	10,200	19,000	8,800	13,400	12,850 ± 4,530

exp. = Experiment.

group of such values. Firstly, the variation between different animals will be considered. The first three rabbits listed in the table were of the 'low-responder' type, while the last three responded well. Several intermediate responses were encountered as exemplified by rabbit No. 8,536. According to our experience, low-responding individuals produced consistently low counts and were, therefore, discarded after a second unsuccessful trial. Secondly, the variation within the same rabbit from one experiment to the other was very small in some animals (e.g., rabbits No. 8,606 and 8,536), but rather large in others (e.g., rabbit No. 8,652).

The reason for the variation between rabbits might be a genetic one, especially as all rabbits of a different breed, namely the yellow-silver breed, gave extremely low values. The variation within a single rabbit, as for No. 8,652 in experiment 1, could be due partly to irregular abrasion of the skin. However, further factors must apply as some animals showed a very regular response pattern throughout all tests. Moreover, an effect on a particular chamber induced by an adjacent chamber containing a different liquid may also be excluded. A series of experiments was conducted in which the distal chamber contained fresh NRS (positive control) and the proximal chamber Gey's solution (negative control), or *vice versa*. The leukocyte emigration clearly proved to be independent of the respective position of either control.

Kinetics of the Localised Leukocyte Mobilisation

Gowland [1964] reported a marked lag period in the course of leukocyte emigration in humans. Chang and Houck [1970], using a millipore filter package applied to abraded rat skin, also observed a time-lag of 3 h before the polymorphs penetrated into the filter. Similar findings were obtained with the rabbit, as shown in figure 3. It represents the kinetics of leukocyte mobilisation in single rabbits induced by Gey's solution and fresh NRS between 1 and 6 h. Up to $2^1/_2$ h little cell migration into the collection chamber was observed, while at 4 h the cell numbers found with NRS were clearly greater than the corresponding negative control values and rapidly increased during the next 2 h. At that time the negative control counts remained considerably below the limit of 1,000 cells/mm³.

The initial slow and subsequent sudden increase of leukocyte mobilisation has already been noted in humans by Senn *et al.* [1969], and Barnhart [1968] reported the same phenomenon in dogs. Cliff [1966] observed the first leukocytes sticking to the endothelium of venules about 1 h after local injection of an immune complex and saw no signs of emigration into the perivascular region before $1^1/_2$ h. He further estimated the mean rate of emigration for leukocytes in tissues to be about 7.4 μm/min. Assuming that the distance from the venules to the abrased skin surface measures between 300 and 500 μm and that a leukocyte migrates in a directional manner only two-thirds of the total distance covered towards the lesion it seems reasonable that the appearance of cells in the ear chamber starts after about 4 h and develops fully in the next few hours.

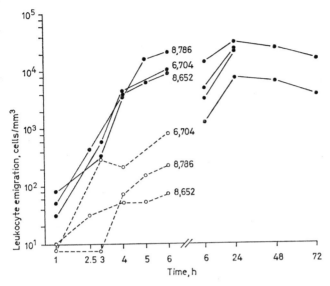

Fig. 3. Kinetics of the localised leukocyte mobilisation as measured by the rabbit collection chamber technique. The solid lines represent values for positive controls (fresh normal rabbit serum [NRS]), while the dotted lines indicate the corresponding values for the negative controls (Gey's solution) during the first 6 h only. The data shown for the 6- to 72-hour periods are derived from rabbits other than those designated by their ear numbers and used in the previous 1- to 6-hour periods.

Moreover, figure 3 demonstrates a further increase of the cell number after 6 h, reaching a peak around 24 h, and diminishing gradually during the next 48 h. At 72 h most cells were dead, but surprisingly no mononuclear cells were encountered, the cell population consisting exclusively of polymorphonuclear leukocytes.

Several attempts were made to induce macrophage emigration into the ear chamber, but none was successful. From *in vitro* experiments [SORKIN *et al.*, 1970] it is known that a post-granular fraction from rabbit liver produces a powerful cytotaxin for macrophages after incubation with fresh NRS, while its effect on neutrophils remained comparatively weak. In the present *in vivo* system this cytotaxin stimulated neutrophil mobilisation to a similar degree as did fresh NRS, but again no mononuclear cells were seen during a 48-hour period. Other substances tested included sodium caseinate (3.5%, as used to elicit peritoneal exudates; Nutritional Biochemicals, Cleveland, Ohio, USA) and a bacterial filtrate from *Escherichia coli* (strain

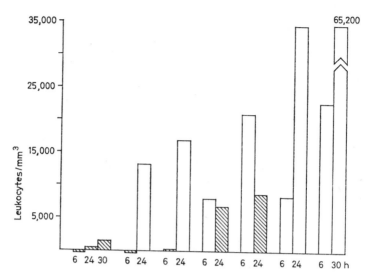

Fig. 4. Effect of changing or renewing the content of collection chambers on the leukocytic emigration. Numbers under columns designate hours when cells were counted and fluids replaced. Hatched columns represent Gey's solution and white columns fresh NRS.

K 12; prepared in our laboratories), both of which affected neither cell type. A commercial endotoxin (Bacto lipopolysaccharide from *E. coli;* Difco, Detroit, Mich., USA) diluted to 15 μg/ml in Gey's solution exerted *per se* a very potent activity for neutrophils, but completely failed to attract macrophages. Replacement of the chamber content with a fresh NRS sample at 6 h resulted in a major increase in the number of granulocytes at 30 h (cf. last pair of columns in fig. 4), but macrophages remained absent. Thus, the mobilisation of mononuclear cells cannot be assessed by the collection chamber technique.

Finally, the effect of changing the content of a chamber on the kinetics of the cellular emigration was investigated. As shown in figure 4, some chambers were filled with Gey's solution for the first 6 h, after which this solution was replaced by fresh NRS for the next 18 h. The reverse procedure was also applied in other chambers which first received fresh NRS and afterwards Gey's solution. A previous application of the negative Gey's solution did not preclude a subsequent influx of cells towards fresh NRS. On the contrary, when Gey's solution followed incubation with NRS, a definite decline in the cell count occurred. The replacement of the first NRS sample by a second fresh NRS sample resulted in an impressive accumulation

Table III. Inhibition by cytochalasin B (CyB) of leukocyte mobilisation *in vivo*

CyB µg/ml	NRS plus CyB from 0–6 h	Replacement by NRS from 6–24 h	NRS plus DMSO (solvent control)	NRS alone (positive control)
1	2,100[1]	nd	12,600	nd
1	220	nd	(DMSO only:	2,460
10	0	5,490	220)	
10	20	nd	8,800	13,400
10	20	30,800	3,200	5,200
10	10	6,890	(DMSO only: 10)	nd

1 Number of cells/mm^3 emigrated into collection chamber at 6 h.
NRS = fresh normal rabbit serum pool, DMSO = dimethylsulphoxide; solvent of CyB (1 mg CyB dissolved in 1 ml DMSO).
nd = Not done.

of granulocytes (as previously mentioned), while in a series of refills with Gey's solution alone no significant leukocyte mobilisation was ever observed.

Reversible Inhibition of the Leukocyte Migration by Cytochalasin B

BECKER *et al.* [1972] and BOREL and STÄHELIN [1972] have demonstrated that cytochalasin B inhibits reversibly the chemotactic migration of neutrophils *in vitro*. The effect of cytochalasin B was also tested in this *in vivo* model using NRS as the active control substance. The data presented in table III indicate complete inhibition of leukocyte migration when cytochalasin B is added to fresh NRS. The neutrophil emigration *in vivo* was clearly inhibited down to a concentration of 1 µg/ml NRS. This inhibition is, however, reversible as replacement of NRS plus cytochalasin B by a sample of fresh NRS alone after 6 h resulted in an unimpeded mobilisation of leukocytes for the next 18 h. The chambers were rinsed 3 times with Gey's solution immediately before replacement and the newly infiltrated cells measured at 24 h. In the last experiment listed in table III a new sample of NRS plus cytochalasin B was again added after 24 h and the cells counted

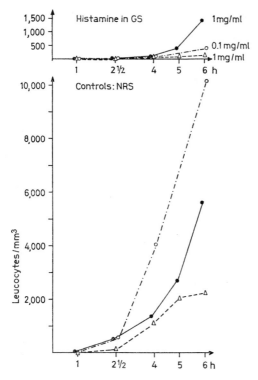

Fig. 5. Effect of histamine applied topically on leukocyte mobilisation. 1.0 or 0.1 mg/ml histamine was added to Gey's solution (GS) which was used to fill the ear chamber (top part), while a positive control (fresh NRS) without histamine was placed on the opposite ear (lower part).

at 30 h (not shown in the table). In these 6 h their number had decreased to 140 cells/mm³, showing that inhibition of leukocyte mobilisation could be induced repeatedly. It seems, therefore, that cytochalasin B affects the motility of rabbit neutrophils not only *in vitro* but also *in vivo*.

Effect of Histamine on the Localised Leukocyte Mobilisation

The use of a vasodilator agent such as histamine should allow evaluation of the extent to which diapedesis of leukocytes is caused by either passive leakage from the venules into the perivascular area or by active migration towards a chemotactic stimulus. HURLEY [1964] noticed that repeated intradermal injections of histamine (100 μg/ml) in rats evoked a

Table IV. Effect of histamine (intravenous application) on leukocyte mobilisation

Rabbit No.	Hist-amine i.v. 2×0.1 mg/kg	Time of injection h	Test fluid	Leukocytes/mm³ at			Control values at 6 h[1]
				2½–3 h	4 h	6 h	
8,606	–½	0[2]	NRS	310	330	2,180	3,450
8,606	0	1	NRS	120	1,070	2,090	3,450
			Gey's	0	0	0	50
6,704	0	1	NRS	nd	7,420	13,400	12,200
			Gey's	nd	330	1,040	810
8,536	0	2	NRS	110	900	2,940	6,620
			Gey's	0	50	20	0
8,534	0	2	NRS	0	220	2,020	3,160
			Gey's	0	0	0	20
8,534	0	3	NRS	180	410	3,120	3,160
6,710	0	3	NRS	260	5,440	10,800	19,500

1 Control values from same rabbits, but tested 3 weeks before or after treatment with histamine.
2 0 h is the time when the collection chamber was applied on the ear and filled.
nd = Not done.

mild delayed emigration of polymorphonuclear cells into the surrounding tissues. GOWLAND [1964] using the 'window-box' method found that local application of a mixture of 300 µg histamine/ml Hanks' solution or physiological saline did not cause any appreciable increase in leukocytic emigration. The present study in which histamine was applied topically confirmed the latter results. The flat curves in the upper part of figure 5 indicate the number of leukocytes having emigrated at 6 h into the collection chamber filled with 100 µg or 1 mg histamine/ml Gey's solution. The steep curves on the lower part represent the control values for the chamber located on the opposite ear and containing fresh NRS alone. Results from the 3 rabbits represented, strongly suggest that vasodilatation produced by local application of histamine could not have induced passive cell leakage into the ear chamber, because the leukocyte counts did not exceed those normally obtained for negative controls with Gey's solution only.

The effect of systemic administration of histamine on localised leukocyte mobilisation was investigated. Rabbits were injected twice intravenously with 100 µg histamine/kg body weight and collection chambers containing

NRS (positive) and Gey's solution (negative control) were applied. The scheme of histamine injection was varied and the emigrated leukocytes counted at 2.5–3, 4 and 6 h. The normal positive and negative control values for each rabbit were determined separately either 3 weeks before or after histamine treatment. The results shown in table IV clearly indicate that repeated intravenous injections of histamine in rabbits did not influence the migratory behaviour of neutrophils.

As both topical and systemic administration of relatively high doses of histamine did not affect the localised leukocyte mobilisation as measured by the collection chamber technique, there is conclusive evidence for considering vasodilatation and leukocyte emigration as two independent phenomena.

Comparison between Localised Leukocyte Mobilisation and Chemotaxis in vitro

There can be no doubt that the *in vitro* test performed in the Boyden chamber measures chemotactic migration. However, it has not yet been conclusively demonstrated that cellular migration, as it occurs in the localised mobilisation test, is truly chemotactic and not a purely random phenomenon.

The results obtained with both methods, the localised leukocyte mobilisation *in vivo* and the chemotaxis test *in vitro*, are represented in table V. Cell numbers of less than 1,000 cells/mm³ at the 6-hour reading *(in vivo)* or less than 30 at the 3-hour reading *(in vitro)* may be considered as negative background values. The following agents therefore proved inactive with both methods: Gey's solution, heat-inactivated NRS, immune complex alone and the liver fraction (S16,000 *g*), the latter fraction showing only questionable activity *in vivo*. Endotoxin was inactive *in vitro*, but caused extensive neutrophil emigration *in vivo*, suggesting that it induces cytotaxin formation when in contact with exuding fluid of the lesion. Besides endotoxin, immune complexes and the liver fraction are also to be regarded as cytotaxigens, but their lack of effect *in vivo* could be due to their enormous particle size which may impair their diffusion into the tissues and thus prevent cytotaxin formation sufficient to promote cell attraction.

NRS activated by immune complexes, endotoxin or a liver fraction (the latter being especially active *in vitro* for macrophages) contained strong cytotaxins for *in vitro* chemotaxis and also effectively stimulated leukocytic emigration *in vivo*. Since heat-inactivated NRS induced extremely weak

Table V. Comparison between leukocytic emigration *in vivo* (plastic collection chamber) and chemotactic migration *in vitro* (Boyden chamber) of rabbit polymorphonuclear leukocytes

Agents tested	Numbers of polymorphonuclear leukocytes	
	per cubic millimetre, 6-hour readings *in vivo*	per microscopic field, 3-hour readings *in vitro*
Gey's solution[1]	171 (7)	0 (4)
Fresh NRS (pool) (100%)	8,850 (31)	–[5]
Heat-inactivated NRS (100%)[2]	802 (4)	25 (41)
Immune complex (human serum albumin [HSA]-anti-HSA) (20%)[2]	73 (3)	1 (11)
Immune complex plus NRS (100%)[2]	7,280 (5)	404 (41)
Endotoxin (15 μg/ml)[1]	9,600 (4)	0 (2)
Endotoxin plus NRS (50%)[3]	10,502 (5)	220 (3)
Liver fraction S 16,300 g (6 mg/ml)[4]	1,300 (3)	1 (23)
Liver fraction plus NRS (100%)[4]	7,586 (5)	71 (14)
Casein (1%)[1]	4,074 (5)[6]	549 (8)

Number of tests in brackets. The figures represent the average cell count of n tests.
1 Gey's solution, endotoxin in Gey's solution and casein were used in identical concentrations for both *in vivo* and *in vitro* experiments.
2 Heat-inactivated NRS, immune complex in Gey's solution and immune complex in NRS were used as indicated for *in vivo* tests, but diluted 1:9 in Gey's solution for *in vitro* tests.
3 Endotoxin with NRS: as indicated for *in vivo* experiments, but only 5% NRS used for *in vitro* tests.
4 Liver fraction in Gey's solution or in NRS was used as indicated for *in vivo* tests. For *in vitro* experiments only 1 mg/ml of the fraction was taken and the NRS diluted to 10% in Gey's solution.
5 Fresh NRS cannot be reliably tested *in vitro*, because it is sometimes suddenly activated by contact with glass or small impurities. Values ranging from negative to highly positive were obtained in such trials.
6 Cell counts with casein (Hammarsten, Merck, Darmstadt, FRG) varied greatly, but the substance was definitely active for leukocytes, though less so than fresh NRS.
N.B. Except for fresh NRS, the liver fraction in Gey's solution and casein, all samples were submitted to the following procedure: incubation for 10 min (endotoxin 30 min) in a water bath at 37 °C followed immediately by a second incubation for 30 min at 56 °C (water bath).

chemotactic migration both *in vivo* and *in vitro* and fresh NRS was highly active in the ear chamber, a heat-labile component or, more likely, a cyto-taxigen reacting with the damaged skin tissue and thus forming a potent cytotaxin is presumed to be present in fresh NRS. Evidence for activation of NRS by various tissues and their extracts has been produced by HURLEY and SPECTOR [1961], who injected such agents into rat skin. BOREL [1970] has shown that endogenous cytotaxins and cytotaxigens were also chemotactic *in vitro*. The abraded skin, however, did not by itself promote chemotactic activity, for Gey's solution was ineffective. Casein alone was very active *in vitro*, but exerted *in vivo* a variable activity. The close agreement between the *in vivo* and the *in vitro* results supports the concept that *in vivo* leukocyte migration is directional.

Discussion

A new method for quantitation of *in vivo* leukocytic emigration in experimental animals has been developed. This involves mainly adaptation to the rabbit ear of the window-box of GOWLAND [1964] or the skin chamber of SENN *et al.* [1969] used on the human forearm. Several attempts to provide a reliable *in vivo* method have been published previously. A straightforward method consists of intradermal injection of material and preparation of histological sections of the inflamed tissue [HURLEY and SPECTOR, 1961; HURLEY, 1964; OZAKI *et al.*, 1971]. This procedure measures leukocyte emigration *in vivo* but does not differentiate between random and directed migration. Another possible technique is to stimulate directly the peritoneum and to quantitate the invading cells. SNYDERMAN *et al.* [1971] have used this technique in C5-deficient mice for evaluating the role of the C5 component in chemotaxis. According to our experience this simple technique was not successful in testing inhibitory substances (i.e., cytochalasin B). OSEBOLD *et al.* [1970] have introduced diffusion chambers into the peritoneal cavity of mice and observed strong chemotaxis of phagocytic cells, but they admittedly created an abnormal *in vivo* situation. CHANG and HOUCK [1970] developed a semi-quantitative *in vivo* technique consisting of a micropore filter package applied to abraded rat skin, and BARNHART [1968] described a skin-window collection chamber for dogs. We have found the two latter methods to be less satisfactory, using the rabbit ear, than the plastic collection chamber which allows direct contact of the test fluid with the lesion. We have found that leukocytes penetrating into filters placed directly over the abraded skin often aggregated and showed extremely irregular migratory patterns.

The major disadvantage of the collection chamber technique in general lies in its failure to promote the escape of mononuclear cells. SENN *et al.* [1969] who made kinetic studies in humans, reported a mononuclear response to occur only if the skin chamber, fixed 24 h previously on the abrasion site, was immediately replaced by a coverslip (Rebuck technique), while in the adjacent skin chamber of concurrent experiments the polymorphonuclear response prevailed. In the collection chamber technique in dogs described by BARNHART [1968], in which a membrane filter is kept in close contact with the lesion, the cell kinetics and granulocyte counts during the first 9 h are comparable to those presented here. In contrast, the appearance of the mononuclear phase was already developing between 4 and 8 h, if the Rebuck technique as modified for dogs was used [RIDDLE and BARNHART, 1965]. However, SOUTHAM and LEVINE [1966], utilising a modified Sykes-Moore tissue culture chamber for collecting cells, have succeeded in obtaining a high percentage of macrophages after 24 h in healthy persons. Except for this case, it appears that collection chamber methods do not reproduce the sequence of cellular events observed with the skin-window technique. The former methods deal predominantly with the acute phase of the inflammatory reaction under nearly physiological conditions, while the latter technique appears to involve the defence reaction against foreign material. The localised leukocyte mobilisation therefore provides a valuable method for quantitating the migration of granulocytes, but not of other cell types. This method may, therefore, prove suitable for the discovery of substances inhibiting granulocyte migration *in vivo* after either systemic or topical application and also help to disclose the mechanism of action of present anti-inflammatory drugs. Furthermore, it seems possible to test, in an autologous system, the chemotactic activity present in biological fluids such as arthritic or gouty synovial exudates.

Histological sections made through the reactive site of rabbit ears on which chambers containing Gey's solution or fresh NRS or NRS activated with the liver fraction had been placed 6 or 24 h before, indicated the presence of mononuclear cells scattered randomly within the tissues. After a 24-hour incubation period with fresh or activated NRS these cells represented approximately 30% of all leukocytes still present in the tissues, which seems a small percentage compared with the large amount of polymorphs having already emigrated out of the lesion into the chamber. However, massive emigration of neutrophils from the venules into the perivascular tissues towards the skin abraison was observed (exept when Gey's solution was used). In the directly underlying tissues of the opposite side of the ear (ventral side) no accumulation of granulocytes was ever detected.

Of interest are the consistent quantitative differences found among individual rabbits of the same breed: animals were found to be either weak, medium or good responders. It seems likely that the differences have a genetic origin, since consistent differences in leukocyte mobilisation were also observed between some breeds.

The assumption that leukocyte mobilisation represents an active process (migration) is strongly supported by the lack of cell invasion in negative controls. Moreover, if the abrasion is excessive with resultant bleeding, the cell count found in negative controls can be quite high; in this case the cells leak through the injured vessels, and erythrocytes are also found in the fluid. Furthermore, when Gey's solution was replaced after 6 h by fresh NRS, the same emigration kinetics was observed as in positive controls. If, however, the NRS was replaced after 6 h by Gey's solution, the cell count dropped significantly in the next 18 h. The reversible inhibition of leukocyte emigration by addition of cytochalasin B to fresh NRS topically in a non-toxic dose also favours the concept of active migration. This idea is furthermore substantiated by the ineffectiveness of histamine to promote the escape of cells after topical as well as systemic application. HURLEY [1964] discovered a total absence of correlation between the time-course and extent of leukocyte emigration and the degree of increased vascular permeability as measured by leakage of dye. OZAKI et al. [1971] isolated from inflamed rat skin a chemotactic factor for polymorphs, and from tumour tissues, two permeability factors which showed independent effects when injected intradermally in rats and mice. It appears, therefore, that in the inflammatory process leukocyte infiltration is dissociable from vasodilatation.

The next question is whether this leukocytic emigration is purely random or directional, that is, chemotactic. There are, as yet, no means to prove or disprove the occurrence of either type of migration in vivo. However, a comparison between the present in vivo data concerning leukocyte stimulation and those derived from in vitro experiments supports the idea that in vivo cell emigration in the present experiments is also chemotactic and mediated by cytotaxins. This concept is further strengthened by the similarity between the migratory behaviour of in vivo and in vitro stimulated leukocytes [KELLER, 1972], by the presence of cytotaxins in inflamed tissues [PHELPS, 1970; WARD and ZVAIFLER, 1971], and by the cell-specificity of the cytotaxins [KELLER and SORKIN, 1969] which explains the cellular events of the inflammatory reaction [BOREL, 1970].

Summary

An *in vivo* technique using a plastic collection chamber to measure the localised mobilisation of leukocytes in the rabbit ear is described. This method deals predominantly with the acute phase of the inflammatory reaction as polymorphonuclear, but not mononuclear cells, migrate through the abrased skin into the collection chamber. Undiluted fresh NRS induced a strong positive cellular response, whilst Gey's solution was used for negative controls. Topical application of cytochalasin B reversibly inhibited the mobilisation of leukocytes. Histamine applied either topically or systemically did not alter the kinetics of the response, thereby indicating that the cell emigration constitutes an active process. A comparison between the results obtained with the collection chamber and with the Boyden chamber *in vitro* strongly suggests that the migration of cells in the present *in vivo* test involves directed, that is chemotactic, cell migration.

Acknowledgements

The constructive criticism of Dr. H. U. KELLER (Department of Pathology, University of Berne) and Dr. H. STÄHELIN (Sandoz Ltd., Basel) as well as the careful correction of the manuscript by Dr. I. HOLMES (Sandoz Ltd., Basel) are gratefully acknowledged. Dr. DOROTHEA WIESINGER (Research Institute Wander, Berne, a Sandoz research unit) initiated this work and her collaboration is much appreciated.

References

BARNHART, M. I.: Role of blood coagulation in acute inflammation; in HOUCK and FORSCHER Chemical biology of inflammation, pp. 205–219 (Pergamon Press, New York 1968).

BECKER, E. L.; DAVIS, A. T.; ESTENSEN, R. D., and QUIE, P. G.: Cytochalasin B. IV. Inhibition and stimulation of chemotaxis of rabbit and human polymorphonuclear leukocytes. J. Immunol. *108:* 396–402 (1972).

BOREL, J. F.: Studies on chemotaxis. Effect of subcellular leukocyte fractions on neutrophils and macrophages. Int. Arch. Allergy *39:* 247–271 (1970).

BOREL, J. F. and STÄHELIN, H.: Effects of cytochalasin B on chemotaxis and immune reactions. Experientia *28:* 745 (1972).

BOYDEN, S.: The chemotactic effect of mixtures of antibody and antigen on polymorphonuclear leucocytes. J. exp. Med. *115:* 453–466 (1962).

BUCKLEY, I. K.: Delayed secondary damage and leucocyte chemotaxis following focal aseptic heat injury *in vivo*. Exp. molec. Path. *2:* 402–417 (1963).

CHANG, D. and HOUCK, J. C.: Demonstration of the chemotactic properties of collagen. Proc. Soc. exp. Biol. Med. *134:* 22–26 (1970).

CLIFF, W. J.: The acute inflammatory reaction in the rabbit ear chamber with particular reference to the phenomenon of leukocytic migration. J. exp. Med. *124:* 543–555 (1966).

GOWLAND, E.: Studies on the emigration of polymorphonuclear leucocytes from skin lesions in man. J. Path. Bact. *87:* 347–352 (1964).

HURLEY, J. V.: Substances promoting leukocyte emigration. Ann. N.Y. Acad. Sci. *116:* 918–935 (1964).

HURLEY, J. V. and SPECTOR, W. G.: Endogenous factors responsible for leukocytic emigration *in vivo.* J. Path. Bact. *82:* 403–420 (1961).

KELLER, H. U.: Chemotaxis and its significance for leucocyte accumulation. Agents Actions *2:* 161–169 (1972).

KELLER, H. U. and SORKIN, E.: Studies on chemotaxis. VI. Specific chemotaxis in rabbit polymorphonuclear leucocytes and mononuclear cells. Int. Arch. Allergy *31:* 575–586 (1967).

KELLER, H. U. and SORKIN, E.: Studies on chemotaxis. XIII. Differences in the chemotactic response of neutrophil and eosinophil polymorphonuclear leucocytes. Int. Arch. Allergy *35:* 279–287 (1969).

OSEBOLD, J. W.; OUTTERIDGE, P. M.; PEARSON, L. D., and DiCAPUA, R. A.: Cellular responses of mice to diffusion chambers. I. Reactions to intraperitoneal diffusion chambers containing *Listeria monocytogenes* and to bacteria-free chambers. Infect. Immun. *2:* 127–131 (1970).

OZAKI, T.; YOSHIDA, K.; USHIJIMA, K., and HAYASHI, H.: Studies on the mechanisms of invasion in cancer. II. *In vivo* effects of a factor chemotactic for cancer cells. Int. J. Cancer *7:* 93–100 (1971).

PHELPS, P.: Appearance of chemotactic activity following intraarticular injection of monosodium urate crystals. Effect of colchicine. J. Lab. clin. Invest. *76:* 622–631 (1970).

REBUCK, J. W. and CROWLEY, J. H.: A method of studying leukocytic functions *in vivo.* Ann. N.Y. Acad. Sci. *59:* 757–794 (1955).

RIDDLE, J. M. and BARNHART, M. I.: The eosinophil as a source for profibrinolysin in acute inflammation. Blood *25:* 776–794 (1965).

SENN, H. J.: Infektabwehr bei Hämoblastosen. Experimentelle Medizin, Pathologie und Klinik, vol. *36* (Springer, Berlin 1972).

SENN, H.; HOLLAND, J. F., and BANERJEE, T.: Kinetic and comparative studies on localized leukocyte mobilization in normal man. J. Lab. clin. Med. *74:* 742–756 (1969).

SNYDERMAN, R.; PHILLIPS, J. K., and MERGENHAGEN, S. E.: Biological activity of complement *in vivo.* Role of C5 in the accumulation of polymorphonuclear leukocytes in inflammatory exudates. J. exp. Med. *134:* 1131–1143 (1971).

SORKIN, E.; BOREL, J. F., and STECHER, V. J.: Chemotaxis of mononuclear and polymorphonuclear phagocytes; in VAN FURTH Mononuclear phagocytes, pp. 397–418 (Blackwell, Oxford 1970).

SOUTHAM, C. M. and LEVINE, A. G.: A quantitative Rebuck technique. Blood *27:* 734–738 (1966).

WARD, P. A. and ZVAIFLER, N. J.: Complement-derived leukotactic factors in inflammatory synovial fluids of humans. J. clin. Invest. *50:* 606–616 (1971).

Authors' address: CAMILLE FEURER and Dr. J. F. BOREL, Biological and Medical Research Division, Sandoz Ltd., *CH-4002 Basel* (Switzerland)

Antibiotics and Chemotherapy, vol. 19, pp. 179–190
(Karger, Basel 1974)

Leucocyte Locomotion and Chemotaxis

W. S. RAMSEY

Biology Department, Kline Biology Tower, Yale University, New Haven, Conn.

Introduction

It is evident that motility is a prerequisite for leucocyte chemotaxis. This feature of chemotaxis is sometimes neglected, thus making it difficult to distinguish between treatments and conditions which affect chemotaxis specifically and those which affect motility in general. The membrane filter technique for determining chemotaxis [BOYDEN, 1962] has achieved almost universal adoption, to the exclusion of other methods, although this technique is not well suited to questions of motility. In fact, direct observations on the behavior of leucocytes at the moment of beginning a chemotactic response have only recently been reported. This paper discusses the nature of leucocyte locomotion, the relationship between chemotaxis and motility, and the inhibition of motility by antimitotic drugs.

The Nature of Leucocyte Locomotion

The method of WRIGHT and COLEBROOK [1921] was used to collect and isolate human polymorphonuclear leucocytes (PMN). In this method a drop of blood is allowed to clot on a microscope slide. After about $1/_2$ h of incubation the clot is washed away, leaving cells (at least 95% PMN) clinging to the glass slide. A thin chamber filled with culture medium was formed above these cells using a supported coverglass as the top of the chamber. Thus, the cells were in contact with glass on one side only, were not compressed in any way, and were bathed with culture medium. Attractant was presented as a clump of *Staphylococcus albus* bacteria or as soluble material

Table I. Extension of lamellipodia. The periods of lamellipodia extension accompanied by concerted cytoplasmic flow into the lamellipodia were measured for a PMN moving in the absence of attractant

Length of extension μm	Duration of extension sec	Time to next extension, sec
11	27	46
5	23	106
3	19	25
5	9	18
11	20	84
15	28	31
4	9	20
14	42	66
8	24	83
10	25	33
13	22	7
11	20	

Table II. Variations in the speed of individual cells. The paths of 4 typical cells undergoing chemotaxis were divided into periods of 4.15 min each and the average speed determined for each cell for each period [Ramsey, 1972a]

Cell	Average speed in each period, μm/min								
a	12.1	5.7	4.1	8.3	7.3	10.2	14.0	10.8	13.3
b	17.8	17.8	11.4	7.0	3.8	7.3	2.6		
c	6.3	8.2	8.2	11.9	12.6	8.2	11.3	13.2	11.6
d	9.2	24.2	23.6	11.4	8.9	8.9	14.0	15.3	14.0

supported in a cylinder of agar. The cells were observed using phase-contrast optics and their behavior was recorded by time-lapse cinemicrography.

The most casual glance at a preparation of motile PMN indicates that the process of locomotion is very spasmodic. At any one time some cells are displacing quickly while others are motionless. This fluctuation is related to the nature of PMN locomotion, with displacement being associated with spreading of a cell against the substratum and immobility with protrusion of the cell up into the medium. A closer examination indicates that locomotion occurs in the following manner [Ramsey, 1972b]. Locomotion in a PMN which is protruding above the substratum and not actively displacing begins when a thin flattened process called a lamellipodium is extended along the substratum. This process is free of visible cytoplasmic granules or vacuoles for some distance from the edge, and is extended flush against the substratum. There is no folding or ruffling activity at the edge, such as accompanies lamellipodia extension in fibroblasts [ABERCROMBIE et al., 1970]. Side views of moving cells suggest that the PMN lamellipodium is a fold of membrane which originates on the upper side of the cell, travels down the cell surface to the substratum and then spreads out over the substratum. Next, the contents of a motile cell (as indicated by cytoplasmic granules, vacuoles and nucleus) flow into and fill the extended lamellipodium, thus causing flattening of a protruding cell and subsequent displacement. The concerted cytoplasmic flow into a lamellipodium leads anew to protrusion of the cell above the substratum and the procedure is repeated.

The spasmodic nature of this process is shown by table I, which indicates the duration and extent of lamellipodia extension-cytoplasmic flow for a cell moving in the absence of attractant. This table does not show the speed of the lamellipodium extension, which may reach 33 μm/min. The alteration of periods of extension and rest does not seem to follow any discernible pattern. If the paths of individual cells moving toward an attractant are divided into portions of quite long duration (4.15 min), and the average speeds determined for each portion, wide fluctuations in speed within single cells are observed (table II). Superimposed on the fluctuations observed in the movement of a single cell is a wide variation in the average speed of a number of cells in a single preparation. In one chemotactic experiment the average speed for all PMN was 10 ± 1.0 μm/min (mean $\pm 95\%$ confidence limits). The average speeds varied from nearly 0 to over 20 μm/min, with a standard deviation of 4.6 μm/min. Comparison of the observed distribution of average speeds with that expected for a normal distribution indicated that a single population, although a broadly distributed one, was present. It seems at least possible that the wide diversity in average cell speeds may have been related to the differences in ages of PMN in the peripheral blood. Pulse-labeling techniques could be used to investigate this point. It might be appropriate to add that the average speed of leucocyte locomotion appears to vary greatly between individual donors. KETCHEL and FAVOUR [1955] reported differences of up to 10 times in the locomotion of leucocytes from different healthy people. Although they used a method which does not allow determination of average speeds of locomotion, it appears that there is wide variation at every level of leucocyte locomotion.

High-magnification microscopic observation of cultured PMN gives a striking impression of incessant movement. At no time may the cell be regarded as quiescent. During periods of protrusion and spreading, cytoplasmic granules, vacuoles and the nucleus may be seen to flow about, seemingly at random. Only during cytoplasmic flow into an extended lamellipodium is there a concerted flow in which every visible particle moves in one direction. Protruding cells also wave about much like a flower on a stem. In addition, side views show the movement of waves down the protruding cell body to the substratum. These waves are produced even when lamellipodial extension is not observed.

Although cell displacement occurs by cytoplasmic flow into a lamellipodium, it does not follow that lamellipodia are produced only on one side of a moving cell. Indeed, several lamellipodia of various sizes may be produced in quick succession all around the circumference of the cell. The bulk of the cytoplasmic flow at any one time, however, is into only one of the several lamellipodia. Thus, some lamellipodia inevitably are unused in the sense that they are not filled by cytoplasmic flow. This observation led to an understanding of the origin of the long fibers which are often seen extending several cell lengths from the end of a moving leucocyte.

Such fibers resemble those seen in fibroblast cultures and are called retraction fibers. The fibroblast adhers to the substratum only at the cell margins [HARRIS, 1971]. These discrete areas of adhesion on lamellipodia tend to move centrifugally away from the cell center and create tension in the cell. When a lamellipodium gives way the cell body is suddenly pulled toward an attached lamellipodium, thus causing locomotion of the cell as a whole. This also causes formation of a retraction fiber, a highly stretched cylinder of cell surface extending from the cell body to the original site of adhesion. A somewhat similar process has been observed in moving PMN. Here, however, the site of adhesion of the retraction fiber is the original adhesion of an extended lamellipodium which was not used in cytoplasmic flow. The lamellipodium appears to be nonetheless adhesive for being unused, however, and as cytoplasmic flow causes displacement of the cell away from this adhesion the cell surface is slowly stretched out to form a retraction fiber. The tail or uropod often seen in moving cells is merely a collection of proximal ends of retraction fibers. Various immunological functions have been suggested for lymphocyte uropods [McFARLAND and HEILMAN, 1965; ROSENSTREICH et al., 1972]. It seems possible that the presence of these structures is more directly related to motility than to other functions.

This view of the origin of retraction fibers and tails makes it unnecessary to postulate alternating adhesion and deadhesion at the cell-substratum level to account for locomotion. One needs only a mechanism to account for adhesion – the deadhesion necessary for locomotion occurs by the physical disruption of the original adhesions or attached retraction fibers.

DE BRUYN described leucocyte locomotion as essentially similar to pseudopod extension in the large amoebae [DE BRUYN, 1944, 1945, 1946]. He reported cytoplasmic streaming of granules adjacent to a region of immobile granules, observations interpreted as the flow of sol in a tube of gel. No cinemicrographs were made, however, and DE BRUYN emphasized the difficulty of making precise observations on the relatively small leucocytes. Frame-by-frame analysis of high magnification cinemicrographs of PMN provides little basis for a gel-sol structure. On the contrary, concerted cytoplasmic flow takes place throughout the entire cross-section of the cell. No regions of nonmoving granules adjacent to streaming granules have been found. It seems, therefore, unnecessary to postulate a gel-sol mechanism for leucocyte locomotion.

It was surprising to observe that the mechanism of locomotion in PMN responding to a chemotactic attractant appears in every way similar to that of a cell moving in the absence of attractant. In each case locomotion is characterized by the apparently random production of lamellipodia on all sides of the cell followed by concerted cytoplasmic flow into one lamellipodium. Since the chemotactic response is oriented displacement, its proximal cause would appear to be a tendency of the concerted cytoplasmic flow to be in a certain direction.

PMN behavior at the moment of initiation of a chemotactic response was recorded by filming an individual cell moving toward attractant held in a micromanipulator. After the cell demonstrated distinct movement toward the attractant, the location of attractant was shifted to behind the cell. When this occurred, the cell immediately stopped moving in the original direction, produced a new lamellipodium at what was previously the tail end, and began moving toward the new location of the attractant. A complete reversal of direction of movement occurred within 45 sec, and the whole process could be repeated several times with one cell. Thus, individual PMN are capable of repeated, rapid and efficient response to changes in the attractant location.

This experiment also tells us something about the PMN sensory mechanism. ADLER [1969] discussed two models for the detection of concentration gradients using formal receptors. In the single receptor model, a motile organism samples the medium and determines attractant concentration, moves a short distance, and samples again. A gradient would be indicated by a change in the concentrations as perceived sequentially by the single receptor. This is sometimes referred to as a time gradient. In the multiple receptor model, the organism would have two or more receptors distributed over its surface and would detect a gradient by simultaneously sampling with its several receptors. This is also called a space gradient. The bacterium *Salmonella typhimurium* responds chemotactically by the one receptor model [McNab and Koshland, 1972]. PMN were capable of a chemotactic response without continuing movement in the original direction, however, so it is clear that of these two models the multiple receptor one applies to these cells. It is difficult to imagine how a receptor mechanism could be sensitive enough to detect a gradient over the length of a cell. One might speculate that receptors are located on the distal ends of retraction fibers, thus greatly increasing the effective cell length.

Chemotaxis and Motility

Preparations of PMN during chemotaxis were filmed and the average speeds for all cells determined [Ramsey, 1972a]. Cells incubated in normal rabbit serum moved with an average speed of $10 \pm 1.0\ \mu m/min$, while those in Gey's salt solution pH 7.2 plus 2% bovine serum albumin (Gey's) moved at an average speed of $6.0 \pm 1.2\ \mu m/min$. Seemingly normal chemotaxis was obtained using either medium; thus, a considerable difference in average speed does not necessarily affect chemotaxis. When cells were incubated in rabbit serum in the absence of attractant, those cells which moved did so at about the same speed as with attractant, although many cells did not move at all. Cells incubated in Gey's in the absence of attractant moved at an average speed of $5.4 \pm 1.2\ \mu m/min$, not significantly different from the chemotactic speed. Some early reports suggested an increase in cell speed just before a cell reached the attractant. No evidence of a change in average speed on reaching the attractant was found in these experiments, however. These results are in agreement with those of Dixon and McCutcheon [1936] who concluded that leucocyte chemotaxis was a factor only of direction of movement. These workers developed the chemotropism index to describe the orientation of the cell path with respect to the attractant. The chemotropism index is the ratio of the displacement of the cell relative to the location of the attractant to the total path length. This index was applied to all cells in a chemotactic experiment with results similar to those of DIXON and MCCUTCHEON. In addition, tracings of the paths were divided by lines parallel to the attractant source into sections corresponding to 100 μm in width, and the chemotropism index determined for each portion of each cell path. The surprising result was that within the 450-μm width of the field, the average chemotropism index did not change. Thus, contrary to expectations, the cell paths did not become progressively oriented as the cells approached attractant. Of equal interest is the fact that the paths were by no means straight, but rather meandered without apparent cause, and even sometimes made loops. The average orientation of paths was such that if movement directly toward attractant were given a value of 0° and that directly away of 180°, the average value would be 50°.

Light Tracings

ALBERT K. HARRIS and I became interested in measuring PMN speed. We felt the following requirements should be met: (1) the method should include all cells of a defined population; (2) the speed of individual cells should be determined, and (3) the answer should be expressed in terms of absolute speed. The procedure we derived involves a novel method of analysis of time-lapse cinemicrographs and will be reported in greater detail elsewhere. Cells were incubated in a perfusion chamber in Gey's medium with about 20 cells per cinemicrographic field. The preparation was illuminated so that the cells appeared as very refractile light images against a relatively dark background and was photographed at 12 exposures/min using reversal film. A few experiments using this method quickly generate an overwhelming amount of experimental data. Projection of the films at normal speed allows one to see drastic alterations in cell speed, but because of the nature of PMN locomotion, small changes cannot be noted with confidence. We have used the light tracings to facilitate extraction of data from such films. The film was divided into sections by projecting about 100 frames (8.3 min in real time) onto unexposed photographic printing paper at 8 frames/sec. Subsequent development of the paper results in a dark streak against a white background for each moving cell. Exposure of separate pieces of paper at the beginning and end of each 100-frame sequence enables one to count the cells with adequate resolution to distinguish between cells and debris in the chamber. The average cell speed was obtained by measuring the entire length of light tracing on one piece of paper, dividing

Table III. Colchicine and average speed. PMN were incubated in various concentrations of colchicine and the average speeds determined from light-tracings

Concentration of colchicine, M	Average speed, μm/min (mean $\pm$ 95% confidence limits); minutes of incubation			
	3	30	50	76
0	7.6 ± 1.7	5.2 ± 1.8	2.4 ± 1.6	5.9 ± 2.6
2×10^{-5}	4.3 ± 1.2	5.8 ± 1.0	4.7 ± 1.4	4.5 ± 1.3
1×10^{-4}	4.6 ± 1.3	4.8 ± 1.6	6.4 ± 1.9	5.0 ± 2.1
1×10^{-3}	3.7 ± 1.1	3.5 ± 1.4	3.1 ± 1.1	2.7 ± 1.4

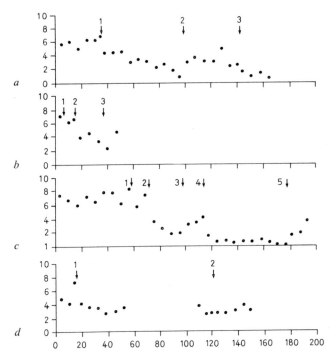

Fig. 1. Antimitotic drugs and average speed. The effects of various drugs on average polymorphonuclear leukocytes (PMN) speed were determined using light-tracings. Abscissa: minutes. Ordinate: average speed, μm/min. *a* Arrow 1, 1×10^{-3} M colchicine; arrow 2, culture medium; arrow 3, 1×10^{-3} M colchicine. *b* Arrow 1, 1×10^{-6} M vinblastin; arrow 2, 1×10^{-5} M vinblastin; arrow 3, 1×10^{-4} M vinblastin. *c* Arrow 1, 0.1 μg/ml cytochalasin B (Cyt B); arrow 2, 0.5 μg/ml Cyt B; arrow 3, culture medium; arrow 4, 1.0 μg/ml Cyt B; arrow 5, culture medium. *d* Culture medium without potassium was used here. Arrow 1, 1×10^{-4} M oubain; arrow 2, culture medium.

by the number of cells involved, subtracting the average cell length, and dividing by the real time involved in the tracing. Confidence limits for the technique were obtained by measuring each cell track separately in a few experiments.

Table III and figure 1a indicate the effect of various concentrations of colchicine on PMN speed. MALAWISTA [1965] reported 10^{-3} M colchicine stopped PMN motility within 45 min. It is clear from our experiments, however, that complete inhibition of motility does not occur under these conditions, and the inhibitory effect is noted only after long incubation in relatively high colchicine concentrations. Likewise, the effects of vinblastin and oubain were relatively slow to take effect (fig. 1b, d). Cytochalasin B (Cyt B), however, inhibited motility quickly, reversibly and at concentrations comparable with its effects in other systems (fig. 1c), suggesting that the effect of this drug is directly on the locomotor apparatus.

The effect of collisions between cells on cell speed was determined for all collisions which occurred in a portion of film of cells moving in the absence of inhibitors or attractants. 21 separate contacts involving 41 cells were examined and the average speeds were as follows: before contact, 14.2 ± 2.0 μm/min; during contact, 11.4 ± 2.1 μm/min; after contact 12.8 ± 2.4 μm/min. The average duration of contact was 1.25 ± 0.26 min, and the average speeds before and after contact were determined over about 1 min. The effect of cell collision, therefore, was to significantly retard motility during the actual contact, although movement continued and the cells eventually separated. Thus contact inhibition of movement, as is found in cultured fibroblasts, does not occur in PMN.

Discussion

The mechanism of leucocyte locomotion is very poorly understood. There is little to indicate the nature of the motile force causing either extension of lamellipodia or cytoplasmic flow. Gross observation of these processes shows they differ in fundamental ways from the extension of pseudopods by large amoebae. Although leucocytes contain microtubules, such structures probably are not directly involved in locomotion. What does seem clear is that chemotaxis can occur as a function of the direction of cell movement only, and that chemotaxis takes place without any apparent alteration of the nature of locomotion.

Any investigation of leucocyte chemotaxis should consider the possibility that conditions and treatments altering chemotaxis may act by altering motility in general. Complete suppression of motility must eliminate chemotaxis – not so evident is the effect on chemotaxis of increasing or decreasing cell speed. This pivotal question has not been answered, although the tools to do so appear to be at hand. This consideration becomes of even greater importance when considering the possibility of *in vivo* chemotaxis, in which the total number of cells arriving at the locus of attraction, which presumably would be affected by cell speed, might be of importance.

The paucity of quantitative data on leucocyte motility is related to the nature of the problem – to determine the average cell speed one must follow a sufficient number of cells for a period of time and divide the distance moved by the time elapsed, a procedure simple to describe but laborious in the extreme. Many reports of leucocyte speeds involve only a few cells and often arbitrarily exclude cells in the field of observation which did not move. One solution to this problem is to separate blood cells by centrifugation in capillary tubes and allow the leucocytes to move up the cell wall. This method is useful for comparisons between treatments but it does not separate the types of leucocytes, and the results cannot be interpreted in terms of absolute speed. Gail and Boone [1970] showed that movement of cultured fibroblasts may be regarded as two-dimensional random walk and thereby characterized by an augmented diffusion constant which is directly related to motility. Peterson and Noble [1972] extended this finding to PMN, a result we have independently confirmed. The use of the augmented diffusion constant eliminates underestimation of total motility due to retracing of paths by the cells. Its application, however, involves determination of the displacement of each cell during a time interval, and does not appear to provide a saving of effort over the more straightforward methods of determining speeds. Although various approaches to analysis of time-lapse cinemicrographs may simplify the problem of determining average speeds, it seems likely that a better method would be to use a tracking microscope, such as invented by Berg and used for following bacteria [Berg, 1971; Berg and Brown, 1972]. This device automatically follows the movement of objects and reports its finding to a computer. Although it can follow only one object at a time, it appears to be the instrument of choice for studying the parameters of leucocyte locomotion.

In a larger sense it will be useful to dissociate motility from chemotaxis, that is, to study the sensing, integrating or response mechanisms without involving directional movement. It may be possible to immobilize a leucocyte and impale it with a micro-electrode. One might then look for changes in resting potential corresponding to delivery of attractant. Another approach would be to isolate and characterize the hypothetical receptors.

Summary

The relationships between *in vitro* granulocyte locomotion and chemotaxis were discussed. Chemotaxis appears to be a function only of the direction of cell movement. High-magnification cinemicrography did not reveal modes of locomotion unique to chemotaxis other than a tendency of cytoplasmic flow to be in the direction leading to the attractant source. The effects of several antimitotic drugs on locomotion were studied using a perfusion chamber and a new method of cinemicrographic analysis. Inhibitors of microtubule synthesis did not appear to have a direct effect on locomotion, although inhibition of locomotion by Cyt B was immediate and reversible.

Acknowledgements

This research was supported by a grant from the National Science Foundation to J. P. TRINKAUS (GB 21240). The author was supported by a United States Public Health Service training grant to the Department of Biology, Yale University (HD 00032). I wish to thank Dr. TRINKAUS for advice and encouragement during the research.

References

ABERCROMBIE, M.; HEAYSMAN, J. E. M., and PEGRUM, S. M.: The locomotion of fibroblasts in culture. II. 'Ruffling'. Exp. Cell Res. *60:* 434–444 (1970).

ADLER, J.: Chemoreceptors in bacteria. Science *166:* 1588–1597 (1969).

BERG, H. C.: How to track bacteria. Rev. Sci. Instr. *42:* 868–871 (1971).

BERG, H. C. and BROWN, D. A.: Chemotaxis in *Escherichia coli* analyzed by three-dimensional tracking. Nature, Lond. *239:* 500–504 (1972).

BOYDEN, S.: The chemotactic effect of mixtures of antibody and antigen on polymorphonuclear leucocytes. J. exp. Med. *115:* 453–466 (1962).

BRUYN, P. P. H. DE: Locomotion of blood cells in tissue cultures. Anatom. Rec. *89:* 43–63 (1944).

BRUYN, P. P. H. DE: The motion of the migrating cells in tissue cultures of lymph nodes. Anat. Rec. *93:* 295–315 (1945).

BRUYN, P. P. H. DE: The amoeboid movement of the mammalian leukocyte in tissue culture. Anat. Rec. *95:* 177–192 (1946).

DIXON, H. M. and McCUTCHEON, M.: Chemotropism of leucocytes in relation to their rate of locomotion. Proc. Soc. exp. Biol. Med. *34:* 173–176 (1936).

GAIL, M. H. and BOONE, C. W.: The locomotion of mouse fibroblasts in tissue culture. Biophys. J. *10:* 980–993 (1970).

Harris, A. K.: The role of adhesion and the cytoskeleton in fibroblast locomotion; PhD dissertation, Yale (1971).

Harris, H.: Chemotaxis of granulocytes. J. Path. Bact. 66: 135–146 (1953).

Ketchel, M. K. and Favour, C. B.: The acceleration and inhibition of migration of human leucocytes in vitro by plasma protein fraction. J. exp. Med. 101: 647–663 (1955).

Lotz, M. and Harris, H.: Factors influencing chemotaxis of the polymorphonuclear leucocyte. Brit. J. exp. Path. 37: 477–480 (1956).

McFarland, W. and Heilman, D. H.: Lymphocyte foot appendage. Its role in lymphocyte function and in immunological reactions. Nature, Lond. 205: 887–888 (1965).

MacNab, R. M. and Koshland, D. E.: The gradient-sensing mechanism in bacterial chemotaxis. Proc. nat. Acad. Sci., Wash. 69: 2509–2512 (1972).

Malawista, S. E.: The action of colchicine in acute gout. Arthrit. Rheumat. 8: 752–756 (1965).

Peterson, S. C. and Noble, P. B.: A two-dimensional random-walk analysis of human granulocyte movement. Biophys. J. 12: 1048–1055 (1972).

Ramsey, W. S.: Analysis of individual leucocyte behavior during chemotaxis. Exp. Cell Res. 70: 129–139 (1972a).

Ramsey, W. S.: Locomotion of human polymorphonuclear leucocytes. Exp. Cell Res. 72: 489–501 (1972b).

Rosenstreich, D. L.; Shevach, E.; Green, I., and Rosenthal, A. S.: The uropod-bearing lymphocyte of the guinea pig. Evidence for thymic origin. J. exp. Med. 135: 1037–1048 (1972).

Wright, A. E. and Colebrook, L.: Technique of the teat and capillary glass tube (Constable, London 1921).

Author's address: Dr. W. Scott Ramsey, Sullivan Park, Corning Glass Works, Corning, NY 14830 (USA)

Antibiotics and Chemotherapy, vol. 19, pp. 191–217
(Karger, Basel 1974)

Mechanism of Movement and Maintenance of Polarity in Leucocytes

A. C. Allison

Clinical Research Centre, Harrow, Middlesex

Two basic requirements for chemotaxis are that cells should be motile and that their movement should be directional rather than random. The latter requires that cells should exhibit polarity. Leukocytes – as well as many other cell types – have long been known to move in an orderly manner, but it is only during the past few years that plausible explanations for this function could be offered. The explanations are based on the identification in the cytoplasm of polymorphs, macrophages, lymphocytes and many other cell types of fibrous protein aggregates. The most important of these are microfilaments, consisting of actin and myosin, and microtubules, consisting of tubulin. There are also other systems, including filaments 10 nm in diameter, the function of which is still unknown.

Microtubules have been found not only in systems that are obviously contractile, such as flagella, cilia and the mitotic spindle, but also in a wide range of dispositions in the cytoplasm of other cells [Porter, 1966; Margulis, in press]. Moreover, evidence is accumulating that actin- and myosin-like proteins (and probably the related regulatory proteins, such as tropomyosin and troponin) are not confined to the several varieties of muscle cell but are present also in other cell types. The number of cell types in which appropriate identification has been performed is still limited, but already various leukocytes, blood platelets, fibroblasts and epithelial cells have been shown to contain microtubules as well as actomyosin microfilaments. Thus, both these systems are probably widespread constituents of living tissues, and either might provide a contractile system in any cell. Probably their relative importance varies from cell to cell and in the same cell at different times, and the analysis of their interactions will require prolonged and detailed investigation.

The first attempts at such an analysis have depended on the use of compounds thought to interact more or less specifically with either microtubules
or microfilaments, namely colchicine and vinca alkaloids on the one hand,
and the cytochalasins on the other. It is, therefore, necessary to review
briefly the evidence for specificity or otherwise of the effects of these compounds. Before this is done, evidence for the presence of actin and myosin
in cells other than muscle will be presented.

Identification of Actin in Cells Other than Muscle

The presence of actin-like proteins in a variety of cell types has been
demonstrated by biochemical and ultrastructural methods. One biochemical
approach has depended upon the ability of crude or purified preparations
to increase the ATPase activity of heavy meromyosin under appropriate
conditions. Actin-like proteins have been obtained from the slime mould
Physarum [HATANO and OOSAWA, 1966], *Acanthamoeba* [WEIHING and KORN,
1969], sea-urchin eggs [HATANO *et al.*, 1969], blood platelets [BETTEX-
GALLAND and LÜSCHER, 1965], the brush border of chicken epithelial cells
[TILNEY and MOOSEKER, 1971], mammalian brain [BERL and PUSZKIN, 1970],
and chick embryo fibroblasts [YANG and PERDUE, 1972].

A second biochemical approach has depended upon the demonstration
in non-muscle cells of a protein of molecular weight about 45,000 with an
actin-like amino acid composition (including the presence of 3-methyl-
histidine) and peptide maps indistinguishable from those of muscle actin
[see POLLARD and KORN, 1972, for references]. A sensitive variant of this
technique is the use of proteins labelled with a radioactive amino acid in
cultures of chick sympathetic nerve cells to demonstrate homology with
muscle actin [FINE and BRAY, 1971]. These authors found that actin peptides
from chick brain and muscle were indistinguishable and conclude that,
although nothing short of the total amino acid sequences of the two proteins
will establish identity, their failure to find any differences in the fingerprints
of actins from brain and muscle suggests that they may be the same protein.
This contrasts with the obvious difference between brain and muscle myosins
discussed below. Actin and tubulin are major bands of nerve cell cytoplasmic
protein submitted to acrylamide gel electrophoresis [FINE and BRAY, 1971],
and are thought to comprise some 12% of the cytoplasmic protein of many
cells [POLLARD and KORN, 1972].

Ultrastructural identification of microfilaments as actin-like has been

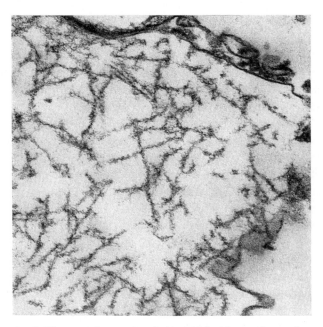

Fig. 1. Electron-micrograph of the peripheral cytoplasm of a mouse peritoneal macrophage, showing the network of microfilaments beneath the plasma membrane. The microfilaments have a thickened and fuzzy appearance because they have been 'decorated' with heavy meromyosin [from ALLISON *et al.*, 1971]. × 59,000.

made possible by the demonstration that such filaments, 50–70 Å in diameter, can bind and become 'decorated' with heavy meromyosin (fig. 1). This procedure, originally used with skeletal muscle F actin [HUXLEY, 1963], has been adapted to other glycerinated cells, including chick embryo cells [ISHIKAWA *et al.*, 1969], the slime mould *Physarum* [NACHMIAS *et al.*, 1970], *Acanthamoeba* [POLLARD *et al.*, 1970], *Amoeba proteus* extracts [POLLARD and KORN, 1971], human blood platelets [BEHNKE *et al.*, 1971a], leukocytes [SENDA *et al.*, 1969], newt eggs [PERRY *et al.*, 1971] and mouse macrophages [ALLISON *et al.*, 1971].

MOORE *et al.* [1970] have made 3-dimensional reconstructions of F actin, thin filaments and thin filaments decorated with heavy meromyosin. These show that the appearance of 'arrowheads' of the correct periodicity depends upon a highly specific interaction of double helical filaments of F-actin with heavy meromyosin, which is very unlikely to occur by chance association with some other material. The presence in non-muscle cells of thin filaments 50–70 Å in diameter which can be decorated with heavy meromyosin is, therefore, acceptable evidence that they are actin-like. The arrowheads are more readily seen in negatively stained preparations than in thin sections, but in the latter thin filaments appear much thicker, with a fuzzy outline, after decoration. Microtubules and the filaments about 100 Å in diameter often seen in cells, and to which no function has yet been assigned, are not decorated with heavy meromyosin.

It appears likely that in at least some cells much of the actin is present in a globular form, and that when suitably stimulated it quickly polymerises into filaments 50–70 Å in diameter. Thus, the circulating blood platelet is disc-shaped and when rapidly fixed shows few microfilaments; in response to various stimuli, including ADP, it is transformed into an amoeboid cell in which many filaments are observed [BEHNKE et al., 1971a]. The cytoplasmic fragments of A. proteus, shown by THOMPSON and WOLPERT [1963] to contract when exposed to ATP and warmed, have in the precursor stage few thin filaments; after stimulation the viscosity of the preparation increases and numerous filaments 50–70 Å in diameter appear, and can be decorated with heavy meromyosin [POLLARD and KORN, 1971]. Later, the filaments aggregate to form birifringent fibrils visible by light microscopy which interact with 60 Å filaments to constitute a contractile system. It seems likely that the actin-like filaments are formed from precursors in the groundplasm; they are unable to contract by themselves, but require the myosin-like thick filaments for contraction.

Thus, part of the actin in at least some cells is in a precursor form which can polymerise when required and part probably remains filamentous. This transition from G to F actin seems to represent, at least in part, the basis for the classical sol-gel transition in cytoplasm, and may be important in locomotion, as discussed below.

There appears to be a close relationship between some of the actin-like filaments and cell membranes. POLLARD and KORN [1972] isolated purified plasma membranes from Acanthamoeba and found typical actin-like filaments regularly associated with them. CRAWFORD [1971] has obtained from blood platelets a fraction consisting of small vesicles and sheet membrane fragments. Actomyosin activity was found in this fraction. Thus, membranes prepared in media that do not dissociate actin may contain these filaments, suggesting that the morphological association may reflect a functional relationship.

In many cell types, such as macrophages [ALLISON et al., 1971], retinal cells [CRAWFORD et al., 1972] and pancreatic β-cells [ORCI et al., 1972] a dense network of microfilaments lies immediately beneath the plasma membrane. These appear to form a cage keeping the nucleus and cytoplasmic organelles (lysosomes, mitochondria and secretory granules) away from the plasma membrane. They may also help to maintain the rounded contour of the cells by a process analogous to muscle tone in whole animals.

Whether actomyosin-like proteins are also present in organelles in which microtubules are prominent, such as the mitotic or meiotic spindle, has been much debated. Peptide mapping shows that tubulin is distinct from actin [STEPHENS, 1971], but the concept that an actomyosin system may exist alongside microtubules receives support from the observations of GAWADI [1971] and of BEHNKE *et al.* [1971 b] on microfilaments which can de decorated with heavy meromyosin amongst the microtubules of mitotic and meiotic spindles of locust and cranefly testis cells.

It is still possible that the microtubules do not themselves contract but form a system of fairly rigid but somewhat flexible rods that can resist compression, while microfilament contraction provides the tension necessary for movement. The microtubules would thus serve a skeletal and directing function, important for polarity of movement of chromosomes and other organelles within cells. Disassembly of microtubules, especially if it can only occur at certain regions, such as one or both ends, would allow controlled shortening of structures such as the spindle. The recent finding that Ca^{2+} brings about disassembly of microtubules [WEISENBERG, 1972], as well as being the trigger for microfilament contraction, implies that the two processes could readily be co-ordinated in cell physiology.

KAMINER and SZONYI [1972] have identified a protein closely resembling tropomyosin in the electric organs of the electric eel and *Torpedo*, so the regulatory proteins may also be found in non-muscle cells.

Identification of Myosin-Like Proteins in Cells Other than Muscle

Three methods have been used to identify myosin-like proteins a variety of cells: biochemical, ultrastructural and immunological. The former depends on the isolation of proteins that have a Mg^{2+}-activated ATPase activity which can be increased by addition of actin. Such materials have been isolated from human and porcine blood platelets [BETTEX-GALLAND and LÜSCHER, 1965; ADELSTEIN *et al.*, 1971] and the slime mould [HATANO and TAZAWA, 1968].

The ultrastructural identification of myosin filaments is at present unsatisfactory because they vary widely in thickness from about 40 to 180 Å, according to the material and method of preparation, and there is no convenient marker such as localisable enzyme activity or heavy meromyosin binding. However, thick filaments have been identified in a variety of preparations, such as human blood platelets [BEHNKE et al., 1971a] and Amoeba cytoplasm [POLLARD and KORN, 1971], and there are functional resaons for believing that these consist of myosin-like proteins.

At the light-microscopical level, useful information about the distribution of myosin in different cell types has come from the use of fluorescent antibodies against human smooth muscle myosin [KEMP et al., 1971; FARROW et al., 1971; ALLISON and BRIGHTON, unpublished information]. These are either prepared by immunisation of rabbits with myosin purified from fresh human uterus or are found as naturally occurring auto-antibodies against smooth muscle in certain human patients with hepatitis. We have investigated the reactions of these antibodies in a variety of fixed and unfixed cells, using the 'sandwich' technique in which, after treatment with antibody and repeated washing, cells are 'stained' with antiglobulins coupled with fluorescein isothiocyanate. Unfixed cells stained after exposure to 0.3 M sodium azide in the cold (to inhibit pinocytosis) show no fluorescence, from which it is clear that the outer surface of the plasma membrane, to which the antibody can gain access, does not contain demonstrable quantities of myosin. Fixed smooth muscle from all parts of the body tested (viscera, blood vessels and bronchi) shows intense cytoplasmic fluorescence, striated muscle (skeletal and cardiac), weak or no fluorescene. Fairly strong fluorescence is observed in the cytoplasm of many fixed cells (fibroblasts, macrophages, lymphocytes, granulocytes, blood platelets, neurons and many different types of tumour cells) (fig. 2–4). Staining of liver cells by this type of antibody has been described by FARROW et al. [1971] and localisation of smooth muscle protein in myoepithelium has been reported by ARCHER et al. [1971]. Most of the staining can be diffuse, as in leukocytes, or in linear arrays along the long axis of cells, as in fibroblasts and some tumour cells; the nucleus remains unstained.

Cells other than muscle are only poorly stained with antibody against skeletal muscle myosin, which stains very intensely striated muscle cells. These antibodies show strong precipitation with homologous myosins in Ouchterlony gels, good reactions with smooth muscle myosins of other mammals, weak reactions with skeletal muscle myosins and no detectable reactions with actin from either source. From these observations it seems that the non-muscle cells so far investigated contain in their peripheral cytoplasm myosin that can be shown immunologically to resemble that of smooth

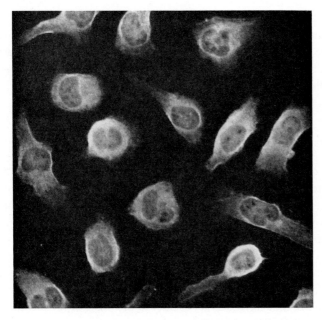

Fig. 2. Culture of mouse peritoneal macrophages, fixed with acetone and treated with rabbit anti-smooth muscle myosin and fluorescein isothiocyanate-coupled porcine anti-rabbit globulin. Fluorescence microscopy shows myosin in the peripheral cytoplasm. × 800.

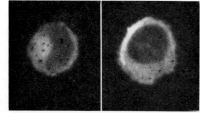

3, 4

Fig. 3. Fluorescence photomicrograph of human peripheral blood monocyte, as described in figure 2 to demonstrate cytoplasmic myosin. × 800.

Fig. 4. Fluorescence photomicrograph of human peripheral blood lymphocyte transformed with phytohaemagglutinin and treated as described in figure 2 to demonstrate cytoplasmic myosin. × 1,000.

muscle rather than that of skeletal muscle. Striated muscle myosins are known to be heterogeneous and it would be premature to conclude that smooth muscle myosins are uniform, whether present in muscle or other cells. This variability of myosin contrasts with the apparent similarity of actin in different cells already discussed.

Membrane Movement

A crucial question when analysing cell motility is whether membranes have any inherent motive power or whether they are moved by the action of extrinsic agents, such as filaments, which are attached to them. During the past few years evidence has accumulated in support of a fluid mosaic model of membranes [SINGER and NICOLSON, 1972], in which the constituents are envisaged as not occupying a fixed relationship in a solid structure but floating in an oily fluid. Proteins are retained within a membrane because they have hydrophobic groups that interact with the hydrocarbon chains of lipids; nevertheless, they have lateral mobility in the plane of the membrane. Examples of such lateral movements are the mixing of membrane antigens following fusion of different cell types [FRYE and EDIDIN, 1970] and the formation of 'patches' and 'caps' in lymphocytes when surface immuno-globulins are cross-linked by bivalent anti-immunoglobulin [TAYLOR et al., 1971]. Patching is not prevented by cytochalasin, and probably follows linkage of membrane protein units brought together by diffusion within the membrane. Formation of such protein 'patches' could, however, result in a distortion of the membrane structure to form small spherical infoldings or outfoldings analogous to the budding of enveloped viruses [ALLISON, 1972b].

An example could be the invagination of membrane of erythrocytes, treated with relatively high concentrations of aminoquinolines to form vesicles within the cell [GINN et al., 1969]. Analogous processes might result in the formation of small pinocytic vacuoles, such as those described below, which is resistant to cytochalasin.

It has also been postulated that energy-driven movement of membrane proteins is involved in specific transport of ions. In liposomes, the concentric lipid bilayers developed by BANGHAM et al. [1965], alteration of the ionic composition of the medium leads to wavy motion of the lamellae in the particles. However, it seems unlikely that any of these limited types of move-ment known to occur within membranes can account for the marked displacement of membranes which is necessary for ruffling, cell movement, phagocytosis, macropinocytosis and the discharge of packed secretions. These appear to require the movement of membrane by some external force.

Colchicine and Microtubules

That colchicine and vinblastine affect microtubular structure by inter-action with their tubulin subunits is well established. The binding of colchi-cine and vinblastine by tubulin [WEISENBERG et al., 1968; WILSON, 1970; WEISENBERG and TIMASHEFF, 1970; OWELLEN et al., 1972] has been studied by many investigators. One molecule of colchicine or vinblastine is bound per 110,000 daltons of the tubulin structural unit, which is a dimer of the in-dividual tubulin proteins. Colchicine and vinblastine bind to different sites on the tubulin molecule, so that neither competes with the other. Binding of radioactive colchicine is widely used as a means of identifying tubulin in fractionation of cell extracts. Moreover, it is generally agreed that in cells treated with colchicine or vinca alkaloids labile cytoplasmic microtubules are disrupted. After vinblastine treatment tubulin is often deposited in crystal-like structures in the cytoplasm. Colchicine does not affect the appearance or function of organised microtubular structures, such as cilia or flagella, but treatment with colchicine prevents their formation [ROSEN-BAUM et al., 1969]. It is thought that colchicine can bind the tubulin dimers, inhibiting their polymerisation, but is prevented from gaining access to binding sites in organised microtubules, possibly by some additional com-ponent of flagella and cilia, such as dynein or nexin [GIBBONS and FRONK, 1972]. Hence, not all microtubule-related functions are impaired by colchicine or vinblastine.

Nevertheless, there is good evidence that colchicine and vinblastine treatment breaks down microtubules in a great variety of cell systems [MAR-GULIS, in press] and presumably functional impairment, such as metaphase arrest, follows. The problem in interpreting observed results of colchicine and vinblastine treatment is that the drugs, particularly when used in high concentrations, may have effects on systems other than microtubules. Thus, colchicine inhibits nucleotide transport across mammalian cell membranes [MIZEL and WILSON, 1972], has a hypocalcaemic action [HEATH et al., 1972], inhibits pancreatic aldose reductase [GABBAY and TZE, 1972] and protein synthesis [CREASEY et al., 1971], and has anti-inflammatory activity which – judging from activity of structurally related compounds – is apparently independent of effects on microtubules [FITZGERALD *et al.*, 1971]. Doubtless other effects of colchicine and vinblastine will be discovered, so that interpretation of the observed results of treatment of cells must be made with due caution. Observation that colchicine and vinblastine leave certain cell functions, such as phagocytosis, intact under conditions when labile cytoplasmic microtubules are lost, show unambiguously that microtubules are not required for these functions. The converse is not necessarily true: observation that colchicine or vinblastine inhibit a function may be due to involvement of labile microtubules, but may be due to other effects of the drugs. It seems that colchicine and vinblastine in concentrations that can disperse labile microtubules do not significantly inhibit actomyosin systems, including contraction of smooth and skeletal muscle, cell movement and ruffle membrane activity, as discussed below.

Cytochalasin and Microfilaments

The evidence that cytochalasin disturbs certain microfilament functions is less direct than that for colchicine effects on microtubules. It is based on observations that cytochalasin inhibits functions of cells which on morphological and other grounds are thought likely to be related to an actomyosin-like microfilament system [SCHROEDER, 1960; WESSELLS *et al.*, 1971]. There are also reports that treatment of cells with cytochalasin disrupts the regular arrangement of microfilaments associated with certain specialised functions. These include the contractile ring of microfilaments thought to be responsible for cleavage after mitosis in the sea-urchin egg [SCHROEDER, 1970, 1972], the bundle of microfilaments observed in epithelial cells during resorption of the tail of ascidian larvae [CLONEY, 1966; LASH *et al.*, 1970] and ordered microfilaments in other situations [WESSELLS *et al.*, 1971]. However, treatment with cytochalasin in low concentrations does not always demonstrably alter microfilament structure [GOLDMAN, 1972]. FORER *et al.* [1972] have reported that cytochalasin does not affect the appearance of actin-like filaments in blood platelets or their capacity to react with heavy meromyosin. Nevertheless, inhibition of microfilament-related functions would not necessarily be accompanied by morphological evidence of disorganisation, as the effects of agents blocking neuromuscular transmission suffice to illustrate.

Evidence in support of a direct effect of cytochalasin B on actomyosin has been published by SPUDICH and LIN [1972]. They reported that cytochalasin decreases the viscosity of the actomyosin complex of rabbit skeletal muscle by at least 60%. Cytochalasin does not affect the viscosity or ATPase activity of heavy meromyosin, suggesting that it interacts with the actin component of the complex, and viscosity measurements showed a strong interaction with actin at nearly stoichiometric concentrations. This work is open to criticism on two grounds: firstly, the concentration of cytochalasin effective (250 μM) is higher than required to inhibit cell movement; secondly, cytochalasin does not affect the contraction of striated muscle *in vivo* or in culture [SANGER *et al.*, 1971]. It would be of interest to repeat the work of SPUDICH and LIN using preparations of actin and myosin from cells that are highly sensitive to cytochalasin. One possibility is that cytochalasin affects purified actin, particularly the transition from G to F actin, but that when in intact striated muscle actin is polymerised and complexed with other materials such as troponin and tropomyosin cytochalasin has little effect.

It has also been shown that cytochalasin inhibits the incorporation of glucosamine into glycoproteins [HOLZER and SANGER, 1972]. This may be related to the inhibition by cytochalasin of the transport of glucose, deoxy-glucose and glucosamine across plasma membranes [KLETZIEN *et al.*, 1972; ZIGMOND and HIRSCH, 1972; ESTENSEN and PLAGEMANN, 1972; MIZEL and WILSON, 1972]. Amino acid transport is not inhibited by cytochalasin, so the effect is selective. A possible explanation is that phosphorylated sugars are removed from the plasma membrane by cytoplasmic protein acceptors [KUNDIG and ROSEMAN, 1971]; if access of these proteins to the plasma membrane is reduced in cytochalasin-treated cells transport of sugars and amino sugars would be inhibited, whereas transport of amino acids and other compounds not requiring cytoplasmic protein acceptors would be unaffected.

The inhibition by cytochalasin of sugar transport has led to speculations that stopping of cellular movements may also follow effects on the plasma membrane rather than on the contractile system itself. This seems unlikely, since the drug inhibits contraction of cortical microfilaments in frog eggs induced by injection of Ca^{2+} ions into the cytoplasm [ASH, quoted by WESSELLS *et al.*, 1971]. Moreover, B. ELFORD and I have found that cyto-chalasin B (10 μg/ml) inhibits the contraction of glycerinated smooth muscle (rabbit *taenia coli*) induced by ATP. Movement of isolated cytoplasm from cut cells of *Chara corralina*, displacement of organelles and rotation of chloroplasts is also inhibited by cytochalasin [WILLIAMSON, 1972]. In none of these situations is plasma membrane permeability a limiting factor. It therefore seems probable that cytochalasin B exerts a direct effect on the contractile system of smooth muscle and the analogous systems of other cells. Effects of cytochalasin on membranes are of interest in long-term results of treatment, such as glycoprotein synthesis, but probably do not explain the marked effects seen in cells exposed to cytochalasin for short periods. Cells treated with cytochalasin (up to 100 μM) for relatively long periods continue to synthetise protein at rates not significantly different from normal [PARKHOUSE and ALLISON, 1972; DAVIES, unpublished information], so it seems unlikely that overall inhibition of metabolism is limiting [see also ESTENSEN and PLAGEMANN, 1972]. Nevertheless, it scarcely needs emphasising that the effects of cytochalasin on contractile and other systems are still poorly defined and that further studies are needed before too much reliance is placed on experiments based solely on the effects of cytochalasin.

Effects of Cytochalasin on Ruffle Membrane Activity and Cell Locomotion

It is generally agreed that when cytochalasin B in low concentrations (0.5–5 μg/ml medium, i.e. 1–11 μM) is added to cells ruffle membrane activity and cell movements cease within a few minutes and that this effect is rapidly reversible when cytochalasin is removed and fresh medium replaced. Such changes have been described in L cells, a permanent line of mouse fibroblasts [CARTER, 1967], elongating nerve cells [WESSELLS et al., 1971], mouse peritoneal macrophages [ALLISON et al., 1971], BHK 21 cells [GOLDMAN, 1972] and the slime mould Dictyostelium discoideum [WIKLUND and ALLISON, 1972]. Apart from a globule of cytoplasm (endoplasm) around the nucleus, containing neraly all cytoplasmic organelles, the cells lie flat on the substratum, suggesting that whatever support there is between upper and lower layers of plasma membrane in normal cells has been lost after cytochalasin treatment. Existing adhesions to the substrate are seldom lost, and they appear to establish the irregular 'spidery' appearance of the cells after treatment with the drug.

The simplest interpretation of these findings is that microfilament-mediated contraction is required for ruffling and for the forward movement of the cell membrane and enclosed cytoplasm which is involved in locomotion. There is no evidence that microtubules are present in the neighbourhood of the ruffles or leading edge, nor that microfilaments can provide a microtubule-like extensible system that can resist compression. At least two models can be considered by which co-ordinated microfilament-mediated contractions can lead to cytoplasmic extension. According to one, a rounded leading edge could be converted to a thin, discoid forward extending edge by contractions bringing the upper and lower parts of the plasma membrane together. This would be analogous to pleating, and if the lower surface were to adhere to the substratum, further contractions could lead to movement of the pleat upwards and backwards (ruffling). The pleating of the membrane of frog eggs following contraction of cortical microfilaments [GINGELL, 1970] illustrates the principle. A second model depends on the observed fluidity of cytoplasm near the site of elongation as compared with other regions. If a high proportion of actin and/or myosin is unpolymerised near the site of elongation, any contractions will tend to squeeze it forwards, so displacing the plasma membrane. Following adhesion to the substratum, a plasma membrane change (perhaps due to apposition of adhesive units) may rapidly trigger polymerisation of actin and/or myosin in the immediate vicinity. The cytoplasm near the point of contact would then become gelated, and subsequent contractions from this site would pull the cell body towards the new site of adhesion. These models are not mutually exclusive. The pattern of birefringence observed supports the frontal contraction theory of amoeboid movement [ALLEN, 1972]. The geometry of the initial forward movement would presumably depend on purely local factors such as prior constraints from adhesion sites, distribution of polymerised actin, and how well activity is controlled. Thus, spreading of ruffling activity, with less polymerised actin filaments around the margin, might convert 'ruffling' into 'blebbing'. This happens at low cytochalasin B concentrations, as well as under unfavourable conditions, and may be due to partial inhibition of the microfilament system.

Effects of Cytochalasin on Endocytosis

There is also general agreement that phagocytosis and macropino-cytosis – the intake of fluid into vacuoles visible by light microscopy, i.e. over 300 nm in diameter – are markedly inhibited by cytochalasin B in the concentrations described above. In contrast, micropinocytosis – the in-vagination of plasma membrane to form small vesicles about 70 nm in diameter visible only by electron-microscopy (which may however contain ferritin, colloidal gold, horseradish peroxidase or other markers added to the extracellular fluid) is not demonstrably inhibited by cytochalasin B [WILLS et al., 1972]. Micropinocytosis is also less temperature-dependent and affected by inhibitors of ATP generation than are phagocytosis and macropinocytosis.

Attachment of opsonised bacteria to polymorphonuclear leukocytes or macrophages is not inhibited by cytochalasin, but electron-micrographs show that the bacteria are never completely engulfed but remain within invaginations of the plasma membrane. Lysosomal granules are seen in close apposition to the plasma membrane, and are discharged into the folds containing the bacteria. Biochemical measurements show that during phagocytosis some lysosomal hydrolases are discharged into the surrounding medium; in cells treated with cytochalasin there is a similar loss, but when cells are exposed to both cytochalasin and bacteria the loss of hydrolases is much greater [DAVIES et al., 1973]. The loss is selective for lysosomal enzymes; the cells remain viable and the cytoplasmic enzyme lactic dehydrogenase is not found in the medium. We conclude that in the intact leukocyte a network of actin-like microfilaments prevents lysosomes from gaining access to the plasma membrane. During phagocytosis this network may well play a role in interiorisation of the membrane. As the phagocytic vacuole is formed, the filament network ceases to be continuous and lysosomes can fuse with the vacuole to form large secondary lysosomes.

Colchicine does not inhibit ruffle membrane activity, amoeboid movement of cells, phagocytosis or macropinocytosis, although the movement of endocytic vacuoles within cells may be affected, as described below. Hence, it is unlikely that the motive power for these activities comes from microtubules.

It has been claimed [RABINOWITCH, 1967] that macrophages require Ca^{2+} ions in the medium for phagocytosis to occur. This observation is based on experiments in which the cells were incubated in medium containing ethylenediamine tetraacetic acid (EDTA) which chelates Mg^{2+} and other ions as well as Ca^{2+} and in our hands soon produces irreversible changes in macrophages. In contrast ethylenebis(oxyethylenenitrilo)tetraacetate (EGTA), an agent which has a much greater affinity for Ca^{2+} than other ionos, des not detectably inhibit phagocytosis of labelled bacteria [DAVIES and ALLISON, unpublished information], so that an exogenous source of Ca^{2+} does not seem to be required for phagocytosis.

A Postulated Mechanism for Phagocytosis

The above observations suggest a mechanism for phagocytosis. The initial stage is attachment of a particle to the plasma membrane. Mammalian phagocytic cells have on their plasma membranes receptors for antibody and complement [LAY and NUSSENZWEIG, 1968], and it is likely that these are involved in the initial attachment of opsonised micro-organisms. Latex particles and formalin-treated erythrocytes are attached in the absence of antibody, and it seems likely that their hydrophobic surface is involved [RABINOWITCH and DI STEFANO, 1970]. In the case of amoeba, positively charged particles are readily attached, suggesting that electrostatic interaction is important [JACQUES, 1969].

We suggest that attachment of a particle traps and immobilises membrane proteins in the vicinity and allows them to form a cluster, thereby increasing the permeability to sodium ions and bringing about membrane depolarisation [ALLISON, 1972b]. Such depolarisation has been reported following attachment of particles to peritoneal exudate cells [KOURI, 1972].

It seems reasonable to postulate that, as in muscle, membrane depolarisation is followed by release of calcium ions from an intracellular site of sequestration, probably smooth membrane vesicles analogous to sarcoplasmic reticulum. Release of calcium ions could trigger contraction of an actomyosin system of microfilaments attached to the plasma membrane, bringing about phagocytosis. Pinocytosis visible by light-microscopy appears to occur by a similar mechanism. In macrophages it can be induced by an anticellular antibody present in calf serum [COHN, 1970]; in tumour cells it can be induced by basic proteins [RYSER, 1967]. In either case these could immobilise membrane proteins as postulated above. Since the cytoplasm is often everted to form a rim over the particle phagocytosed, activity at the periphery of the particle as well as below it is required.

Microtubules and Cellular Polarity

Many cells show polarity of structure, of movement over a substratum and of movement of endocytic vesicles within their cytoplasm. Polarity is most obvious in cells such as those in epithelia, which have luminal and basal borders as well as lateral contacts with other epithelial cells. Even fibroblasts and leukocytes in culture show polarity, with ruffle membrane activity often confined to one or two poles (fig. 5). Movement occurs towards the area of major ruffle membrane activity [ABERCROMBIE et al., 1970] and pinocytosis visible by light-microscopy is usually confined to this region.

Persistence of movement in one direction is readily demonstrable by phase-contrast time-lapse cine-photomicrography in fibroblasts [VASILIEV

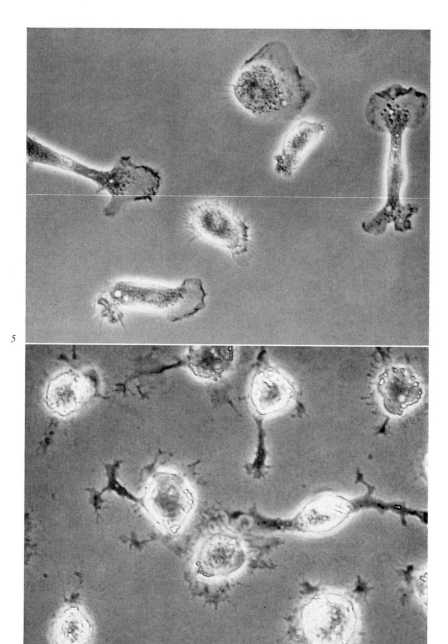

5

6

et al., 1970] and in macrophages (fig. 7) [ALLISON, 1971b]. This appears to be a property of individual cells, and may be referred to as intrinsic polarity of movement, in contrast to extrinsic polarity of movement which is imposed from the environment, e.g. by having a substratum of nearly parallel fibres or in the presence of a chemotactic influence.

MISZURSKI [1969] reported that chick heart fibroblasts 'assume a rounded or polygonal shape' when treated with colcemid. VASILIEV *et al.* [1970] found that colcemid retards the migration of confluent human diploid cells into spaces created by scraping away part of the confluent layer, and affects the directional movement of fibroblasts. In macrophages [ALLISON, 1971a] and fibroblasts [GAIL and BOONE, 1971] treated with low concentrations of colcemid or colchicine, ruffle membrane activity, instead of being confined to certain areas, spreads over the entire cell surface. Directional gliding movements are replaced by random amoeboid movements (fig. 8). Fibroblasts (BHK 13) treated with colchicine after cell division assume a rounded, epithelial-like appearance [GOLDMAN, 1971]. After treatment with colchicine HeLa cells showed randomisation of the distribution of lysosomal and Golgi elements [ROBBINS and GONATAS, 1964], and in fibroblasts there was loss of radial organisation of acid phosphatase-reactive granules around the centrosome and of directed saltatory movements of endocytic vacuoles towards the centrosomal region [FREED and LEBOWITZ, 1970]. Loss of polarity may play a role in the inhibition of leukocyte chemotaxis by colchicine [CANER, 1965], which may limit the number of inflammatory cells entering gouty joints.

An important role of microtubules in the maintenance of membrane structure has been recently shown. Normally phagocytosis by polymorphonuclear leukocytes (PMN) is not accompanied by internalisation of membrane sites involved in specific transport of amino acids [TSAN and BERLIN, 1971]. However, treatment of PMN with 1 μM of colchicine or vinblastine leads to the internalisation of large areas of membrane involved in active transport

---

Fig. 5. Phase-contrast photomicrograph of normal mouse peritoneal macrophages in culture, showing cytoplasmic extensions and ruffled membranes at poles. × 600.

Fig. 6. Phase-contrast photomicrograph of mouse peritoneal macrophages in culture treated for 15 min. with cytochalasin B (10 μg/ml). The cells are immobile and show irregular remnants of pseudopodia and very flattened areas of peripheral cytoplasm, with a globular mass of cytoplasm around the nucleus. × 845.

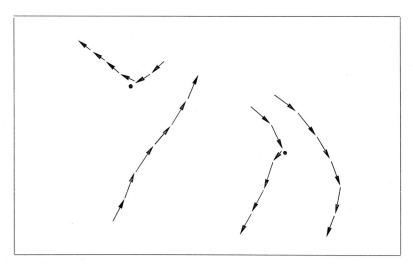

Fig. 7. Tracing of the movement of normal mouse peritoneal macrophages, taken from time-lapse cine-micrography. Note that cells tend to maintain gliding movements in the same direction unless they meet an obstacle, in which case they deviate and then continue moving in that direction (intrinsic polarity) [fig. 5, 6, ALLISON, 1971b].

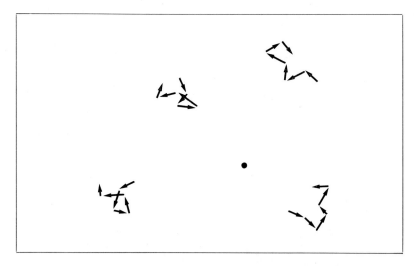

Fig. 8. Tracing of the movements of mouse peritoneal macrophages treated with colchicine, showing random amoeboid movement (no intrinsic polarity).

[UKENA and BERLIN, 1972]. This suggests a loss of the mosaic-type structure normally existing in the PMN, with separation of phagocytic and specific transport sites mediated by intact microtubules.

All these observations suggest that microtubules play a major role in maintaining the shape of cells, the disposition of organelles within them, and polarity of movement of cells in relation to the substratum as well as of endocytic vacuoles and other organelles within cells. When microtubules are laid down after cell division, they tend to be parallel, and this determines the subsequent polarity of the cell during interphase.

Inhibition of the Actomyosin System by Cyclic AMP

Evidence is accumulating that hormones and drugs are first bound to specific receptors in the plasma membranes of target cells. Indirect – and more recently direct – measurements of agonist binding show that it is reversible, with association constants in the range 10^{-8}–10^{-11} l/mol [PASTAN and PERLMAN, 1971]. Elsewhere [ALLISON, 1972a] it has been suggested that attachment of agonists to plasma membranes has one of two main consequences. Firstly, it can bring about depolarisation of the membrane, thereby activating an actomyosin system and triggering contraction of smooth muscle, endocytosis and discharge of packaged secretions. This we have termed the *contractile or A effector system*, A because it depends on actomyosin filaments. The essential feature of this system is that the result is *movement* of some kind.

The second major system, the *B or metabolic effector system*, involves changes in the *metabolism* of target cells. In this case attachment of an agonist to the target cell plasma membrane is followed by activation of adenyl cyclase or an analogous membrane enzyme, increase in the level of cyclic AMP or another cyclic nucleotide and stimulation of protein kinase activity. As a result substrates such as the enzyme phosphorylase are phosphorylated. This system has been extensively discussed [ROBISON et al., 1968; PASTAN and PERLMAN, 1971]. Hormones shown to raise intracellular levels of cyclic AMP in target tissues include catecholamines (through β-receptors), glucagon adenocorticotropic hormone (ACTH), melanocyte-stimulating hormone (MSH), luteinizing hormone (LH), vasopressin, parathyroid hormone, prostaglandins and thyrocalcitonin.

What is relevant from the point of view of the present discussion is the fact that raising intracellular levels of cyclic AMP often antagonizes the contractile system. In smooth muscle, catecholamine α-receptor stimulation activates the A effector system and brings about contraction whereas stimulation of catecholamine β-receptors activates the B system, increases cyclic AMP levels and antagonizes histamine-mediated contraction of bronchial smooth muscle [AXELSSON, 1971] and uterine smooth muscle [MITZNEGG et al., 1971]. As shown by the latter authors, this effect is specific for cyclic AMP; cyclic GMP and other nucleotides will not substitute for it. Granule release in mast cells and platelets is probably triggered by microfilament contraction and is antagonised by increasing the levels in the target cells of cyclic AMP [BOURNE et al., 1972; DEYKIN et al., 1972]. There is also evidence that the movement of fibroblasts [JOHNSON et al., 1972] and macrophages [PICK, 1972] is inhibited by treatments which are known to increase levels of cyclic AMP.

General Discussion

The number of organelles and biochemical processes which cells have at their disposal is limited, and often a system is adapted for various purposes in the course of evolution. It is now clear that the distinction between leukocytes, fibroblasts and smooth muscle cells is one of degree rather than qualitative: that all these cells have in their cytoplasm a contractile system consisting of actin and myosin filaments. The movement of cells, as well as functions such as phagocytosis, macropinocytosis and the discharge of packaged secretions, probably depend on this system. All these cells have, in addition, cytoplasmic microtubules. At the simplest level of analysis, microfilaments provide the motive power for cell movements while microtubules provide not only a framework or cytoskeleton but also polarity in the cytoplasm. This polarity is reflected in the orderly movement of endocytic vacuoles within the cell and in the plasma membrane itself. The latter is suggested by observations that colchicine treatment allows ruffle membrane activity to spread over the whole plasma membrane, instead of being confined to certain areas, and abolishes the distinction between sites involved in endocytosis and in amino acid transport. The directional movement of leukocytes appears to result from an adaptation of the intrinsic polarity of the cell, in the maintenance of which cytoplasmic microtubules play a major part, to chemotactic influences exerting extrinsic polarity. The inter-relationship between the plasma membrane and colchicine-sensitive functions, presumably microtubular, may well be pertinent to this adaptation. Increases in calcium ions in the cytoplasm could also favour contraction of the microfilament system and disaggreation of microtubules.

Not only has the primitive contractile system been used for cell movement, endocytosis and exocytosis; the primitive control system based on cyclic AMP [which goes back to slime moulds and even prokaryotes – see PASTAN and PERLMAN, 1971] has been used to antagonise and so modulate the contractile activity. These concepts may be applicable to understanding the effects of chemotactic agents at the cellular level. It seems likely that such agents become attached to specific sites on the plasma membrane of the target leukocytes and there initiate either one of two reactions: they depolarise the membrane, as a result of which calcium ions are released from an intracellular reservoir and trigger contraction of microfilaments; or they inhibit adenyl cyclase activity, so encouraging movement. Conceivably both processes occur simultaneously. Thus, ruffling of the membrane, and consequent cell elongation, is more likely to occur towards the course of the chemotactic stimulus than at other sites. These hypotheses are testable: for example, on the first hypothesis chemotactic factors should promote ruffle membrane activity whereas on the second they should antagonise leukocyte adenyl cyclase activation. With regard to the first, our recent observation of increased cation flux through plasma membranes associated with clustering of the membrane proteins [ALLISON *et al.*, in press] suggests that chemotactic factors could bring about depolarisation by clustering proteins in the membrane of the target leukocytes. Further observations are required to test these hypotheses. Recent studies showing that cyclic GMP sometimes antagonizes the action of cyclic AMP suggest that increased levels of cyclic GMP might favour leukocyte movement, and effects of agents promoting chemotaxis on cyclic GMP levels in leukocytes should also be investigated.

Summary

1. All types of leukocytes contain microtubules and actomyosin microfilaments. Actin-like microfilaments in the peripheral cytoplasm can be identified in electron micrographs because of their capacity to bind heavy meromyosin. Myosin resembling that in smooth muscle is found in leukocytes by immunofluorescence. Both proteins are demonstrable in leukocytes by biochemical studies.

2. Cytochalasin B in small doses reversibly inhibits ruffle membrane activity, locomotion, phagocytosis and pinocytosis visible by light-microscopy. Micropinocytosis visible only by electron-microscopy is not inhibited. Reasons are given why these effects of

cytochalasin are thought to be due to inhibition of the microfilament-associated contractile system rather than membrane transport.

3. Colchicine, in concentrations which break down cytoplasmic microtubules, does not inhibit ruffle membrane activity, cytoplasmic movement, phagocytosis or pinocytosis. However, the orderly locomotion of cells and movement of particles within cells is abolished.

4. It is concluded that the actomyosin contractile system with membrane attachment provides the motive power for cell locomotion, phagocytosis and macropinocytosis, whereas microtubules subserve a cytoskeletal function and maintain the polarity of cells, which is required for orderly locomotion and movement of the plasma membrane and of organelles within the cytoplasm. The relationship of these processes to chemotaxis is discussed.

References

ABERCROMBIE, M.; HEAYSMAN, J. E. M., and PEGRUM, S. M.: The locomotion of fibroblasts in culture. I. Movements of the leading edge. Exp. Cell Res. *59:* 393–398 (1970).

ADELSTEIN, R. S.; POLLARD, T. D., and KUEHL, W. M.: Isolation and characterization of myosin and two myosin fragments from human blood platelets. Proc. nat. Acad. Sci., Wash. *68:* 2703–2707 (1971).

ALLEN, R. D.: Pattern of birefringence in the giant amoeba, *Chaos carolinensis.* Exp. Cell Res. *72:* 34–45 (1972).

ALLISON, A. C.: The role of membranes in the replication of animal viruses; in RICHTER and EPSTEIN International revue of experimental pathology, vol. 10, pp. 181–242 (Academic Press, New York 1971a).

ALLISON, A. C.: Discussion; in UHR and LANDY Immunologic intervention, pp. 294–298 (Academic Press, New York 1971b).

ALLISON, A. C.: Role of membranes in effector systems for hormones and drug action. Chem. Phys. Lipids *7:* 118–129 (1972a).

ALLISON, A. C.: Analogies between triggering mechanisms in immune and other cellular reactions; in SILVESTRI Cell interactions. 3rd Lepetit Coll., pp. 156–161 (North-Holland, Amsterdam 1972b).

ALLISON, A. C.; BÄCHI, T., and HAMMOND, V.: Observations on the mechanism of haemolysis by Sendai virus. Nature, Lond. (in press).

ALLISON, A. C. and BRIGHTON, W. D.: Myosin in non-muscle cells (in preparation).

ALLISON, A. C.; DAVIES, P., and PETRIS, S. DE: Role of contractile microfilaments in movement and endocytosis. Nature New Biol. *232:* 153–155 (1971).

ARCHER, F. L.; BECK, J. S., and MELVIN, J. M. O.: Localisation of smooth muscle protein in myoepithelium by immunofluorescence. Amer. J. Path. *63:* 109–118 (1971).

AXELSSON, J.: Catecholamine functions. Annu. Rev. Physiol. *33:* 1–30 (1971).

BANGHAM, A. D.; STANDISH, M. M., and WATKINS, J. C.: Diffusion of univalent ions across the lamellae of swollen phospholipids. J. molec. Biol. *13:* 238–252 (1965).

BEHNKE, O.; FORER, A., and EMMERSEN, J.: Actin in sperm tails and meiotic spindles. Nature, Lond. *234:* 408–410 (1971b).

BEHNKE, O.; KRISTENSEN, B. I., and NIELSEN, L. E.: Electron microscopical observations on actinoid and myosinoid filaments in blood platelets. J. Ultrastruc. Res. *37:* 351–369 (1971a).

BERL, S. and PUSZKIN, S.: Mg^{2+}–Ca^{2+}-activated adenosine triphosphatase system isolated from mammalian brain. Biochemistry *9:* 2058–2067 (1970).

BETTEX-GALLAND, M. and LÜSCHER, E. F.: Thrombosthenin. The contractile protein from blood platelets and its relation to other contractile proteins. Adv. Protein Chem. *20:* 1–35 (1965).

BOURNE, H. R.; LICHTENSTEIN, L. M., and MELMON, K. L.: Pharmacologic control of allergic histamine release *in vitro*. Evidence for an inhibitory role of 3'5'-adenosine monophosphate in human leukocytes. J. Immunol. *108:* 695–705 (1972).

CANER, J. E. Z.: Colchicine inhibition of chemotaxis. Arth. Rheum. *8:* 757–764 (1965).

CARTER, S.: Effects of cytochalasins on mammalian cells. Nature, Lond. *213:* 261–26 (1967).

CLONEY, R. A.: Cytoplasmic filaments and cell movements. Epidermal cells during ascidian metamorphosis. J. Ultrastruct. Res. *14:* 300–328 (1966).

COHN, Z.: Endocytosis and intracellular digestion; in VAN FURTH Mononuclear phagocytes, pp. 121–132 (Blackwell, Oxford 1970).

CRAWFORD, N.: The presence of contractile proteins in platelet microparticles isolated from human and animal platelet-free plasma. Brit. J. Haemat. *21:* 53–69 (1971).

CRAWFORD, B.; CLONEY, R. A., and CAHN, R. D.: Cloned pigmented retinal cells. The effects of cytochalasin B on ultrastructure and behaviour. Z. Zellforsch., Abt. Histochem. *130:* 135–151 (1972).

CREASEY, W. A.; BENSCH, K. G., and MALAWISTA, S. E.: Colchicine, vinblastine and griseofulvin. Pharmacological studies with human leukocytes. Biochem. Pharmacol. *20:* 1579–1588 (1971).

DAVIES, P.; FOX, R. I.; POLYZONIS, M.; ALLISON, A. C., and HASWELL, A. D.: The inhibition of phagocytosis and facilitation of exocytosis in rabbit PMN by cytochalasin B. Lab. Invest. *28:* 16–22 (1973).

DEYKIN, D.; PARRIS, E. W., and SALZMAN, E. W.: Cyclic AMP and platelet function. New Engl. J. Med. *286:* 358–363 (1972).

ESTENSEN, R. D. and PLAGEMANN, P. G. W.: Cytochalasin B: inhibition of glucose and glucosamine transport. Proc. nat. Acad. Sci., Wash. *69:* 1430–1434 (1972).

FARROW, L. J.; HOLBOROW, E. J., and BRIGHTON, W. D.: Reaction of human smooth muscle antibody with liver cells. Nature New Biol. *232:* 196–187 (1971).

FINE, R. E. and BRAY, D.: Actin in growing nerve cells. Nature New Biol. *234:* 115–118 (1971).

FITZGERALD, T. J.; WILLIAMS, B., and UYEKI, E. M.: Colchicine on sodium urate-induced paw swelling in mice. Structure-activity relationship of colchicine derivatives. Proc. Soc. exp. Biol. Med. *136:* 115–120 (1971).

FORER, A.; EMMERSEN, J., and BEHNKE, O.: Cytochalasin B. Does it affect actin-like filaments? Science *175:* 774–776 (1972).

FREED, J. J. and LEBOWITZ, M. M.: The association of a class of saltatory movements with microtubules in cultured cells. J. Cell Biol. *45:* 334–354 (1970).

FRYE, L. D. and EDIDIN, M.: The rapid intermixing of cell surface antigens after formation of mouse-human heterokaryons. J. Cell Sci. *7:* 319–335 (1970).

GABBAY, K. H. and TZE, W. J.: Cytochalasin B-sensitive emiocytosis in the beta-cell. Diabetes, N.Y. *21:* suppl., p. 327 (1972).

GAIL, M. H. and BOONE, C. W.: Effect of colcemid on fibroblast motility. Exp. Cell Res. *65:* 221–227 (1971).

GAWADI, N.: Actin in the mitotic spindle. Nature, Lond. *234:* 410 (1971).

GIBBONS, I. R. and FRONK, E.: Some properties of bound and soluble dynein from sea urchin sperm flagella. J. Cell Biol. *54:* 365–381 (1972).

GINGELL, D.: Contractile responses at the surface of an amphibian egg. J. Embryol. exp. Morph. *23:* 583–609 (1970).

GINN, F. L.; HOCHSTEIN, P., and TRUMP, B.: Membrane alterations in hemolysis. Internalization of plasmalemma induced by primaquine. Science *164:* 843–845 (1969).

HATANO, S.; KONDO, H., and MIKI-NOUMURA, T.: Purification of sea urchin egg actin. Exp. Cell Res. *55:* 275–277 (1969).

HATANO, S. and OOSAWA, F.: Isolation and characterization of plasmodium actin. Biochim. biophys. Acta *127:* 488–498 (1966).

HATANO, S. and TAZAWA, M.: Isolation, purification and characterization of myosin B from myxomycete plasmodium. Biochim. biophys. Acta *154:* 507–519 (1968).

HEATH, D. A.; PALMER, J. S., and AURBACH, G. D.: The hypocalcemic action of colchicine. Endocrinology *90:* 1589–1593 (1972).

HOLZER, H. and SANGER, J. W.: Cytochalasin B: microfilaments, cell movement and what else? Develop. Biol. *27:* 444–446 (1972).

HUXLEY, H. E.: Electron microscope studies on the structure of natural and synthetic protein filaments from striated muscle. J. molec. Biol. *7:* 281–308 (1963).

ISHIKAWA, H.; BISCHOFF, R., and HOLTZER, H.: Formation of arrowhead complexes with heavy meromyosin in a variety of cell types. J. Cell Biol. *43:* 312–328 (1969).

JACQUES, P.: Endocytosis; in DINGLE and FELL Lysosomes in biology and pathology, vol. 2, pp. 395–420 (North-Holland, Amsterdam 1969).

JOHNSON, G. S.; MORGAN, W. D., and PASTAN, I.: Regulation of cell motility by cyclic AMP. Nature, Lond. *235:* 54–56 (1972).

KAMINER, B. and SZONYI, E.: Tropomyosin in electric organs of eel and *Torpedo*. Abstract 257. J. Cell Biol. *55:* (1972).

KEMP, R. B.; JONES, B. M., and GROSCHEL-STEWART, U.: Aggregative behaviour of embryonic chick cells in the presence of antibodies directed against actomyosins. J. Cell Sci. *9:* 103–122 (1971).

KLETZIEN, R. F.; PERDUE, J., and SPRINGER, A.: Cytochalasin A and B. Inhibition of sugar uptake in cultured cells. J. biol. Chem. *247:* 2964–2966 (1972).

KOURI, J.: Personal communication (1972).

KUNDIG, W. and ROSEMAN, S.: Sugar transport. II. Characterization of constitutive membrane-bound enzymes II of the *Escherichia coli* phosphotransferase system. J. biol. Chem. *246:* 1407–1418 (1971).

LASH, J.; CLONEY, R. A., and MINOR, R. R.: Tail adsorption in Ascidians. Effects of cytochalasin B. Biol. Bull. *139:* 427–428 (1970).

LAY, W. H. and NUSSENZWEIG, V.: Receptors for complement on leukocytes. J. exp. Med. *128:* 991–1007 (1968).

MARGULIS, L.: Microtubules. Int. Rev. Cytol. (in press).

MISZURSKI, B.: Effects of colchicine on resting cells in tissue cultures. Exp. Cell Res. suppl. 1, pp. 450–451 (1969).

MITZNEGG, P.; HACH, B., and HEIM, F.: The influence of guanosine 3'5'monophosphate and other cyclic nucleotides on contractile responses induced by oxytocin in isolated rat uterus. Life Sci. 10: 1285–1289 (1971).

MIZEL, S. B. and WILSON, L.: Nucleotide transport in mammalian cells. Inhibition by colchicine. Biochemistry 11: 2573–2578 (1972).

MOORE, P. B.; HUXLEY, H. E., and DEROSIER, D. J.: Three dimensional reconstruction of F-actin, thin filaments and decorated thin filaments. J. molec. Biol. 50: 279–295 (1970).

NACHMIAS, V. T.; HUXLEY, H. E., and KESSLER, D.: Electron microscope observations on actomyosin and actin preparations from Physarum polycephalum and on their inter-action with heavy meromyosin subfragment I from muscle myosin. J. molec. Biol. 50: 83–90 (1970).

ORCI, L.; GABBAY, K. H., and MALAISSE, W. J.: Pancreatic β-cell web. Its possible role in insulin secretion. Science 175: 1128–1130 (1972).

OWELLEN, R. J.; OWENS, A. H., and DONIGIAN, C.: The binding of vincristine, vinblastine and colchicine to tubulin. Biochem. biophys. Res. Commun. 47: 685–691 (1972).

PARKHOUSE, R. M. E. and ALLISON, A. C.: Failure of cytochalasin or colchicine to inhibit secretion of immunoglobulins. Nature New Biol. 235: 220–222 (1972).

PASTAN, I. and PERLMAN, R. L.: Cyclic AMP in metabolism. Nature New Biol. 229: 5–7 (1971).

PERRY, M. M.; JOHN, H. A., and THOMAS, N. S. P.: Actin-like filaments in the cleavage furrow of newt egg. Exp. Cell Res. 65: 249–253 (1971).

PICK, E.: Cyclic AMP affects macrophage migration. Nature New Biol. 238: 176–177 (1972).

POLLARD, T. D. and KORN, E. D.: Filaments of Amoeba proteus. II. Binding of heavy meromyosin by thin filaments in motile cytoplasmic extracts. J. Cell Biol. 48: 216–219 (1971).

POLLARD, T. D. and KORN, E. D.: The 'contractile' proteins of Acanthamoeba castellani. Cold Spr. Harb. Symp. quant. Biol. 37: 573–584 (1972).

POLLARD, T. D.; SHELTON, E.; WEIHING, R. R., and KORN, E. D.: Ultrastructural charac-terization of F-actin isolated from Acanthamoeba castellani and identification of cytoplasmic filaments as F-actin by reaction with rabbit heavy meromyosin. J. molec. Biol. 50: 91–97 (1970).

PORTER, K. R.: Cytoplasmic microtubules and their functions. Ciba Foundation Sym-posium on Principles of Biomolecular Organization, pp. 308–356 (Churchill, London 1966).

RABINOVITCH, M.: The dissociation of the attachment and ingestion phases of phago-cytosis by macrophages. Exp. Cell Res. 46: 19–28 (1967).

RABINOVITCH, M. and STEFANO, M. DI: Interactions of red cells with phagocytes of the wax-moth (Galleria mellonella, L) and mouse. Exp. Cell Res. 59: 272–282 (1970).

ROBBINS, E. and GONATAS, N. K.: Histochemical and ultrastructural studies on HeLa cell cultures exposed to spindle inhibitors with special reference to the interphase cell. J. histochem. Cytochem. 12: 704–711 (1964).

ROBISON, G. A.; BUTCHER, R. W., and SUTHERLAND, E. W.: Cyclic AMP. Annu. Rev. Biochem. 37: 149–174 (1968).

ROSENBAUM, J. L.; MOULDER, J. E., and RINGO, D. L.: Flagellar elongation and shortening in *Chlamydomonas*. The use of cycloheximide and colchicine to study the synthesis and assembly of flagella proteins. J. Cell Biol. *41:* 601–619 (1969).

RYSER, H. J. P.: Studies on protein uptake by isolated tumor cells. 3. Apparent stimulations due to pH, hypertonicity, polycations, or dehydration and their relation to the enhanced penetration of infectious nucleic acids. J. Cell Biol. *32:* 737–750 (1967).

SCHROEDER, T. E.: The contractile ring. I. Fine structure of dividing mammalian (HeLa) cells and the effects of cytochalasin B. Z. Zellforsch. *109:* 431–449 (1970).

SCHROEDER, T. E.: The contractile ring. II. Determining its brief existence, volumetric changes and vital role in cleaving *Arbacia* eggs. J. Cell Biol. *53:* 419–434 (1972).

SENDA, N.; SHIBATA, N.; TATSUMI, N.; KONDO, K., and HAMADA, K.: A contractile protein from leucocytes. Its extraction and some of its properties. Biochim. biophys. Acta *181:* 191–200 (1969).

SINGER, S. J. and NICOLSON, G. L.: The fluid mosaic model of the structure of cell membranes. Science *175:* 720–731 (1972).

SPUDICH, J. A. and LIN, S.: Cytochalasin B, its interaction with actin and actomyosin from muscle. Proc. nat. Acad. Sci., Wash. *69:* 442–446 (1972).

STEPHENS, R. E.: Microtubules; in TIMASHEFF and FASMAN Biological macromolecules, vol. 5, chap. 8, pp. 355–391 (Dekker, New York 1971).

TAYLOR, R. B.; DUFFUS, W. P. H.; RAFF, M. C., and PETRIS, S. DE: Redistribution and pinocytosis of lymphocyte surface immunoglobulin molecules induced by anti-immunoglobulin antibody. Nature, Lond. *233:* 225–229 (1971).

THOMPSON, C. M. and WOLPERT, L.: The isolation of motile cytoplasm from *Amoeba proteus*. Exp. Cell Res. *32:* 156–160 (1963).

TILNEY, L. G. and MOOSEKER, M.: Actin in the brush-border of epithelial cells of the chicken intestine. Proc. nat. Acad. Sci., Wash. *68:* 2611–2615 (1971).

TSAN, M. F. and BERLIN, R. D.: Effect of phagocytosis on membrane transport of electrolytes. J. exp. Med. *134:* 1016–1035 (1971).

UKENA, T. E. and BERLIN, R. D.: Effect of colchicine and vinblastine on the topographical separation of membrane functions. J. exp. Med. *136:* 1–7 (1972).

VASILIEV, J. M.; GELFAND, I. M.; DOMNINA, L. V.; IVANOVA, O. Y.; KOMM, S. G., and OLSHEVSKAJA, L. V.: Effect of colcemid on the locomotory behaviour of fibroblasts. J. Embryol. exp. Morph. *24:* 625–640 (1970).

WEIHING, R. R. and KORN, E. D.: Amoeba actin. The presence of 3-methylhistidine. Biochem. biophys. Res. Commun. *35:* 906–912 (1969).

WEISENBERG, R. C.: Changes in the organization of tubulin during meiosis in the eggs of the surf clam, *Spisula solidissima*. J. Cell Biol. *54:* 266–278 (1972).

WEISENBERG, R. C. and TIMASHEFF, S. N.: Aggregation of microtubule subunit protein. The effects of divalent cations, colchicine and vinblastine. Biochemistry *9:* 4110–4116 (1970).

WESSELLS, N. K.; SPOONER, D. S.; ASH, J. F.; BRADLEY, M. O.; LUDUENA, M. A.; TAYLOR E. L.; WRENN, J. T., and YAMADA, K. M.: Microfilaments in cellular and developmental processes. Science *171:* 135–143 (1971).

WIKLUND, R. A. and ALLISON, A. C.: The effects of anaesthetics on the motility of *Dictyostelium discoideum*. Evidence for a possible mechanism of anaesthesia. Nature New Biol. *239:* 221–222 (1972).

WILLIAMSON, R. E.: A light-microscope study of the action of cytochalasin B on the cells
and isolated cytoplasm of the Characeae. J. Cell Sci. *10:* 811–819 (1972).

WILLS, E. J.; DAVIES, P.; ALLISON, A. C., and HASWELL, A. D.: Cytochalasin B fails to
inhibit pinocytosis by macrophages. Nature New Biol. *240:* 58–60 (1972).

WILSON, L.: Properties of colchicine binding protein from chick embryo brain. Interactions
with vinca alkaloids and podophyllotoxin. Biochemistry *9:* 4999–5007 (1970).

YANG, Y-Y. and PERDUE, J. F.: Contractile proteins of cultured cells. 1. The isolation
and characterization of an actin-like protein from cultured chick embryo fibroblasts.
J. biol. Chem. *247:* 4503–4509 (1972).

ZIGMOND, S. H. and HIRSCH, J. G.: Cytochalasin B. Inhibition of D-2-deoxyglucose
transport into leukocytes and fibroblasts. Science *176:* 1432–1434 (1972).

Author's address: Dr. A. C. ALLISON, Clinical Research Centre, Watford Road,
Harrow, Middlesex (England)

Antibiotics and Chemotherapy, vol. 19, pp. 218–232
(Karger, Basel 1974)

Active Site Chemotactic Factors and the Regulation of the Human Neutrophil Chemotactic Response

E. J. Goetzl and K. F. Austen

Department of Medicine, Harvard Medical School, and Department of Medicine, Robert B. Brigham Hospital, Boston, Mass.

The magnitude and composition of the *in vitro* chemotactic response of human leukocytes are determined both by the inherent activity and specificity of the chemotactic stimulus and by the presence of other principles which inhibit the response through an influence on either the activity of the stimulus or the migration of the leukocytes. Complexes and fragments of complement components such as C $\overline{567}$[15] and C5a [11, 24], an eosinophil chemotactic factor of anaphylaxis (ECF-A) [12, 13] and the serum enzymes kallikrein and plasminogen activator [8, 10] are each capable of attracting both neutrophils and eosinophils. However, with comparable leukocyte pools *in vitro* ECF-A is preferentially chemotactic for eosinophils [13] and kallikrein for human neutrophils [8]. The chemotactic factor inactivator (CFI) [25], a heterogeneous serum protein, irreversibly inhibits diverse chemotactic stimuli. More selective action is seen with the inhibitor of the activated first component of complement, $\overline{CI}$NH, which suppresses the chemotactic activity of kallikrein, and α_2-macroglobulin which blocks the chemotactic activity of both plasminogen activator and kallikrein, in both instances by inhibition of the respective active enzymatic site [6a]. Recently we have found a peptide inhibitor of leukocyte mobility, the neutrophil immobilizing factor (NIF), which irreversibly suppresses the human polymorphonuclear leukocyte response to diverse chemotactic factors without altering either the *in vitro* phagocytic or metabolic capabilities of these leukocytes or the chemotactic response of mononuclear leukocytes [5, 6]. This section will examine regulation of the chemotactic response of human leukocytes by native active site inhibitors and by NIF.

Human Plasma Active Site Chemotactic Factors

Human plasma kallikrein, which cleaves the nonapeptide bradykinin from kininogen, and plasminogen activator, which converts plasminogen to plasmin thus initiating fibrinolysis, are both present in native plasma as inactive proenzymes termed prekallikrein and plasminogen proactivator [7, 9]. Interaction of these proenzymes either as mixtures in crude chromatographic fractions derived from plasma or during each subsequent purification step with intact active Hageman factor or Hageman factor fragments unmasks the kinin-generating and plasminogen-activating sites with concomitant appearance of neutrophil chemotactic activity [8, 10]. As illustrated in figure 1, chemotactic activity was demonstrated after the addition of Hageman factor fragments to the Sephadex G-200 fractions obtained from whole plasma which was collected by a method [20] which minimizes spontaneous appearance of the active enzymes. Chemotactic activity assayed by a modified Boyden method [1, 5] was contained in a single broad peak coinciding with the peak of kinin-generating activity (fig. 1) and plasminogen-activating activity (not shown) between the IgG and bovine serum albumin (BSA) markers. The moderate kinin-generating, plasminogen-activating and chemotactic activities of the unactivated fractions were each increased at least 2-fold by the addition of Hageman factor fragments which alone demonstrated neither function. Further, Hageman factor activation of the plasminogen proactivator in a quaternary ammonium ethyl (QAE) Sephadex effluent of prekallikrein-deficient (Fletcher factor deficient) plasma generated approximately 50% of the chemotactic activity derived from normal plasma containing both prekallikrein and plasminogen proactivator [10]. Finally, the two proenzymes in plasma and the active enzymes in serum have been purified by serial chromatography on QAE-Sephadex, sulphopropyl (SP) Sephadex, and Sephadex G-150 so that IgG was the sole contaminant recognized on disc gel electrophoresis or isoelectric focusing of functionally pure (pre)kallikrein or plasminogen (pro)activator [8, 10]. Two discrete peaks of chemotactic activity were obtained with chromatographic fractions of either serum or subsequently activated plasma, one superimposed on the kinin-generating peak and the other on the plasminogen-activating peak in the eluate fractions from SP-Sephadex, and each of these peaks was further purified by Sephadex G-150 gel filtration to remove minor contamination with the other active site chemotactic factor. Removal of the contaminating IgG by application to an anti-IgG immunoadsorbent column has yielded a plasminogen proactivator preparation giving a single band on isoelectric

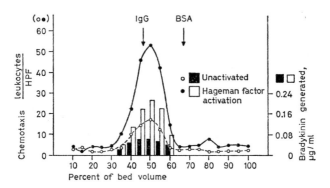

Fig. 1. Fractionation of human plasma chemotactic activity on Sephadex G-200. Three milliliters of human plasma from blood collected in plastic tubes with 9 mg of sodium ethylenediaminetetraacetate and 3.6 mg of hexadimethrine/100 ml blood was applied to a 15 × 850 mm Sephadex G-200 column equilibrated with Hanks' balanced salt solution. Fractions of 3.75 ml were collected and 1.0 ml of each fraction was assayed for chemotactic activity as the mean number of leukocytes in 10 high power fields (HPF) in duplicate modified Boyden chambers, both directly and after activation with Hageman factor fragments. Kinin-generating activity was bioassayed in each fraction directly and after Hageman factor activation by adding 0.1 of the fraction to 0.2 ml heat-treated plasma as a source of prekallikrein. Bovine serum albumin (BSA) and immunoglobulin G (IgG) were employed as markers.

focusing and revealing both plasminogen activating and chemotactic activity upon elution from unstained gels and activation with Hageman factor fragments [GOETZL and KAPLAN, unpublished observations].

The essential role in neutrophil chemotaxis of the serine esterase site of kallikrein and plasminogen activator was demonstrated by inhibition with diisopropylfluorophosphate (DFP) or soybean trypsin inhibitor (SBTI) of the chemotactic activity in the initial effluent from a QAE-Sephadex column (fig. 2) or after extensive purification of the enzymes [8, 10]. Suppression of chemotactic activity with DFP or SBTI always resulted in corresponding decrements in kinin-generating (fig. 2) and plasminogen-activating (not shown) activities. DFP treatment of the precursor molecules did not inhibit the appearance of their chemotactic or other activities upon removal of DFP and activation with Hageman factor, indicating that the active site is protected in the proenzymes [8, 10]. That not all serine esterases are chemotactic is indicated by studies with plasma thromboplastin antecedent (PTA) which also has a Hageman factor activatable serine esterase but no detectable neutrophil chemotactic activity [10].

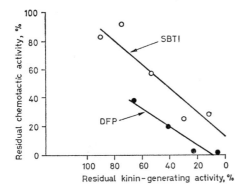

Fig. 2. Inhibition of active site chemotactic activity by diisopropylfluorophosphate (DFP) and soybean trypsin inhibitor (SBTI). SBTI in final concentrations of 0.1–1.0 μg/ml was incubated with 0.25 ml aliquots of the concentrated effluent from a quaternary ammonium ethyl (QAE) Sephadex column chromatographic fractionation of human serum containing both kallikrein and plasminogen activator; DFP was similarly interacted with this mixture of enzymes at final concentrations ranging from 10^{-7} to 10^{-3} M and the mixture was dialyzed free of DFP prior to assaying residual activity. Kinin-generating activity in 20 μl of solution was assayed as in figure 1 and the maximum activity of 1.6 μg bradykinin/ml in the untreated control was set at 100%. Chemotaxis was assayed by a radiochemotactic method [4] using a 10-fold dilution of the remainder of each solution in duplicate chambers; maximum activity representing 9.8 net percent radioactivity (table II) was designated 100%.

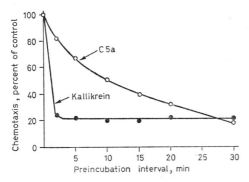

Fig. 3. Deactivation of the human neutrophil chemotactic response by purified chemotactic factors. Each point represents the residual chemotactic activity assayed by the modified Boyden technique of neutrophils either untreated (100%) or incubated with a chemotactic factor for a specific interval, washed twice with buffer, and then examined for the response to the original stimulus. The concentration of kallikrein employed had a chemotactic potency of 11.0 net percent radioactivity and that of C5a a chemotactic potency of 4.6 net percent radioactivity (table I).

Table I. Chemotactic deactivation of human neutrophils by purified chemotactic factors

Deactivating factor	Chemotactic potency[1]; net radioactivity, %	Deactivation of neutrophils; suppression, %[2]	
		C5a stimulus	kallikrein stimulus
C5a	2.0	62	80
	4.6	74	83
	8.4	87	86
Kallikrein	3.3	−7	20
	5.6	27	49
	11.0	84	81

1 Radiochemotactic determinations of the chemotactic potency of deactivating doses of C5a and kallikrein were carried out using ^{51}Cr-labeled mixed human leukocytes in a chamber with a layer of two micropore filters. Net percent radioactivity was calculated as $\dfrac{Rs - Rc}{Rt - Rc} \times 100\%$, where Rs = the radioactivity in the bottom filter of stimulated chambers, Rc = the radioactivity in the bottom filter of control chambers without a stimulus, and Rt = the total radioactivity in the initial leukocyte suspension added to each chamber.

2 The deactivation effect of prior exposure of leukocytes to a chemotactic stimulus is expressed as the percent suppression calculated by the formula $\left(1 - \dfrac{AD}{AC}\right) \times 100\%$, where AD is the radiochemotactic response of deactivated leukocytes and AC the radiochemotactic response of control leukocytes to the same test stimulus. Deactivation conditions were 25 min and 37 °C.

The chemotactically active enzymes, but not their precursors, prekallikrein and plasminogen proactivator are also capable of deactivating human neutrophils to further chemotactic stimulation (fig. 3, table I). Exposure of peripheral neutrophils to active kallikrein for as brief a period as 2 min followed by extensive washing of the leukocytes results in a state of deactivation in which the neutrophils are unresponsive to a second dose of kallikrein (fig. 3). C5a prepared by tryptic digestion of purified C5 followed by inhibition of residual trypsin with SBTI [24] deactivates neutrophils to a homologous stimulus but with a slower time course of action (fig. 3).

Kallikrein and C5a neutrophil cross-deactivation were studied using three concentrations of each factor, the chemotactic potencies of which were assessed with a radiochemotactic assay employing ^{51}Cr-labeled leukocytes and a chamber equipped with a layer of two micropore filters [4]. Both kallikrein and C5a deactivation produced parallel suppression when assessed with the homologous and heterologous stimuli (table I). C5a was more potent in leukocyte deactivation than kallikrein at a chemotactically equivalent level; the lowest concentration of C5a gave two-thirds of maximum deactivation whereas the lowest concentration of kallikrein showed only a minor deactivation.

Inhibition of Plasma Hageman Factor –
Activatable Enzyme Chemotactic Factors by Human Serum Proteins

Both $\overline{\text{CIINH}}$ and α_2-macroglobulin from human sera [18, 19] suppress the neutrophil chemotactic activity of mixtures of kallikrein and plasminogen activator in the QAE-Sephadex effluent of human serum. Studies utilizing concentrations of each inhibitor sufficient to obliterate all detectable kinin-generating activity revealed that only α_2-macroglobulin inhibited all chemotactic activity. It had previously been demonstrated that $\overline{\text{CIINH}}$ was incapable of inhibiting the plasminogen-activating function of the plasminogen activator [19]. Since α_2-macroglobulin inhibits plasmin, and therefore interferes with the assay for plasminogen activator, two other approaches were necessary to study the differential capacities of these two inhibitors.

A QAE-Sephadex effluent from 12 ml of human serum, containing both chemotactic enzymes, was concentrated to 3 ml and divided into three equal portions which were treated respectively with buffer alone (control), α_2-macroglobulin or $\overline{\text{CIINH}}$, so that detectable kallikrein activity was absent from the mixtures with inhibitor. Each mixture was rechromatographed on QAE-Sephadex to remove free $\overline{\text{CIINH}}$ or α_2-macroglobulin [19]. Neither inhibitor-mixture effluent upon concentration to initial volume had kinin-generating activity and the effluent from the α_2-macroglobulin mixture was also free of detectable plasminogen-activating activity. The control showed a mean chemotactic activity of 26 neutrophils per high power field (HPF) in the modified Boyden system. The α_2-macroglobulin-inhibited effluent attracted only 3 neutrophils/HPF whereas the $\overline{\text{CIINH}}$-inhibited effluent stimulated 13 neutrophils/HPF. These data reveal that α_2-macroglobulin but not $\overline{\text{CIINH}}$ blocks both the chemotactic and the plasminogen-activating

Table II. Suppression of active site chemotactic factors by purified serum protein inhibitors

Chemotactic factor	Chemotactic potency[1]; net radioactivity, %	Suppression of chemotaxis, %[2]	
		CĪINH	α_2-macroglobulin
Kallikrein	4.6	73	70
Plasminogen activator	6.1	6	46

1 Radiochemotactic assessment of the chemotactic potency of kallikrein and plasminogen activator were carried out as described in table I.

2 Suppression of chemotaxis was calculated by the formula $\left(1 - \dfrac{AI}{AC}\right) \times 100\%$, where AI is the radiochemotactic response of neutrophils to a mixture of a stimulus and an inhibitor and AC the response of the same number of neutrophils to the same concentration of the stimulus without an inhibitor. Inhibitors were preincubated with the chemotactic stimuli for 60 min at 4°C.

activity of plasminogen activator. The residual concentration of free α_2-macroglobulin after QAE-Sephadex chromatography of the enzyme-inhibitor mixtures was not sufficient to interfere with the assay of plasmin, thereby establishing the specificity of α_2-macroglobulin inhibition for the plasminogen activator.

Since the QAE-Sephadex column may also have removed plasminogen activator-α_2-macroglobulin complexes with possible chemotactic activity, studies were also undertaken with kallikrein and plasminogen activator purified free of detectable activity of the other enzyme (table II). The addition of CĪINH or α_2-macroglobulin to aliquots of kallikrein in amounts sufficient to reduce the kinin-generating activity by 80%, as measured by addition to 0.2 ml heat-inactivated plasma, suppressed the neutrophil chemotactic activity of kallikrein by 70% as assessed in the radiochemotactic assay. Addition of the same quantity of each inhibitor to aliquots of isolated plasminogen activator revealed approximately 50% suppression of both neutrophil chemotactic and plasminogen-activating activities with α_2-macroglobulin and no effect with CĪINH. Purified α_1-antitrypsin affected neither enzyme in terms of chemotaxis or its other functions. Therefore, CĪINH represents a probe capable of distinguishing the active site of kallikrein from that of plasminogen activator.

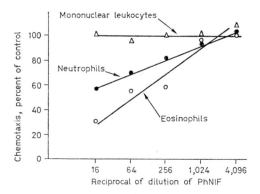

Fig. 4. Neutrophil immobilizing factor (NIF) effect on chemotaxis of purified human leukocytes. Each point represents the residual chemotactic activity, assayed by the modified Boyden technique, of leukocytes incubated in a specified dilution of partially purified phagocytosis NIF (PhNIF) relative to the activity of leukocytes incubated in the same dilution of control for NIF (100%). Neutrophils and mononuclear leukocytes purified on Ficoll-Hypaque cushions were stimulated with kallikrein and C5a, respectively. Eosinophils purified on metrizoate cushions were stimulated with human eosinophil chemotactic factor of anaphylaxis (ECF-A).

A Leukocyte-Derived NIF

Incubation of mononuclear or polymorphonuclear leukocytes from human peripheral blood in acidic medium (ANIF), with endotoxin (ENIF) or with particles undergoing phagocytosis (PhNIF) releases a soluble inhibitor of chemotaxis [5]. This inhibitor, termed the NIF, irreversibly blocks the response of human neutrophils and eosinophils [6] but not mononuclear leukocytes to diverse chemotactic stimuli without impairing their viability as assessed by dye exclusion. Only trivial inhibitory activity is found in the supernatant of unstimulated leukocytes which serves as the control for NIF [5]. The inhibitory activity of NIF is separable from coexistent chemotactic activity present in the leukocyte supernatants by heating at 56 °C for 1 h followed by chromatography on Sephadex G-25. NIF inhibitory activity elutes with an approximate molecular weight of 4,000–5,000. The chemotactic inhibitory activity of column-purified NIF obtained from a phagocytosis supernatant (PhNIF) is shown in figure 4. Exposure of leukocytes to partially purified PhNIF at varying dilutions prior to assessment of their chemotactic activity in the modified Boyden assay results in marked suppression of the directed migration of purified neutrophils to

Table III. Regulation of the chemotactic response of human leukocytes

Leukocyte[1]	Source of chemotactic factor (I)[2]	Chemotactic factor (II)	Chemotactic factor inhibitor (III)	Leukocyte inhibitor (IV)	
				CF deactivation	NIF
Neutrophil	activation of Hageman factor	kallikrein	$\overline{C1}$INH, α_2-macroglobulin	+	+
		plasminogen activator	α_2-macroglobulin	+	+
	classical or alternate	C3a	AI, CFI	+	+
	complement pathway	C5a	AI, CFI	+	+
		$\overline{C567}$	CFI	+	+
Eosinophil	classical or alternate pathway	C5a	AI, CFI	+	+
	anaphylactic reactions; mast cells	ECF-A	ND	+	+
Monocyte	classical or alternate pathway	C5a	AI, CFI	ND	–

+ = Effect demonstrated; – = no effect; ND = experiment not done; AI = anaphylatoxin inactivator; CFI = chemotactic factor inactivator; ECF-A = eosinophil chemotactic factor of anaphylaxis.

1 Basophils and lymphocytes are not included because the basophils have not been adequately separated from other leukocytes and human lymphocytes do not appear to be capable of chemotaxis in Boyden chambers.

2 Lymphocyte-derived chemotactic factors are excluded since they are inadequately purified at present.

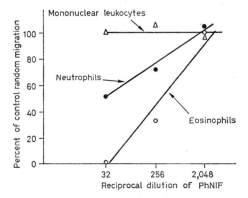

Fig. 5. NIF effect on random migration of purified human leukocytes. Each point represents the residual random migration, assayed in a modified Boyden system without a chemotactic stimulus, of leukocytes incubated in a specified dilution of phagocytosis NIF (PhNIF) relative to leukocytes incubated in the same dilution of control (100%) for NIF. Leukocytes were purified as in figure 4.

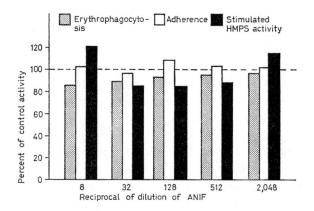

Fig. 6. NIF effect on erythrophagocytosis, surface adherence and stimulation of the hexose monophosphate shunt (HMPS) activity by kallikrein of purified neutrophils. Each bar represents the residual activity of Ficoll-Hypaque cushion-purified human neutrophils in a specified dilution of acidic medium NIF (ANIF) relative to the same activity of neutrophils in an identical dilution of control (100%) for NIF. Erythrophagocytosis was assessed as described [3] using antibody-sensitized erythrocytes coated with C1, C4, C2 and C3. Surface adherence was assayed by measuring the residual cellular DNA dissolved in 3% sodium lauryl sulfate from plastic Petri dishes incubated with $3–4 \times 10^6$ neutrophils and then washed three times to remove nonadherent cells [6]. Hexose monophosphate shunt activity was assessed by measuring the radioactivity of $^{14}CO_2$ generated from $1-^{14}C$-glucose metabolism by adherent neutrophils in sealed Petri dishes [6]. ANIF = acid medium NIF.

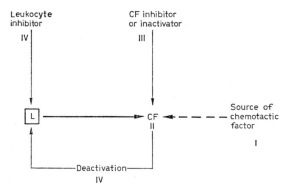

Fig. 7. Regulation of the chemotactic response of human leukocytes. Symbols used are: heavy arrow = leukocyte response, dashed arrow = generation of a factor, thin arrow = inactivation of leukocytes or chemotactic factors; CF = chemotactic factor, L = leukocyte. The specific factors for each type of leukocyte are shown in table III.

kallikrein or C5a (not shown) and purified eosinophils to ECF-A or C5a (not shown) without altering the response of mononuclear leukocytes to C5a. NIF similarly suppresses the random migration of neutrophils and eosinophils, but not mononuclear leukocytes, as assessed in the modified Boyden system but without a chemotactic stimulus (fig. 5).

The noncytolytic action of NIF and its functional selectivity is shown by its inability to significantly suppress the phagocytosis by suspensions of purified neutrophils of antibody-sensitized erythrocytes coated with the first four components of complement (fig. 6). NIF is also unable to suppress the adherence of purified neutrophils to plastic Petri dishes, the phagocytic function of adherent neutrophils, or the 2- to 5-fold increase in the activity of the hexose monophosphate shunt (HMPS) of adherent purified neutrophils following the introduction of a chemotactic factor (fig. 6) [6].

Conclusions

The chemotactic response of human polymorphonuclear leukocytes is modulated by humoral factors through four basic pathways (fig. 7). Since the chemotactic factors are generated during an inflammatory reaction, each mechanism of generation represents an initial level of modulation determined by the nature and location of the inflammatory reaction: activation of the classical or alternate complement pathways by immune complexes or microbial cell wall materials leads to the formation of C3a, C5a, and $\overline{C567}$ [16, 17]; activation of Hageman factor by an interaction with basement membranes or collagen leads to the generation of kallikrein and plasminogen activator (fig. 1) [2, 7]; and antigen challenge of mast cells sensitized with IgE results in the release of preformed ECF-A along with histamine [12, 27]. At a second level, chemotactic factors have unique structural features which determine their cellular specificity. The active site chemotactic factors kallikrein and plasminogen activator selectively attract neutrophils by a mechanism which requires their serine esterase function (fig. 2), but this alone is not sufficient for chemotaxis since PTA which also possesses a Hageman factor activatable esteratic site is not chemotactic for neutrophils [10]. The fragment of the fifth complement component, C5a, attracts monocytes as well as polymorphonuclear leukocytes [21], while the trimolecular complex $\overline{C567}$ has appreciable chemotactic activity only for polymorphonuclear leukocytes [15]. ECF-A, a 500 molecular weight peptide [12, 13, 27] which preferentially attracts eosinophils [12], and C5a, a 12,000 molecular weight peptide [24] which also appears selective for eosinophils [11], both demonstrate their eosinophilotactic specificity when the eosinophils represent 10% or more of the leukocyte pool and act synergistically in attracting eosinophils [14].

A third pathway of modulation (fig. 7) involves inhibitors or inactivators of chemotactic factors and includes the anaphylatoxin inactivator (AI) which inhibits complement-stimulated chemotaxis by destroying the anaphylatoxins, C3a and C5a [26]. The CFI, a heterogeneous protein found in high concentrations in sera of patients with Hodgkin's disease or agamma globulinemia [25] irreversibly inactivates diverse chemotactic factors including those derived from complement. The exact structural and functional relationships between AI and CFI are yet to be established, but preliminary studies indicate that they are separate inhibitors [26]. The active site chemotactic factors are selectively modulated by specific serum proteins which inhibit their active site functions. The $\overline{CI}$INH blocks both the kinin-generating and neutrophil chemotactic activities of kallikrein without influencing the functions of the plasminogen activator (table II) [18], while α_2-macroglobulin inhibits both kallikrein and the plasminogen activator (table II) [19] without influencing the chemotactic activity of complement-derived stimuli [6a].

Finally, the chemotactic response is regulated by factors which influence the reactivity of leukocytes (fig. 7). Both deactivation by chemotactic factors and inhibition by NIF suppress chemotaxis by a direct effect on leukocytes, and their inhibitory effect is independent of the specific test chemotactic stimulus (table I) [5, 23]. The polymorphonuclear leukocyte inhibition by NIF, which has no chemotactic activity, extends to random migration (fig. 5) [6] and is polymorphonuclear leukocyte specific. NIF irreversibly and selectively suppresses both neutrophilic and eosinophilic leukocyte directed and random migration (fig. 4, 5) [6] without cytotoxicity or suppression of such other leukocyte functions as phagocytosis, surface adherence or stimulated HMPS activity (fig. 6). An *in vitro* biologic role of NIF then may be the trapping of neutrophils in early inflammatory foci without affecting their other functions. Higher concentrations of NIF released from both mononuclear and polymorphonuclear leukocytes could subsequently provide a feedback control by limiting the influx of neutrophils but not mononuclear leukocytes. NIF together with the mononuclear leukocyte-selective chemotactic factor present in polymorphonuclear leukocyte lysates [22] would provide a leukocyte-directed transition from neutrophils to mononuclear leukocytes in those lesions in which there is early polymorphonuclear leukocyte predominance.

Summary

The activation of purified prekallikrein or plasminogen proactivator by fragments of activated Hageman factor, themselves devoid of activity, gave rise to chemotactic activity for human neutrophils. This neutrophil chemotactic activity was associated with the active enzymes kallikrein and plasminogen activator during their serial chromatographic separations and purification. Chemotactic activity required an intact serine esterase active site since it was destroyed by treating the enzymes with DFP; such treatment also resulted in a concomitant suppression of kinin-generating activity or plasminogen-activating activity, respectively. The chemotactic activity of both enzymes was inhibited by α_2-macroglobulin and the response to kallikrein, but not to plasminogen activator, was suppressed by preincubation with the inhibitor of the first component of complement (CĪINH), in both cases by an effect on the factors themselves. The response to these factors was also suppressed by preincubating the neutrophils with either active site chemotactic factor, a process termed deactivation, or by exposure to the NIF, which suppresses the random or directed mobility of polymorphonuclear leukocytes by a direct action on the cells.

Acknowledgements

The authors are grateful for the collaborative efforts of Dr. ALLEN P. KAPLAN, Dr. IRMA GIGLI, Dr. ALAN D. SCHREIBER, Dr. ELBERT MAGOON, and Dr. BRUCE U. WINTROUB. Ms. JANET WOODS and IRMA FINKS provided valuable technical assistance. This work was supported by grants AI-07722 and AI-10356 from the National Institutes of Health and a grant from the John A. Hartford Foundation, Inc.

References

1 BOYDEN, S.: The chemotactic effect of mixtures of antibody and antigen on poly-morphonuclear leukocytes. J. exp. Med. *115:* 453–466 (1962).

2 COCHRANE, C. G.; REVAK, S. D.; AIKIN, B. S., and WUEPPER, K. D.: The structural characteristics and activation of Hageman factor; in LEPOW and WARD Inflammation. Mechanisms and control, pp. 119–138 (Academic Press, New York 1972).

3 GIGLI, I. and NELSON, R. A., jr.: Complement dependent immune phagocytosis. I. Requirements for C1, C4, C2, C3. Exp. Cell Res. *51:* 45–67 (1968).

4. GOETZL, E. J. and AUSTEN, K. F.: A method for assessing the *in vitro* chemotactic response of neutrophils utilizing ^{51}Cr-labeled human leukocytes. Immunol. Commun. *1:* 421–430 (1972).

5 GOETZL, E. J. and AUSTEN, K. F.: A neutrophil immobilizing factor derived from human leukocytes. I. Generation and partial characterization. J. exp. Med. *136:* 1564–1580 (1972).

6 GOETZL, E. J.; GIGLI, I.; WASSERMAN, S. I., and AUSTEN, K. F.: A neutrophil im-mobilizing factor derived from human leukocytes. II. Specificity of action on poly-morphonuclear leukocyte mobility. J. Immunol. *111:* 938–945 (1973).

6a GOETZL, E. J.; SCHREIBER, A. D., and AUSTEN, K. F.: Modulation of active site chemotactic factors by serum protein inhibitors. (in preparation).

7 KAPLAN, A. P. and AUSTEN, K. F.: A prealbumin activator of prekallikrein. J. Im-munol. *105:* 802–811 (1970).

8 KAPLAN, A. P.; KAY, A. B., and AUSTEN, K. F.: A prealbumin activator of pre-kallikrein. III. Appearance of chemotactic activity for human neutrophils by the conversion of human prekallikrein to kallikrein. J. exp. Med. *135:* 81–97 (1972).

9 KAPLAN, A. P. and AUSTEN, K. F.: The fibrinolytic pathway of human plasma. J. exp. Med. *136:* 1378–1393 (1972).

10 KAPLAN, A. P.; GOETZL, E. J., and AUSTEN, K. F.: The fibrinolytic pathway of human plasma. II. The generation of chemotactic activity by activation of plasmino-gen proactivator. J. clin. Invest. *52:* 2591–2595 (1973).

11 KAY, A. B.: Studies on eosinophil leukocyte migration. II. Factors specifically chemotactic for eosinophils and neutrophils generated from guinea pig serum by antigen-antibody complexes. Clin. exp. Immunol. *7:* 723–737 (1970).

12 KAY, A. B. and AUSTEN, K. F.: The IgE-mediated release of an eosinophil leukocyte chemotactic factor from human lung. J. Immunol. *107:* 899–902 (1971).

13 KAY, A. B.; STECHSCHULTE, D. J., and AUSTEN, K. F.: An eosinophil leukocyte chemotactic factor of anaphylaxis. J. exp. Med. *133:* 602–619 (1971).

14 KAY, A. B.; SHIN, H. S., and AUSTEN, K. F.: Selective attraction of eosinophils and synergism between eosinophil chemotactic factor of anaphylaxis (ECF-A) and a fragment cleaved from the fifth component of complement (C5a). Immunology, Lond. *24:* 969–976 (1973).

15 LACHMANN, P. J.; KAY, A. B., and THOMPSON, R. A.: The chemotactic activity for neutrophil and eosinophil leukocytes of the trimolecular complex of the fifth, sixth, and seventh components of human complement (C$\overline{567}$) prepared in free solution by the 'reactive lysis' procedure. Immunology, Lond. *19:* 895–899 (1970).

16 MÜLLER-EBERHARD, H. J.: Chemistry and reaction mechanisms of complement. Adv. Immunol. *8:* 1–80 (1968).

17 RUDDY, S.; GIGLI, I., and AUSTEN, K. F.: The complement system of man. New Engl. J. Med. *287:* 489–495, 545–549, 592–596, 642–646 (1972).

18 SCHREIBER, A. D.; KAPLAN, A. P., and AUSTEN, K. F.: Inhibition by C$\overline{\text{I}}$INH of Hageman factor fragment activation of coagulation, fibrinolysis, and kinin generation. J. clin. Invest. *52:* 1402 (1973).

19 SCHREIBER, A. D.; KAPLAN, A. P., and AUSTEN, K. F.: Plasma inhibitors of the fibrinolytic pathway in man. J. clin. Invest. *52:* 1394 (1973).

20 SPRAGG, J.; TALAMO, R. C.; WINTROUB, B. U.; HABER, E., and AUSTEN, K. F.: Immunoassays for bradykinin and kininogen; in CHASE and WILLIAMS Methods in immunology and immunochemistry, vol. V (in press).

21 SNYDERMAN, R.; ALTMAN, L. C.; HAUSMAN, M. S., and MERGENHAGEN, S. E.: Human mononuclear leukocyte chemotaxis. A quantitative assay for humoral and cellular chemotactic factors. J. Immunol. *108:* 857–860 (1972).

22 WARD, P. A.: Chemotaxis of mononuclear cells. J. exp. Med. *128:* 1201–1221 (1968).

23 WARD, P. A. and BECKER, E. L.: The deactivation of rabbit neutrophils by chemotactic factor and the nature of the activatable esterase. J. exp. Med. *127:* 693–709 (1968).

24 WARD, P. A. and NEWMAN, L. J.: A neutrophil chemotactic factor from human C5. J. Immunol. *102:* 93–99 (1969).

25 WARD, P. A.: Natural and synthetic inhibitors of leukotaxis; in LEPOW and WARD Inflammation. Mechanisms and control, pp. 301–310 (Academic Press, New York 1972).

26 WARD, P. A.: Personal commun. (1973).

27 WASSERMAN, S. I.; GOETZL, E. J., and AUSTEN, K. F.: Preformed eosinophil chemotactic factor of anaphylaxis (ECF-A). J. Immunol. *112:* 351–358 (1974).

Authors' address: EDWARD J. GOETZL, MD and K. FRANK AUSTEN, MD, Department of Medicine, Harvard Medical School, and Department of Medicine, Robert B. Brigham Hospital, *Boston, MA 02120* (USA)

Antibiotics and Chemotherapy, vol. 19, pp. 233–270
(Karger, Basel 1974)

Substances that Attract Eosinophils *in vitro* and *in vivo*, and that Elicit Blood Eosinophilia

W. E. PARISH

Lister Institute of Preventive Medicine, Elstree, Hertfordshire

The substances that attract eosinophils *in vitro* may be considered as belonging to one of 4 groups according to their properties and origin. Some substances attract both neutrophils and eosinophils; others are selective for eosinophils. The groups of substances are as follows.

1. Anaphylatoxin-associated substances generated by incubating antigen-antibody complexes, or other agents, in serum-containing complement [KELLER and SORKIN, 1969; WARD, 1969; KAY, 1970; LACHMANN et al., 1970; SORKIN et al., 1970; WISSLER et al., 1972].

2. Substances released from anaphylactic lung, or anaphylactic basophils and mast cells [KAY et al., 1971; KAY and AUSTEN, 1971; PARISH, 1972a].

3. Substances formed when damaged neutrophils or their components are incubated in serum-containing complement [BOREL et al., 1969; SORKIN et al., 1970; PARISH, 1972 a, b].

4. The substance formed when antigen-stimulated guinea-pig lymph node cell supernatant medium is incubated with immune complexes of homologous antibody with the same antigen used to stimulate the lymph-node cells [COHEN and WARD, 1971].

The *in vitro* tests detect only the substances that attract eosinophils. The *in vivo* events are more complex, and stimuli other than attraction may elicit eosinophilia in tissues. 'Eosinophilia' is an imprecise term for an increase above the normal number of eosinophils, though histologists use the term more strictly to describe tissue changes in which the eosinophils are the predominant infiltrating cells. Eosinophilia designates the accumulation of eosinophils selectively attracted to an organ or lesion, e.g. the nasal mucosa in hay-fever or a parasitic granuloma, the increased numbers in the blood as occurs in anaphylaxis or asthma and the increased maturation in the bone marrow as occurs in some allergic disorders. Moreover, eosinophilia of the non-haematogenous tissues, e.g. the skin or lung, may occur independently of eosinophilia of the blood, which may be independent of that of the bone marrow. Three examples illustrate the independence of skin, blood and bone marrow eosinophilia. Eosinophils accumulate in anaphylactic or antigen-antibody complex treated skin, probably by selective retention of cells randomly infiltrating the site, without any increase in the number in the blood [PARISH, 1972 b]. The rapid increase in the number of eosinophils in blood during anaphylaxis or after treatment with extracts of anaphylactic tissues (see below) occurs without any increase in the number in the bone marrow. Such rapid eosinophilia probably reflects the 'mobilisation of eosinophils from the marginated pool' as described by SPRY [1971 a]; that is, the release of mature eosinophils from reserves in the spleen, bone marrow or other tissues. This differs from the blood eosinophilia accompanied by increased formation in the bone marrow occurring several days after para-sitising rats, and which is mediated by lymphocytes [BASTEN and BEESON, 1970; SPRY, 1971 b].

This chapter considers the results of our studies on substances attracting eosinophils *in vitro* and *in vivo* and, when known, correlates this activity with blood and bone marrow eosinophilia.

Distinctions between Eosinophil Retention, Attraction and Mobilisation

Histamine as a Substance Retaining Eosinophils in Tissues

It has been variously claimed that histamine attracts eosinophils into tissues, and elicits a haematogenous eosinophilia. Our tests [PARISH, 1970a] failed to show that histamine was chemotactic for eosinophils, but showed that histamine in tissues could retain eosinophils for a longer time than neutrophils, giving the appearance of a selective attraction for eosinophils.

Histamine did not attract eosinophils or elicit eosinophilia, on the evidence (1) anti-histamine drugs modifying acute anaphylaxis in guinea-pigs failed to reduce the eosinophilia of blood or lungs; (2) intravenous histamine in guinea-pigs induced eosinopenia in 30 min to 3 h, and the slight to moderate eosinophilia at 24 h probably resulted from release of eosinophils sequestered in the tissues, and (3) eosinophils were not attracted to histamine *in vitro* as monolayers in glass chambers, or in later tests in Boyden-type chambers when a gradient of increasing concentration of histamine was perfused

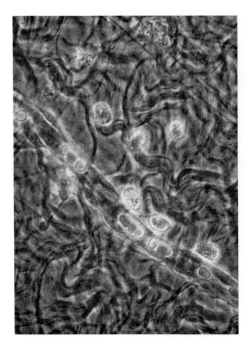

Fig. 1. Guinea-pig mesentery 45 min after application of 5 μg histamine. Eosinophils with refractile granules remain near the vessel. Two neutrophils migrate away (top) one is shedding granules. A third neutrophil to the left disappears into the depth of the tissue. Phase contrast [from PARISH, 1970a].

through the bottom, test, chamber. This confirmed the results of the *in vitro* test of KELLER and SORKIN [1969].

However, when histamine was applied *in vivo* to tissues which were observed continuously, there was a short period of perivascular eosinophilia which could have been erroneously interpreted as selective eosinophil attraction.

The mesenteries of normal or of anaphylactically sensitised, anaesthetised guinea-pigs were exposed and mounted intact between thin siliconed glass slides at 37 °C and examined continuously under phase contrast illumination. Despite the several technical limitations that may have depressed some changes and accelerated others, it was possible to observe events for 1–2 h. Animals with a natural eosinophilia of 300–400 cells/mm³ had histamine applied to the mesentery as 0.02 ml containing 2.5 or 5 μg histamine base/ml.

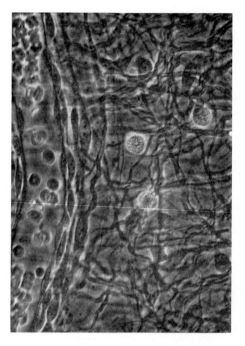

Fig. 2. Guinea-pig mesentery 90 min after application of 5 µg histamine. Three eosinophils begin to migrate from the vessel; the granules do not enter the translucent edge of the cytoplasm as they do in neutrophils. One neutrophil remains, bottom right. Phase contrast [from PARISH, 1970a].

Small arterioles contracted, but there was no noticeable change in character of small vessels, though they became congested with a sluggish flow. Both mononuclear and polymorphonuclear leucocytes adhered to the endothelium, but while some of the polymorphonuclear cells emigrated in 10–15 min, the mononuclear cells either remained adherent or were swept away. In 30–50 min the neutrophils either migrated further from the vessel or wandered near the point of their emergence; a few discharged their granules and presumably began to degenerate. The eosinophils, recognisable by the size and number of their granules and by the more obvious translucent membrane, were less active, and those that emigrated tended to round up and stay near the vessel, where they remained as the predominant cells when the more numerous neutrophils moved away (fig. 1, 2).

Histamine, therefore, did not selectively induce eosinophil emigration, because more neutrophils than eosinophils infiltrated the tissues, but as the eosinophils stayed by the vessel when the neutrophils migrated further, histamine appeared to selectively retain the eosinophils. Such eosinophil retention is probably of short duration, because sections of skin treated with histamine show no persistent perivascular eosinophils.

Different Activities of IgG2 Complexes which Attract, and of IgG1 Complexes which Attract Eosinophils and Elicit Blood Eosinophilia

Complexes formed of guinea-pig IgG2 with antibody and complement, and of the anaphylactic IgG1 which has two forms, IgG1a and IgG1b [PARISH, 1970b; OVARY and WARNER, 1972], also with antigen and complement, attract eosinophils *in vitro*, though the IgG2-containing complexes attracted more eosinophils than those containing IgG1 (table I). Our findings [PARISH, 1970a; 1972b] that complexes containing IgG1 antibody were less able to attract eosinophils than those containing IgG2 appear to conflict with those of KAY [1970] who found both types of antibody to be equally effective in forming eosinophil-attracting complexes. The difference, however, may be due to the methods of preparing the complexes. The complexes used in our tests were formed by adding IgG1 fractions to antigen in the presence of complement, whereas those of KAY were formed by mixing IgG1 and antigen and subsequently incubating the complexes in complement.

Complexes containing IgG2 were less effective than those containing IgG1 in attracting eosinophils *in vivo*, and failed to elicit blood or peritoneal eosinophilia. Table II [modified from PARISH, 1970a] shows that in guinea-pigs with less than 25 eosinophils/mm³ blood, intradermal injection of IgG2 complexes did not attract many eosinophils into the skin 18 h later, though they stimulated eosinophil infiltration into the draining lymph-node. When, however, IgG2 complexes were injected intradermally into guinea-pigs whose blood had 100 or more eosinophils/mm³, eosinophils were attracted into the skin in proportion to the numbers in the blood. The effect of the complexes was not selective for eosinophils, because there were also numerous neutrophils and some mononuclear cells.

Injection of the IgG2 complexes intraperitoneally attracted only small numbers of eosinophils into the peritoneal fluid, and did not elicit any increase in the number of eosinophils in the blood, at 18–24 h, or in the bone marrow at 1 or 6 days (table II).

In contrast to the changes induced by the IgG2 complexes, the IgG1 complexes not only attracted eosinophils into injected skin sites but also

Table I. Summary of the number of eosinophils attracted to antigen-antibody complexes in Boyden-type chambers

Test substance in medium with complement	Eosinophils, 100 fields
Complexes with IgG2 and IgG1a in whole serum	244
Complexes with IgG2	298
Complexes with IgG1a	68
Complexes with IgG1b	51
Antigen with 10% normal serum	8
10% normal serum	10
Crushed guinea-pig neutrophils	271

Complexes were formed of guinea-pig anti-picryl-chloride antibodies with picryl-chloride guinea-pig albumin.
[From PARISH, 1972b, by permission of the editor of Immunology, Lond.]

Table II. Eosinophilia elicited by complexes of guinea-pig IgG1a and IgG2 antibodies with antigen

Complexes of	Intradermal injection		Intraperitoneal injection			
	skin	lymph-node	peri-toneal cavity	periph-eral blood	bone marrow	
					1 day	6 days
IgG1a anti-BSA plus BSA	+	+	+	+	–	–
IgG2 anti-BSA plus BSA	–	+	±	–	–	–
IgG2 anti-tetanus toxoid plus tetanus toxoid	–	+	–	–	–	–

Guinea-pigs used had less than 25 eosinophils/mm³ blood before test.
BSA = bovine serum albumin.
+ skin = >7 eosinophils/516 field.
+ lymph-node = >20 eosinophils/516 × field.
+ peritoneal fluid = >500 eosinophils/mm³.
+ blood = >2-fold increase/mm³.
[Modified from PARISH, 1970a, by kind permission of the editor of Brit. J. Derm.]

Table III. Number of eosinophils in skin injected with IgG1a anti-picryl-chloride (control) and with IgG1a anti-BSA (test), 18 h after intravenous challenge with BSA. Each concentration of antibody was tested on 9 guinea-pigs in 3 groups of 3 according to the number of eosinophils in the peripheral blood

Antibody, μg	Antibody class	Group	Number of eosinophils in skin sites; recipients eosinophils/mm³ blood in range of		
			0–25	100–250	1,125–2,000
0.5	IgG1	control	3	6	31
	IgG2	control	1	6	29
	IgG1	test	6	18	83
	IgG2	test	1	5	44
5	IgG1	control	3	13	54
	IgG2	control	2	15	48
	IgG1	test	16	82	652
	IgG2	test	6	19	233
50	IgG1	control	4	22	66
	IgG2	control	4	19	71
	IgG1	test	38	101	788
	IgG2	test	10	64	594

Guinea-pigs passively sensitised for 4 h before challenge.
[From Parish, 1972b, by permission of the editor of Immunology, Lond.]

elicited eosinophilia in the peritoneum and blood when injected intra-peritoneally. The increased numbers in the blood were not accompanied by increased numbers in the bone marrow (table II).

Complexes formed of both types of antibody induced eosinophilia in the draining lymph-nodes, but this phenomenon is distinct from that occurring in skin or blood, because antigen alone or immunoglobulin alone, intra-cutaneously, induced regional lymph-node eosinophilia but no infiltration of the injected skin.

Anaphylactic Eosinophilia in Skin Sensitised by IgG1 Antibody

Differences in the ability of IgG1 and IgG2 antibodies to prepare tissues for eosinophilia were more apparent after passive sensitisation and challenge *in vivo* than after injection of preformed complexes. When preparations of each type of antibody were injected intracutaneously, and challenged intravenously 4 h later, eosinophils were attracted into sites prepared with IgG1 antibody, but few into those prepared with IgG2. Moreover, the number of eosinophils infiltrating the test site was influenced by the amount of antibody injected and the number of circulating eosinophils (table III). Passive cutaneous anaphylaxis was not accompanied by increased numbers of haematogenous eosinophils.

When samples of the anaphylactic skin were examined at intervals after challenge, eosinophil infiltration was first apparent at the 4th to 8th hours and the greatest number was present at 12 h and persisted until 24 h, whereas the greatest number of neutrophils occurred at 4 h, and thereafter the numbers were reduced as they migrated away. This evidence suggests that the cutaneous eosinophilia results from a selective retention of eosinophils randomly infiltrating the skin rather than release of a substance selectively attracting eosinophils from the blood, whereas the neutrophils are not selectively retained in passive anaphylaxis tests (as they are in the presence of complexes in Arthus reactions) and migrate away.

Summary of Tests on Retention and Attraction
of Eosinophils in Tissues

This account of our tests based on further experimental data [PARISH, 1970a, 1972b, c] is interpreted as follows.

1. Histamine does not attract eosinophils *in vitro*, or into tissues *in vivo*, or elicit blood eosinophilia. When applied to mesentery, it retains emigrated eosinophils longer than the neutrophils, resulting in a short-term selective retention of eosinophils.

2. Complexes of IgG2 and of IgG1 antibody with antigen and complement attract eosinophils *in vitro;* the IgG2 complexes being more effective than those containing IgG1. When injected into skin, the IgG2 complexes were less effective in attracting eosinophils than those containing IgG1. Furthermore, the IgG2 complexes when injected intraperitoneally attracted few eosinophils and did not elicit blood eosinophilia, whereas the IgG1 complexes mediated both phenomena.

3. After intracutaneous injection of preformed complexes, or after passive cutaneous anaphylaxis, the numbers of eosinophils infiltrating the site progressively in 12 h is proportional to the numbers in the blood, and no

increase occurs in the number in the blood. This is thought to show that eosinophils randomly infiltrate the site from the blood, and are selectively retained there, rather than they are selectively attracted to the site.

Studies on Anaphylactic Substances Eliciting Blood Eosinophilia or Selectively Attracting Eosinophils

Anaphylaxis is associated with eosinophilia of the shock organs and of the blood. It is apparent that the eosinophilia is not induced directly by complexes of anaphylactic antibody with antigen, but by substances released from the anaphylactic tissues. The following summaries describe our attempts to determine the origin and properties of substances released from anaphylactic tissues.

Substances in Anaphylactic Guinea-Pig Tissues Eliciting Eosinophilia or Attracting Eosinophils

SAMTER *et al.* [1953] contributed much to our understanding of eosinophilia in anaphylaxis, when they showed that pieces of guinea-pig anaphylactic lung implanted into the peritoneal cavity of a normal guinea-pig induced eosinophilia in the blood of the recipient 24 h later. When confirming this property of anaphylactic lung, LITT [1960] and PARISH and COOMBS [1968] showed that the eosinophilia was induced by a soluble substance that could be extracted from the lung.

The soluble substance extracted from anaphylactic lung passed through 0.8 μm filter pores but did not pass, or completely pass, through dialysis membranes [PARISH and COOMBS, 1968], though subsequently the eosinophil-eliciting substance was sometimes found to diffuse through dialysis membranes leaving activity inside and outside the sac. The variable results may be due to differences in samples of the Visking dialysis tubing and to differences in osmotic pressure in each preparation.

At the time that these tests were made, LITT's assertion [1961, 1964] that antigen-antibody complexes attracted eosinophils and elicited eosinophilia was generally accepted. In order to see whether or not the eosinophilia-inducing substance in extracts of anaphylactic lung was an antigen-antibody complex, extracts were treated to remove antigen, antibody and the C3 complement component before being tested in guinea-pigs.

The lungs of guinea-pigs actively sensitised to β-lactoglobulin or to bovine serum albumin were perfused with Hank's or Tyrode's solutions. The lungs were challenged with 250 μg antigen, either by adding it to the fluid perfusing the whole lung, or to the fluid perfusing lung which had been perfused, chopped and rewashed before challenge. The post-challenge cell-free perfusates were treated with insoluble immuno-adsorbents to remove the antigen and all IgG. Three extracts were also treated to remove guinea-pig complement C3. The adsorptions were effective in removing the complexes because antigens labelled with ^{131}I before challenge were completely removed from the perfusate and, in a control test in which normal lung was passively sensitised with ^{131}I-labelled IgG1, this globulin was completely removed from the perfusate. The adsorbed perfusates were as effective as the unadsorbed samples in eliciting haematogenous and peritoneal eosinophilia in normal guinea-pigs, but were less effective in attracting eosinophils to the site of a cutaneous injection. The attraction of fewer eosinophils to skin injected with adsorbed than with unadsorbed perfusates was not due to removal of complexes, but to non-specific adsorption by the immuno-adsorbent, because adsorbents prepared with unrelated substances, e.g. ovalbumin, also reduced the ability of the extract to attract eosinophils into skin.

These tests showed that perfusates of anaphylactic guinea-pig lung contained one or more substances that induced eosinophilia in the blood and peritoneum of normal guinea-pigs, and attracted eosinophils when injected into the skin. The eosinophil-eliciting and attracting activity was not due to antigen-antibody complexes, and though the participation of the trimolecular complement components $C\overline{567}$ was not excluded, they were unlikely to be formed in tissue perfused free of blood before antigen challenge.

The first detailed report of a substance from anaphylactic guinea-pig lung that selectively attracted eosinophils in chambers *in vitro*, was presented by KAY et al. [1971]. They showed that the substance was released from anaphylactic lung concomitantly with histamine and slow reacting substance (SRS-A), requiring the same divalent cations and equally independant of complement as is release of histamine. It has a molecular weight of between 500 and 1,000, and is destroyed by boiling in alkaline solution. This substance was designated eosinophil chemotactic factor (ECF-A). It appears to be the same eosinophil-attracting substance examined in the *in vivo* tests of LITT [1960] and of PARISH and COOMBS [1968].

In further tests on the perfusates of whole lung, or chopped fragments of lung, from actively or passively sensitised guinea-pigs, the eosinophil-attracting substance was assayed *in vitro* in Boyden-type chambers, similar to the techniques used by KAY *et al.* [1971]. We found it to have the following properties.

1. The substance is released from anaphylactic lung concomitantly with histamine and SRS-A, but whereas release of histamine virtually ceases within 30 min, release of the eosinophil-attracting substance continues for 60 min, though in much smaller amounts than in the first 15 min.

2. The substance is stable for 4 days at 0–4 °C, may be frozen to –25 °C and thawed, or heated at 56 °C for 1 h without losing more than 15–20% activity. The properties resemble those reported by LITT [1960] after *in vivo* tests.

3. The substance passes through Diaflo membranes retaining substances greater than 10,000 molecular weight.

Much activity is retained within a Visking dialysis sac, but the variable results of dialysis against saline or Tyrode's solution may be due to differences between samples of the tubing, osmotic pressure, polymerisation of the substance, or aggregation with other substances.

4. The substance could be separated from many other components of the perfusate by filtration on columns of Sephadex G25 and G15.

5. The substance, partially purified by Sephadex gel filtration, withstands treatment with trypsin or with chymotrypsin. Commercial purified enzymes, 0.01% and 0.1% concentrations at pH 5 and 7, were incubated at 30 °C for 1 h, and then inactivated by heating at 56 °C for 30 min, or removed by filtration through membranes retaining substances of 10,000 molecular weight or greater. In a further test, both enzymes were bound to Enzacryl particles (Koch Light Ltd.) to facilitate their removal from the perfusate after digestion. It was important to completely remove or inactivate the enzymes before test. Treatment with these enzymes did not destroy the activity for eosinophils.

This substance appears identical with the ECF-A of KAY *et al.* [1971] and KAY [this volume, p. 271].

Release of Eosinophil-Attracting Substance from
Tissues Other than Lung

Anaphylactic lung is not the only organ to release the eosinophil-eliciting substance when tested in normal recipients. LITT [1960] found that peritoneum from actively sensitised guinea-pigs, together with the appropriate

Table IV. Eosinophilia-eliciting or eosinophil-attracting activities from various tissues of anaphylactically sensitised guinea-pigs, challenged *in vitro*

Tissue	Organ[1], number of eosinophils in blood	Extract *in vivo*, number of eosinophils			Extract *in vitro*[4]	
		blood	peritoneum[2]	skin[3]	eosinophils	neutrophils
Lung	221	134	954	18	14	6
Diaphragm	52	36	424	13	11	0
Peritoneum	36	29	402	12	7	4
Mesentery	41	62	616	14	12	0
Skin	101	40	733	13	18	31
Liver	9	19	88	5	0	2
Kidney	11	14	93	6	1	3
Tyrode's solution	18	16	96	4	0	1
Nil, pre-test	15	17	268	2	not applicable	

1 Average number of eosinophils per cubic millimetre in blood of recipient 24 h after tissue was implanted intraperitoneally.
2 Average number of eosinophils per cubic millimetre in blood or peritoneal fluid of recipient 9 or 12 h after intraperitoneal injection of anaphylactic tissue extract.
3 Average number of eosinophils per 66 fields in skin, 16 h after injection of 0.1 ml tissue extract.
4 Average number of eosinophils or neutrophils per field on membranes of Boyden-type chambers after tests with tissue extracts.
The pre-test *in vivo* eosinophil counts are omitted to simplify the table: the Tyrode's solution and nil test are suitable control numbers.

antigen, elicited an eosinophilia in the peritoneal fluid of the recipient 24 h later. Similarly, PARISH and COOMBS [1968] showed that pieces of diaphragm taken from anaphylactic guinea-pigs, or from actively sensitised guinea-pigs and challenged *in vitro*, induced haematogenous eosinophilia 24 h after intraperitoneal implantation into normal animals.

Tissues from anaphylactically sensitised guinea-pigs were perfused, or chopped and washed as appropriate. Portions of the tissues were treated with antigen and immediately implanted into the peritoneal cavity of normal animals or were challenged after implantation, and 24 h later the number of eosinophils in the blood was compared to that present before the implantation. Attempts to count the eosinophils in the peritoneal cavity, as described by Litt [1960], were stopped because the peritoneal cells usually aggregated, or adhered to the implanted tissue. Sensitised tissues were also treated with antigen to obtain post-challenge perfusates, which were injected intraperitoneally into normal animals whose blood and peritoneal eosinophils were counted before, and 9 or 12 h after injection. The extracts were also injected intradermally, and the skin sites excised 16 h later and examined histologically. The ability of the extracts to attract 95% pure suspensions of neutrophils and eosinophils *in vitro* was examined in Boyden-type chambers.

The results (table IV) show that anaphylactic lung, skin, diaphragm, peritoneum and mesentery released substances that elicited eosinophilia of the blood, attracted eosinophils into the peritoneal cavity and skin of the recipients, and preferentially attracted eosinophils *in vitro*. Perfusates of skin were an exception in that they also contained considerable ability to attract neutrophils *in vitro*. A possible explanation for this activity attracting neutrophils is offered in the later section on tests on human skin which was examined in more detail. Perfusates of the liver and kidney had no ability to elicit blood eosinophilia or to attract eosinophils.

It is, therefore, evident that the anaphylactic 'shock' tissues of the guinea-pig, which release histamine on challenge, which are infiltrated by eosinophils after *in vivo* anaphylaxis and which, incidentally, also contain numerous mast cells, are the tissues which release one or more substances during anaphylaxis that elicit blood eosinophilia and attract eosinophils.

A Substance Eliciting Haematogenous Eosinophilia Distinct from Eosinophil-Attracting Substance (or ECF-A) in Perfusates from Anaphylactic Lungs

The tests reported above showed that some sensitised tissues, perfused or washed free of blood, and subsequently treated with antigen, released a substance or substances that attracted eosinophils *in vitro* and elicited eosinophilia *in vivo*. However, when the tests were repeated on anaphylactic lung perfusates separated by gel filtration to concentrate the fractions with activity for eosinophils, it appeared that the substance attracting eosinophils differed from that eliciting eosinophilia in the blood of normal recipients.

The anaphylactic perfusates from lung were separated on Sephadex G-25 or G-15 columns. As reported by Kay *et al.* [1971], the substance attracting eosinophils *in vitro* was fully recovered in a late peak of the material eluted from the column. The pooled fractions containing the eosinophil-attracting

Table V. Perfusates of anaphylactic guinea-pig lungs and of the fractions with *in vitro* eosinophil-attracting substance, tested for eosinophil-attraction *in vitro* and *in vivo* and eosinophil-eliciting activity

Test substance	Attract		Elicit, in blood[3], per mm^3	
	in vitro[1], number per field	in skin[2], number per 66 fields	pre	post
Whole perfusate	9	42	8	121
G-25 fraction	9	37	11	48
Tyrode's solution	0	6	9	7
Whole perfusate	7	21	3	98
G-15 fraction	8	29	6	15
Tyrode's solution	0	0	4	9

1 Mean of tests on two eosinophil samples; each test on 3 membranes.
2 Three guinea-pigs for each of the two perfusates. Each animal injected twice with 0.1 ml of each substance. Three sections of each site taken at 24 h, and 66 fields counted per section [PARISH, 1972b].
3 Blood counts. Mean of three guinea-pigs for each substance 9 h after intraperitoneal injection of 2 ml. The 6- and 24-hour counts not reported.

substance, attracted as many eosinophils per field as the whole, unseparated perfusate, but they no longer elicited eosinophilia of the blood of normal recipients (table V). The other fractions from the gel filtration were shown not to attract eosinophils *in vitro*, but have not yet been tested for their ability to induce eosinophilia *in vivo*.

The probable explanation of the results (table V) is that there are at least two substances with activity for eosinophils in the perfusate of anaphylactic guinea-pig lung. One selectively attracts eosinophils *in vitro*, and attracts and retains eosinophils when injected into skin; the other elicits an increase in the number of eosinophils in the blood, and probably does not selectively attract them since all the *in vitro* attracting activity could be accounted for in the fraction containing the first substance described. An alternative explanation is that the substance that attracts eosinophils will also elicit a generalised eosinophilia *in vivo* but the eliciting activity either requires a co-factor in the perfusate, or is unstable when separated from the whole perfusate. Other evidence tends to support the concept that an eosinophil-eliciting substance exists independently of the attracting substance, because the supernatant fluid from rat anaphylactic mast cells strongly attracts eosinophils, but elicits only a slight increase in the number of eosinophils in the blood (see below).

Substance Released from Anaphylactic Human Basophils Selectively Attracting Eosinophils

Anaphylactic human lung releases an eosinophil-attracting substance concomitantly with histamine and SRS-A similar to that released from guinea-pig lung [KAY and AUSTEN, 1971]. Release occurs when lung sensitised by IgE is challenged with antigen or with anti-IgE.

Some of this substance is probably released from mast cells because isolated human basophils, sensitised by IgE antibody, also release a substance selectively attracting eosinophils, concomitantly with histamine when the basophils are treated with antigen or with anti-IgE [PARISH, 1972a]. The eosinophil-attracting substance is a product of the basophils because IgE, aggregated chemically or with antigen as complexes, did not generate *in vitro* chemotropic activity in serum-containing complement.

Human basophils were isolated as suspensions 92% or greater purity by the method of DAY [1972], and human eosinophils 90–95% pure with 3–6% neutrophils were prepared by the method of DAY [1970].

Undamaged or slightly damaged basophils in Eagle's medium containing 0.5% human albumin did not attract neutrophils or eosinophils in Boyden-type chambers. There were a few exceptions in which basophils from atopic persons spontaneously released eosinophil-attracting substances. If, however, the basophils were disrupted by distilled water and incubated in serum-containing complement, a substance was found that attracted both neutrophils and eosinophils (table VI). Thus, in this respect, basophils behave like neutrophils, because damaged neutrophils incubated in whole serum release or form a substance attracting neutrophils [BOREL *et al.*, 1969; SORKIN *et al.*, 1970] or eosinophils [PARISH, 1972b] (table I).

Basophils treated with non-anaphylactic human IgG and washed before the addition of antigen, released no leucocyte-attracting activity, whereas basophils treated with IgE antibody and washed before challenge with antigen released a substance attracting eosinophils but not neutrophils (table VI). In further tests, the anaphylactic release of the eosinophil-attracting substance was correlated with release of histamine. Human and monkey basophils were sensitised with human IgE anti-β-lactoglobulin and then washed free of

Table VI. Chemotaxis elicited by basophils and neutrophils, untreated, or treated with antibody, antigen and normal serum

State or treatment of test basophils or neutrophils	Addition of β-lacto-globulin	Chamber diluent	Number of neutrophils or eosinophils attracted to treated			
			basophils		neutrophils	
			neutro-phils	eosino-phils	neutro-phils	eosino-phils
Undamaged by isolation	−	medium	2	0	0	2
Slightly damaged	+	medium	2	0	3	0
Disrupted by distilled water	−	serum	88	74	136	128
Treated with IgG Ab, washed	+	medium	4	2	18	11
Sensitised with IgE Ab, washed	+	medium	8	48	5	2

IgG and IgE antibodies to β-lactoglobulin.
[Modified from PARISH, 1972a, by permission of the editor of Clin. Allergy.]

unbound serum proteins before challenge with antigen, anti-IgE or anti-IgG. Release of eosinophil-attracting substance and of histamine occurred concomitantly from sensitised basophils treated with antigen or with anti-IgE (table VII).

Anaphylactic basophils are, therefore, one source of eosinophil-attracting substance (EAS-A basophil). If mast cells are similar to basophils in this respect they probably contribute much, if not all, of the eosinophil-attracting substance released from human lung. In order to examine this possibility, mast cells were obtained from monkey tracheal and bronchial mucosa as suspensions 64 and 71% pure; most of the contaminants being small squames or cuboidal cells. Samples of these suspensions were treated with human serum containing IgE anti-β-lactoglobulin, unheated or heated at 56 °C, washed to remove unfixed serum proteins and challenged with β-lactoglobulin or with rabbit anti-IgE. In the *in vitro* chemotaxis chambers, monkey eosinophils were tested because human eosinophils may have been attracted by non-anaphylactic substances in the monkey mast cells.

Table VII. Concomitant release of eosinophil-attracting substance and histamine from sensitised basophils tested with antigen or with anti-IgE

Origin of basophils	Reaginic serum treatment	Sensitised cell challenge	Post-challenge supernatant fluid activity		
			chemotaxis		histamine release, % total
			neutro-phils	eosino-phils	
Man	unheated	antigen	5	32	76
Man	heated	antigen	3	1	12
Man	unheated	anti-IgE	1	18	85
Man	unheated	anti-IgG	4	0	14
Monkey	unheated	antigen	7	22	33
Monkey	heated	antigen	7	3	5

Results of antigenic challenge of human and monkey basophils are a mean of two tests.
Results of antiglobulin challenge are from one test.
[Modified from PARISH, 1972a, by permission of the editor of Clin. Allergy.]

Monkey mast cells challenged with antigen or with anti-human IgE released both histamine and a monkey eosinophil-attracting substance. Unfortunately the pre- and post-challenge supernatant fluids from the cell suspensions also attracted monkey neutrophils, so the results did not provide as definite evidence of release of a substance selectively attracting eosinophils as did the human basophils. It is possible that the neutrophil-attracting activity was generated by the serum reacting with damaged cuboidal cells or with bacteria from the bronchial flora.

The eosinophil-attracting substance released from human basophils appears similar to, or identical with, that released from human lung, as described by KAY and AUSTEN [1971]. It shares the properties of being released from the anaphylactic cells concomitantly with histamine; it separates on a G-15 or G-25 Sephadex column before histamine; some of the activity passes a Visking dialysis membrane, and it is destroyed by boiling in 0.05 N NaOH.

Substances from Human Skin Attracting Neutrophils and
Eosinophils *in vitro*

Human skin is particularly susceptible to anaphylaxis, and was therefore
examined for release of substances attracting eosinophils similar to those
released from lung.

Fresh skin, usually from mastectomy or abdominal operations, was
trimmed of all the subcutaneous tissue, and finely chopped into thin sections.
These sections were washed 5 times in Tyrode's solution, and then further
washed in a slow current of this solution for 1 h.

The washed sections were sensitised with IgE-containing sera, and
washed again to remove unbound serum proteins before challenge with
antigen or with rabbit IgG anti-human IgE. Sensitised skin treated with
antigen or with anti-IgE released into the supernatant fluid histamine and
one or more substances attracting neutrophils and eosinophils *in vitro*.
Normal, unsensitised, skin releases small amounts of histamine spontaneously but, after incubation in sera from non-allergic or allergic persons, and
washed, more histamine was released and substances weakly attracting both
types of polymorphonuclear leucocytes. The amount of polymorphonuclear
leucocyte-attracting substance released could be reduced by (1) removing
complement from the sensitising serum by absorbing it with complexes at
4 °C, and (2) removing most of the stratum corneum by adhesive tape.

The supernatant fluid from suspensions of challenged, sensitised, skin
attracted both neutrophils and eosinophils. The fluid was separated on a
Sephadex G-25 column and, in preliminary tests, showed one fraction
attracting neutrophils only, another attracting both neutrophils and eosinophils and a third which eluted slowly, a little before the histamine-containing
fraction, and which attracts mainly eosinophils. This third substance resists
boiling in acid solution, but its activity for eosinophils is destroyed when
boiled in alkaline solution. These are properties of the eosinophil-attracting
substance from human basophils (above) and of the similar substance from
human lung [KAY and AUSTEN, 1971].

More discriminating results were obtained when the skin of actively
sensitised persons was tested with antigen *in vitro* without incubation with
serum. In tests on two atopic persons, one sensitive to egg, the other to
pollen, the appropriate antigens released histamine and substance(s) attracting eosinophils (table VIII). The supernatant fluid from the pollen-sensitive
skin was very selective for eosinophils without Sephadex gel filtration; the
lack of neutrophil attraction may have been due to the absence of direct

Table VIII. In vitro release from human skin of histamine and substances attracting neutrophils and eosinophils

Allergic state of skin donor	TCT[1] donor serum	Antigenic challenge	Histamine, % release	Leucocytes attracted per field[2]			
				neutrophils		eosinophils	
				control	test	control	test
Asthma, egg	64	egg	19	2	11	0	5
Asthma, egg	–	nil	7	3	5	0	0
Hay-fever, timothy	0	pollen	26	0	4	0	12
Hay-fever, pollen		nil	9	1	2	0	0
Normal donor	2	pollen	7	2	5	0	0
Normal donor		nil	5	1	5	0	0

1 Tanned cell agglutination titre (TCT) for antigen tested on skin.
2 Number of cells per ×40 field in control (Eagle's +0.5% human albumin) medium and test (plus supernatant fluid from skin, treated or not treated with antigen).

agglutinating (IgG or IgM) antibody in the serum as judged by the tanned cell agglutination test. The serum of the egg-sensitive person contained agglutinating antibody, and the antigen-treated skin attracted some neutrophils (table VIII).

Anaphylactic human skin therefore releases an eosinophil-attracting substance. This, however, is one of several skin-associated substances attracting polymorphonuclear leucocytes. The stratum corneum incubated with complement-containing serum generates activity for both neutrophils and eosinophils, the nature of which is unknown. It may be an anaphylatoxin-associated substance as described by WISSLER *et al.* [1972], because stratum granulosum and stratum corneum may aggregate IgG and IgM by their Fc portions, and fix complement (C3) [COWAN and PARISH, unpublished information].

Some Causes of Variations in Human Leucocyte *in vitro* Activity

We found much variation in the separation and *in vitro* activity of human polymorphonuclear leucocytes obtained from different donors or, on different occasions, from the same donor. Guinea-pigs and rats, though usually sampled only once, were much more consistent.

Separation of suspensions of 90%, or greater, purity of each type of leucocyte, particularly eosinophils and basophils by the techniques of DAY [1970, 1972] was more difficult to achieve from recently vaccinated or infected persons, menstruating women, or men who had taken alcohol the previous evening. Fewer cells, relative to the number in the blood, were isolated in pure suspension from such people, whereas men who had not taken alcohol, or women in the inter-menstrual period, gave fairly consistent yields of cells. It is possible that the reduced numbers sometimes obtained reflected a change in the cell membrane, or the weight of the cell, so that the specific gravity of the separating solutions was no longer optimal.

Variations between donors were also observed in the *in vitro* mobility of the neutrophils and eosinophils. It is difficult to compare mobilities, or responses to chemotropic substances, of different samples of leucocytes because leucocyte separation and the Boyden chamber test are not exact procedures, and the number of neutrophils or eosinophils responding to a test substance in the morning may differ from that obtained when a fresh sample of cells from the same person are tested in the afternoon. Nevertheless, when, in a sample from one person, the number of polymorphonuclear leucocytes attracted to a control test substance, e.g. complexes, is much lower than that of cells from other normal persons examined on the same day, it may be concluded that they are less responsive.

Reduced chemotaxis was observed in leucocytes from persons who had been vaccinated against smallpox in the previous 3 weeks, and in men who had taken alcohol the previous evening. The leucocytes of other persons, as in women or during chronic bacterial infection, though sometimes difficult to separate from blood, behaved normally in chemotaxis tests.

The effect of alcohol is interesting. The men reported that they had taken only a 'convivial' amount, which was considerably less than that reported in two other studies, in one of which leucocyte emigration into skin was reduced after taking alcohol [BRAYTON *et al.*, 1970]; in the other, human leucocytes treated with alcohol *in vitro* were less responsive to chemotactic stimuli [KLEPSER and NUNGESTER, 1939]. Our findings are not the result of a planned experiment, but are incidental observations made when carrying out chemotaxis tests. They indicate, however, that alcohol in moderate quantities taken by healthy sportsmen the night before the samples were taken, may still influence leucocyte behaviour the following morning.

These observations, that changes in cell physiology resulting from infection or ingested substances may alter the chemotactic response, account for some discrepancies in reports on human leucocyte activity.

Anaphylactic Mast Cells Attracting Eosinophils in vivo and in vitro

Eosinophils accumulate in tissues in which mast cells show anaphylactic change and, as described above, anaphylactic tissues containing many mast cells release a substance attracting eosinophils. Moreover, many eosinophils are found in tissues rich in mast cells after the mast cell histamine has been released by 48/80 or similar chemical inducers [RILEY, 1956, 1959], or when mast cell damage is prolonged, as occurs when repeatedly washing out the peritoneal cavity of rats with saline [ARCHER and HIRSCH, 1963; PARISH, 1970a].

In order to determine the importance of mast cells for anaphylactic eosinophilia, the *in vivo* attraction of changed mast cells for eosinophils was examined in anaphylactic guinea-pig mesentery and subcutaneous tissue, and also the *in vitro* and *in vivo* attraction of rat anaphylactic mast cells for rat eosinophils.

In vivo Attraction of Eosinophils to Changed Mast Cells in Anaphylactic Guinea-Pigs

In mesenteric spreads from guinea-pigs dying of anaphylaxis after intravenous antigen, or killed after intraperitoneal injection of a very dilute solution of antigen, eosinophils tended to accumulate round changed mast cells, leaving a surrounding zone devoid of eosinophils though they were diffusely spread elsewhere in the tissue [PARISH, 1970a]. About 8% of the mast cells had attracted eosinophils. Some of the eosinophils appeared to enter the damaged mast cells. They lay within the dispersed metachromatic material surrounding the mast cell, and also on the mast cell membrane. The eosinophils shed some of their own granules and were vacuolated, and occasionally 2–4 dark blue granules were seen within the vacuoles. It is possible that the blue granules were mast cell substances, but microscopic examination is insufficient to discriminate between mast cell granules and other basophilic material possibly ingested by eosinophils. When the anaphylactic changes were severe, neutrophils would also accumulate round the

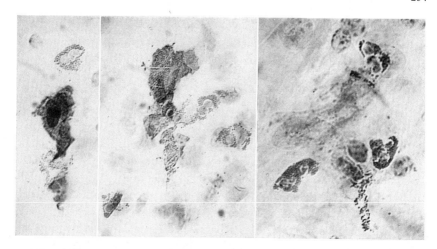

Fig. 3. Three fields of subcutaneous tissue at passive cutaneous anaphylaxis (PCA) sites. Two at 20 min and one at 30 min after intravenous challenge. At 20 min the mast cells are enlarged and homogeneous blue or metachromatic material hides the granules and begins to leave the cell. At 30 min nearly all the mast cell substance is released. Eosinophils adhere to the mast cells, usually shedding granules and becoming vacuolated. Occasional vacuoles contain metachromatic (? mast) cell substance [from PARISH, 1972b]. Peroxidase-Giemsa. × 1,800.

changed mast cells, so the phenomenon was not selective for one type of polymorphonuclear leucocyte.

In subcutaneous connective tissue spreads, however, the anaphylactically changed mast cells in passive cutaneous anaphylaxis attracted eosinophils only [PARISH, 1972b]. Nearly all the mast cells become changed 30 min after intravenous challenge. They enlarge, become round and contain homogeneous cytoplasm with a few intact granules staining metachromatically. Some of the homogeneous material leaks from the cell. The first eosinophils infiltrating the site are attracted towards the mast cells until 1–3 eosinophils surround each cell, and may enter or overlie it (fig. 3). A few eosinophils shed their granules or become vacuolated, as seen in the mesentery. By 4 h after challenge, though few eosinophils remain close to mast cells, the majority are diffusely dispersed in the tissues.

It is believed that these changes represent the early anaphylactic release from mast cells of a substance attracting eosinophils (and also neutrophils in severely affected mesentery which may contain antigen-antibody complexes). The first eosinophils to enter the tissue in 30–60 min are attracted directly to the mast cells where there is the greatest concentration of the attracting substance. Thereafter, the attracting substance diffuses away, or may become inactivated, so that by 4 h the eosinophils accumulate diffusely in the tissue and are not attracted directly to the mast cells.

In vitro and *in vivo* Attraction of Eosinophils to Rat Anaphylactic Mast Cells

The observations that anaphylactic guinea-pig mast cells in tissues attract eosinophils, and that anaphylactic human basophils release a substance selectively attracting eosinophils *in vitro*, suggest that mast cells and basophils may be the major source of substances eliciting eosinophilia in anaphylaxis. However, our tests on actively and passively sensitised rat mast cells, showed that substances released from these cells attracted fewer eosinophils to skin sites, or elicited fewer eosinophils in the blood than occurred on antigenic challenge of actively sensitised rats [PARISH, 1972a]. The tests were made on rats because they are susceptible to anaphylaxis and their mast cells could be prepared in pure suspension.

Suspensions of peritoneal intact mast cells, about 96% pure, were sensitised *in vitro* with a serum fraction containing rat IgE-like antibody, or in serum containing IgE-like and some IgG2 antibody, or treated with normal serum. The cells were then washed and challenged with antigen in Eagle's medium or Tyrode's salt solution containing bovine serum albumin, and the cell-free supernatant fluid separated for testing. Cells sensitised with the IgE-like antibody and challenged, released a substance selectively attracting rat eosinophils in Boyden-type chambers. More eosinophils were attracted to the supernatant fluid of passively sensitised than of actively sensitised mast cells, and to the mast cell supernatant fluids than to the fluids from mesentery, lung, skin, kidney or blood buffy coat.

When the supernatant medium from the anaphylactic mast cells was injected intraperitoneally into rats, it elicited eosinophilia of the blood and peritoneal fluid. The increase in the number of eosinophils, however, was about the same as occurred in rats passively sensitised with sera containing IgG2 or with IgE-like antibodies and challenged *in vivo*, and was always much less than the number of eosinophils found after active sensitisation and challenge *in vivo*, even though at the time of challenge the actively sensitised rats had low levels of circulating antibody and few blood eosinophils.

It is concluded that, in the rat, the anaphylactic mast cell-derived substance is probably responsible for most of the eosinophilia of passively sensitised rats, but other mediators, e.g. from lymphocytes, probably contribute to the eosinophilia of active anaphylaxis. Another feature probably contributing to the greater eosinophilia in actively than passively sensitised rats, is that actively sensitised rats tend to have greater numbers of mature eosinophils stored in the tissues and which are readily mobilised, as occurs in guinea-pigs which have been primed with an unrelated antigen before test [LITT, 1960; PARISH, 1972c]. This feature is being examined in anaphylaxis in passively sensitised rats.

Comparison of Eosinophilia in Actively and Passively Sensitised Guinea-Pigs

In tests on rats, described above, the eosinophilia resulting from injection of anaphylactic mast cell substances or after challenging passively sensitised animals, was less than that following challenge after active sensitisation. One possible explanation for this phenomenon is that actively sensitised animals have more anaphylactic antibody than passively sensitised recipients.

This possibility was examined in guinea-pigs weighing 200 g. Sufficient, concentrated IgG1 anti-bovine plasma albumin antibody was injected intravenously at 3-hour intervals into normal guinea-pigs to achieve a serum passive cutaneous anaphylaxis (PCA) titre of 1:8 to 1:16, 16 h after the last injection. The eosinophilia after intraperitoneal challenge of the passively sensitised animals was less than that of animals actively sensitised subcutaneously or intramuscularly, and whose serum contained the same titre of PCA antibody as in the passively sensitised animals [PARISH, 1970c].

It thus appears that the greater eosinophilia in actively sensitised animals is not due to the amount of anaphylactic sensitising antibody in the serum, but is more likely to be due to a greater number of readily mobilised eosinophils and possibly to lymphocyte substances.

Attraction of Eosinophils to Lymph-Nodes Draining Sites of Injected Substances

Eosinophils infiltrate lymph-nodes draining skin injected with antigens, with antigen-antibody complexes, or even with non-antigenic particles. The infiltration starts within 15 min of injection and is at its maximum after 12 h.

The rate of eosinophil infiltration, the areas of the nodes to which the eosinophils are attracted and their subsequent emigration, have been fully described by LITT [1964] and by COHEN *et al.* [1966]. This eosinophil infiltration occurs in normal animals injected for the first time with the antigen and also with non-antigenic particles [KOSTAGE *et al.*, 1967]. Moreover, the transient eosinophilia of lymph-nodes is independent of anaphylactic activity of injected complexes, because there is no eosinophilia of guinea-pig skin injected with complexes of non-anaphylactic horse antibody with tetanus toxoid, though many eosinophils infiltrate the regional draining lymph-nodes [PARISH, 1970a].

We believe that the eosinophils are attracted into the lymph-nodes by the activity of the stimulated T lymphocytes or possibly of other, larger, lymphoid cells, which may synthesise nucleic acids or some other substance within minutes of stimulation. These T lymphocyte substances then attract, and for some hours retain eosinophils in the lymph-node. Though this contention is still to be investigated, we have used the eosinophilia of draining lymph-nodes as a system representing lymphocyte-mediated eosinophilia, when distinguishing between lymphocyte-mediated eosinophilia and anaphylactic mobilisation of eosinophils.

Differentiation between Anaphylactic (? Mast Cell) Eosinophilia,
and Lymphocyte-Induced Eosinophilia,
by Treatment with Anti-Lymphocyte Globulin or with Puromycin

The events described in the studies reported above are those occurring during anaphylaxis when there is an increase in the number of eosinophils within hours of exposure to the antigen, and which returns to the pre-test level within 2 or 3 days. Another system mediating eosinophilia has been described by Prof. BEESON and his colleagues. Rats injected for the first time with *Trichinella* larvae have a blood eosinophilia appearing about the 6th day and persisting several weeks, and is accompanied by increased numbers of eosinophils in the bone marrow [BASTEN *et al.*, 1970; SPRY, 1971b]. The eosinophilia occurs only when a local cellular response is provoked by the antigen [WALLS and BEESON, 1972] and it appears to be mediated by the T lymphocytes [BASTEN and BEESON, 1970; WALLS *et al.*, 1971].

We have used anti-lymphocyte globulin and puromycin to distinguish between the rapidly occurring blood eosinophilia of anaphylaxis and the blood eosinophilia mediated by lymphocytes stimulated by helminth particles, or the eosinophilia of draining lymph-nodes which we believe to be an expression of lymphocyte activity.

Failure of Anti-Lymphocyte Globulin to Modify Anaphylactic Eosinophils

The results of experiments modifying eosinophilia with anti-lymphocyte serum by PARISH [1970a] and by BASTEN and BEESON [1970] differentiate between the anaphylactic eosinophilia elicited by substances which attract mature 'stored' eosinophils, and the more persistent eosinophilia elicited by lymphocytes (substances) that induce eosinophil multiplication in, and release from, the bone marrow in actively sensitised animals.

A rabbit anti-guinea-pig lymphocyte globulin (ALG), prepared from guinea-pig thymocytes, was absorbed with suspensions of separated erythrocytes, neutrophils and eosinophils. This anti-globulin greatly depressed the cutaneous delayed hypersensitivity response to tuberculin or to dinitrochlorobenzene. It did not affect the numbers of eosinophils appearing in the blood or peritoneal fluid 16–24 h after the animals were passively sensitised with guinea-pig IgG1 or with rabbit antiserum and subsequently challenged with antigen, irrespective of whether the animals were treated with ALG intraperitoneally or subcutaneously for 1 or up to 6 days [PARISH, 1970a]. Therefore, the rapidly occurring eosinophilia of anaphylaxis which results from the mobilisation of stored eosinophils, is not affected by anti-lymphocyte globulin and is therefore almost certainly not controlled by the T lymphocyte system.

It was shown by BASTEN and BEESON [1970] that anti-lymphocyte serum given for 6 days, starting 1 day before parasite antigenic stimulation, suppressed the lymphocyte-mediated eosinophilia, but the anticipated eosinophilia occurred when the ALG was first given 72 h after antigen exposure. We have since confirmed these findings when injecting *Ascaris* worm fragments intravenously into guinea-pigs.

Another feature of eosinophilia is that lymph-nodes draining sites of antigen injection attract eosinophils, already mentioned in table II. The lymph-node eosinophilia occurs even after injection of non-antigenic particles [KOSTAGE *et al.*, 1967]. Attraction of eosinophils to draining lymph-nodes was previously reported to be partially depressed by treatment with ALG [PARISH, 1970a]. We have since shown that if sufficient ALG is injected, that the lymph-nodes draining sites of antigen or particles injected on the first occasion do not attract eosinophils. It is, therefore, believed that the attraction of stimulated lymph-nodes for eosinophils is a property of T cells, probably reflecting synthesis of nucleic acids, and is not due to rapid formation of antibody as suggested by LITT [1964, 1972].

Modification of Eosinophilia with Puromycin

Puromycin, a drug inhibiting synthesis of protein, was used by LITT [1972] to inhibit the attraction of lymph-nodes draining sites of antigen for eosinophils. When the drug was given just before the first injection of haemocyanin, eosinophil infiltration into the lymph-node did not occur. If the animal had been immunised previously against haemocyanin, treatment with puromycin did not prevent the eosinophil infiltration following a further injection of the antigen.

We had previously found that the intraperitoneal injection of a large dose of puromycin into sensitised guinea-pigs did not prevent the anaphylactic eosinophilia on intraperitoneal or intradermal challenge. This was thought to show that protein synthesis was probably not required for the rapid infiltration of eosinophils into the blood, peritoneal cavity or skin during anaphylaxis, and therefore eosinophil mobilisation was not controlled by a newly synthetised protein.

The experiments of LITT [1972] were confirmed that treatment with puromycin into the footpad of a guinea-pig prevented the lymph-node eosinophilia on first exposure to diphtheria toxoid, but not if the antigen was being given for the third time. This was similar to our findings with ALG also injected into the footpad. However, if non-antigenic particles (e.g. latex or bentonite) were injected three times, the ALG or puromycin prevented any significant lymph-node eosinophilia. Table IX shows a provisional summary of our present tests, in which it appears that ALG or puromycin prevent eosinophilia occurring in the draining lymph-nodes on the first treatment with antigenic or non-antigenic substances, but in animals treated thrice with antigens, neither ALG nor puromycin prevented the lymph-node eosinophilia. Both drugs, however, prevented the eosinophilia occurring in animals treated thrice with non-antigenic substances.

It is possible that the lymph-nodes attract eosinophils when any substance is deposited in the appropriate skin site for the first time, because the T lymphocytes are stimulated to synthetise protein or nucleic acid, which attracts eosinophils. In immunised animals, the lymph-nodes attract eosino-

Table IX. The presence or absence of eosinophils in the popliteal lymph-nodes after 1 or 3 injections of the test substances into the footpad of animals untreated, or treated in the footpad with anti-lymphocyte globulin (ALG) or puromycin before the test substance

Test substance injected into footpad	Untreated		ALG		Puromycin	
	1st exp.	3rd exp.	1st exp.	3rd exp.	1st exp.	3rd exp.
IgG2 + β-lactoglobulin (complex)	+	+	−	+	−	+
Aggregated IgG1	+	+	−	±	−	±
Soluble diphtheria toxoid	+	+	−	+	−	+
Soluble guinea-pig albumin	+	+	−	−	−	−
Latex particles	+	+	−	−	−	−
Bentonite particles	+	+	−	−	−	−
Sephadex G-200	+	+	−	±	−	+

Guinea-pigs injected thrice had a 1-week interval between the first and second doses, and 6 weeks between the second and third. The ALG or puromycin was given once only, before the third injection; exp. = exposure.

phils partly because the T lymphocytes are already primed, or because the B lymphocytes are activated. The attraction of eosinophils into the lymph-node is unlikely to be due to the rapid formation of antigen-antibody complexes, because the complexes should also attract neutrophils. The ALG and puromycin can inhibit activation of the T lymphocytes occurring on first exposure to the antigen, or on subsequent exposure to non-antigenic particles; they cannot inhibit the attraction of eosinophils into the sensitised lymph-node re-exposed to antigen, either because the T lymphocytes are already primed and further synthesis of some substance no longer occurs, or there is activation of the B lymphocytes which are not susceptible to the action of ALG or puromycin.

Attraction between Eosinophils and Inflammatory
'Lymphoid' Cells *in vivo*

As ALG reduces or inhibits eosinophilia of lymph-nodes draining sites of the first injection of antigen but does not do so after repeated injections,

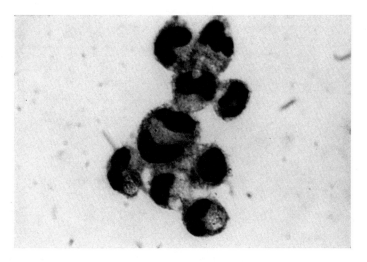

Fig. 4. Peritoneal exudate of an anaphylactic guinea-pig. A mitotic or binuclear cell is surrounded by 7 eosinophils and one neutrophil. May-Grunwald Giemsa.

i.e. when the animal is sensitised, it appears that sensitised lymphoid tissue releases more than one substance attracting eosinophils.

The lymph-node cells attracting the eosinophils have not been identified, though there are several descriptions of enlarged vacuolated mononuclear cells and of large 'lymphoblast-like' cells that may be implicated. The vacuolated mononuclear cells do not attract eosinophils selectively, but the lymphoblast-like cells appear to do so. After intraperitoneal antigenic injections of sensitised mice, swollen, vacuolated mononuclear cells become the centre of rosettes which, depending upon the duration of the response, sometimes comprise eosinophils [SPEIRS and SPEIRS, 1963; SPEIRS, 1964]. It is noteworthy that COTTIER *et al.* [1964] reported an increase in the number of karyorrhetic cells that preceded the eosinophil infiltration in mouse popliteal lymph-nodes draining antigen. Eosinophils are attracted selectively to large basophilic cells in the mesenteric milk spots of anaphylactic guinea-pigs [PARISH, 1970a]; it is unlikely that the attracting cells are macrophages and the attraction due to non-selective adherence between phagocytes, because macrophages containing antigen, within and without the milk spots, do not attract significant numbers of polymorphonuclear leucocytes. Moreover, the granulomata resulting from injection of *Ascaris* particles in mice had an outer ring of large basophilic mononuclear cells which attracted eosinophils

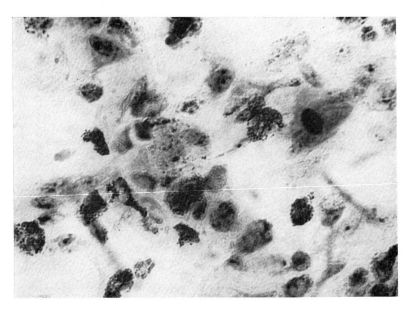

Fig. 5. Human peribronchial connective tissue spread from asthmatic lung. Eosino-phils accumulate round a karyorrhetic mononuclear cell. There is a changed mast cell (right). May-Grunwald Giemsa.

and resembled the changes occurring in the lymphoid milk spots [PARISH, 1970a]. Similar peri-granulomatous eosinophilia has been observed in rats infected with helminth larvae [BOYER *et al.*, 1971, quoted by WALLS and BEESON, 1972], and rosettes of large mononuclear cells surrounded by eosino-phils after peritoneal challenge of sensitised rats [WALLS and BEESON, 1972].

The identity of the cells attracting the eosinophils and the nature of the attraction are still to be determined (fig. 4, 5). However, these phenomena are probably very similar to those occurring in lymph-nodes draining antigen. It is conceivable that T lymphocytes stimulated by antigen form large baso-philic lymphoblast cells that selectively attract eosinophils; if, or when, they degenerate, the karyorrhetic residues also attract other leucoytes. Another possibility that we are examining is the existence of a large lymphoid cell, distinct from the T and B lymphocytes, that is stimulated by antigen to release eosinophil-attracting substances or to release substances which interact with antigen-antibody complexes to form eosinophil-attracting substances, as described by COHEN and WARD [1971].

Some Properties of Substances Retaining or
Attracting Eosinophils in vivo

Despite the several recent reports on substances selectively attracting eosinophils, it has not been ascertained that the substances attracting eosinophils *in vitro* are those that attract eosinophils into tissues. It is probable that some substances that selectively attract eosinophils *in vitro*, though capable of eliciting local eosinophilia *in vivo* when injected into tissues, are of little importance in the eosinophilia of disease.

In the studies reported above, it is evident that there are at least two systems controlling eosinophilia. A provisional summary of some properties of their mediators is presented in table X. The first system comprises the group of substances associated with anaphylaxis or anaphylactoid reactions, in which eosinophils accumulate in affected tissues within hours, usually associated with a blood eosinophilia occurring for 1–2 days, when mature, stored eosinophils are released or mobilised. There is no increase in the number of eosinophils in the bone marrow after one anaphylactic episode. Moreover, animals may be anaphylactically sensitised and show blood eosinophilia on challenge without any appreciable increase in the number in the bone marrow, though sensitisation by repeated injections, or by antigens adsorbed to alum results in bone marrow eosinophilia, and greater numbers in the blood. Several examples of anaphylactic substances attracting eosinophils or eliciting blood eosinophilia are described in this chapter. They are released from anaphylactic guinea-pig lung, skin, mesentery and peritoneum, rat mast cells and human basophils and skin.

Anaphylactoid reactions are also associated with generation of substances selectively attracting eosinophils, as shown by the meticulous separation of anaphylatoxin from cocytotaxin by WISSLER *et al.* [1972, see also this volume, p. 442]. We have also separated eosinophil from neutrophil *in vitro* attracting activities generated in guinea-pig serum, but the substances were not of the purity achieved by WISSLER.

The second system controlling eosinophilia is associated with T lymphocyte activity, as investigated in parasitised rats [BASTEN *et al.*, 1970; BASTEN and BEESON, 1970; SPRY, 1971 a, b; WALLS and BEESON, 1972]. A single injection of particulate antigen induces eosinophilia of the blood and bone marrow about 6 days later. Eosinophils are attracted to and accumulate round the antigen-containing granulomata. This may resemble the eosinophilia of lymph-nodes or mesenteric milk spots stimulated by antigen described above, and the eosinophilia occurring in sites of delayed sensitivity

Table X. Summary of some properties of mediators of eosinophilia occurring in two systems

System	Substance	In vitro attraction	In vivo attraction or retention	elicit in blood 1st to 2nd day	elicit in blood 6th day et sequentes	bone marrow 1st day	6th day
Anaphylaxis and anaphylactoid reactions	histamine	–	±	–	–	–	–
	guinea-pig IgG2+Ag+C	+	+	–	–	–	–
	guinea-pig IgG1+Ag+C	+	+	+	–	–	–
	anaphylactic lung substance	+	+	+	–	–	–
	purified ECF-A[1]	+	+	– ?	–	–	–
	anaphylactic rat mast cells	+	+	+	–	–	–
	anaphylatoxin[2] associated substances	+	±				
Lymphocyte-mediated	T lymphocytes[3] parasitised rats		+ (granulomata)	–	+	–	+
	lymph-node substances[4] +Ag–Ab complexes	+	+				

Data from studies described in this chapter, also:
1 KAY, this volume.
2 WISSLER et al. [1972] and this volume.
3 See references of BASTEN and BEESON [1970], SPRY [1971a, b] and WALLS and BEESON [1972].
4 COHEN and WARD [1971].

responses when rechallenged with antigen [ARNASON and WAKSMAN, 1963]. This attraction may be mediated, in part, by the lymphocyte substance modified by antigen-antibody complexes described by COHEN and WARD [1971].

If any one type of leucocyte predominates in inflamed tissue, it has either been selectively retained at, or selectively attracted to, the site. Histamine appears to be an example of a substance that may retain eosinophils in a tissue for a longer period than neutrophils, so that after a single application of histamine there is a short-term accumulation of eosinophils.

The accumulation of eosinophils at sites of passive anaphylaxis may also result from a selective retention rather than a selective attraction. Eosinophils infiltrate the site progressively, reaching their greatest number after 12 h, and the number is proportional to that in the blood (table III). The neutrophils which are most numerous at 4 h are not selectively retained, and emigrate. It therefore appears that local tissue eosinophilia results from the random infiltration of eosinophils from the blood, which are selectively retained at the site, and not selectively attracted to it.

Though substances are released from anaphylactic guinea-pig, rat or human tissues that selectively attract eosinophils *in vitro*, the evidence that they attract eosinophils *in vivo* is inconclusive. The amounts injected experimentally are probably very much greater than that released from a tissue site. Moreover, in order to attract eosinophils *in vivo*, any substance must diffuse some distance from the site of release, and not be inactivated by leucocytes or humoral antagonists. It is more likely that a slow release of the eosinophilotropic substances occurs in the tissue, and eosinophils randomly infiltrating the site, with other leucocytes, are selectively retained. They may, however, be retained in a tissue because they are attracted to mast cells (fig. 3), to basophilic lymphoid cells or to certain degenerate (vacuolated) mononuclear cells (fig. 4, 5). This retention of eosinophils in a tissue may result from local cell attraction. The *in vitro* tests to examine this phenomenon appear to be an exaggeration of the *in vivo* events.

If the numbers of eosinophils infiltrating anaphylactic or granulomatous tissue are influenced by the numbers in the blood, it therefore follows that more eosinophils will infiltrate a damaged tissue if there is at the same time an increased number in the blood, as occurs in the rapid mobilisation of eosinophils in anaphylaxis, and the greatly increased formation of eosinophils mediated by lymphocytes in the bone marrow of parasitised animals. It is very probable that the eosinophil has a beneficial function in inflammation; therefore the concomitant activity of substances eliciting increased numbers in the blood, with substances retaining eosinophils in the damaged tissue, ensures a rapid accumulation of these beneficial cells.

Summary

Experiments are reported which show that substances attracting eosinophils *in vitro*, e.g. antigen-antibody-complement complexes, may elicit little or no eosinophilia *in vivo*. Eosinophilia of organ tissues and blood appears to be mediated mainly by two systems: by substances released from anaphylactic tissues, probably from mast cells, and by the activity of leucocytes, including but not exclusively, the T cells.

Complexes of guinea-pig IgG2 antibody and complement strongly attract eosinophils *in vitro*, but attract few eosinophils *in vivo* when injected into skin unless the recipient has many blood eosinophils and much antibody is injected. They do not elicit a blood or peritoneal eosinophilia in normal guinea-pigs. In contrast, complexes of IgG1a or IgG1b antibodies attract fewer eosinophils *in vitro* than do those containing IgG2, but attract more eosinophils into skin and elicit a blood or peritoneal eosinophilia. The *in vitro* attraction of IgG2 complexes is probably mediated by one or more anaphylatoxin-associated substances. The *in vivo* eosinophilia induced by IgG1 appears mediated by anaphylactic substances released from the tissues.

In an examination of guinea-pig anaphylactic tissues, lung, skin, diaphragm, peritoneum and mesentery, when implanted into the peritoneal cavity of normal guinea-pigs, induced eosinophilia of the blood of the recipients 24 h later. Cell-free perfusates of these anaphylactic tissues induced eosinophilia of the blood and peritoneum 9 or 12 h after intraperitoneal injection, and of the skin 16 h after intradermal injection. The same perfusates selectively attracted eosinophils *in vitro*, with the exception of skin perfusates which also attracted neutrophils. Antigen-treated liver and kidney from anaphylactically sensitised guinea-pigs did not induce eosinophilia when implanted into normal recipients, and did not release substances that induced eosinophilia or attracted eosinophils. It is evident that the anaphylactic 'shock' organs or tissues of guinea-pigs, which contain numerous mast cells and release histamine during anaphylaxis, also release one or more substances that elicit blood eosinophilia *in vivo* and attract eosinophils *in vitro*.

The anaphylactic substance that selectively attracts eosinophils *in vitro* is released concomitantly with histamine and SRS-A and appears to be preformed. It withstands freezing at $-25\,°C$ but loses about 20% of its activity when heated at $56\,°C$ for 1 h. It has a molecular weight of about 1,000, and can be separated from most other substances in the anaphylactic perfusate by gel filtration. It is not degraded by trypsin or chymotrypsin. Preliminary Sephadex gel separation tests show that fractions of the perfusate containing the partially purified active substance retain all the ability of the whole perfusate to attract eosinophils *in vitro*, but are less effective in inducing blood eosinophilia *in vivo*. Therefore, mobilisation of eosinophils into the blood may be due to another substance in the perfusates, or may require a co-factor.

Human anaphylactic basophils release a substance selectively attracting human eosinophils *in vitro*, which appears similar to that released from the guinea-pig tissues.

The eosinophil-attracting substance may be released from anaphylactically sensitised baso-phils by antigen or by anti-IgE, but not by anti-IgG. Impure preparations of monkey mast cells, passively sensitised with human reagin, on challenge release a substance attracting monkey eosinophils.

Human skin, passively sensitised with reaginic serum and challenged with antigen or with anti-IgE releases a substance attracting eosinophils *in vitro*, but the activity of this substance is masked by other leucocyte-attracting substances released from skin. How-ever, skin from atopic persons may be challenged by antigen to release a substance selectively attracting eosinophils.

Evidence is presented that the main, if not the only sources, of these anaphylactic eosinophil-attracting substances are the mast cells and basophils. Rat mast cells passively sensitised *in vitro* with rat IgE-like antibody and challenged with antigen, release a low molecular weight substance selectively attracting rat eosinophils *in vitro*. When this substance is injected into normal rats it elicits eosinophilia of the blood and peritoneum similar to that occurring in rats passively sensitised with antibody, but less than that occurring in actively sensitised rats. It appears that eosinophilia of passively sensitised rats is mediated by mast cell substances, but in actively sensitised and challenged rats other substances contribute to the eosinophilia.

Eosinophilia of the blood and tissues is also mediated by another non-anaphylactic system, controlled by T lymphocytes or possibly by another lymphoid cell. Many injected particulate or soluble antigens, though not attractive to eosinophils, elicit a transient eosinophilia of the draining lymph-node, possibly due to stimulation of the T cells. More-over, rats infected once with helminths, or guinea-pigs injected once with helminth particles, have a blood eosinophilia starting after 6 days, accompanied by increased numbers in the bone marrow. It was shown that anti-lymphocyte globulin or puromycin had no effect on the rapid eosinophilia of anaphylaxis, but both substances inhibited the eosino-philia of draining lymph-nodes when first exposed to antigen. Neither substance prevented eosinophilia of the lymph-nodes after repeated treatment with the antigen. Moreover anti-lymphocyte globulin suppressed the blood eosinophilia induced by the first injection of helminth fragments into guinea-pigs, differentiating this eosinophilia from that of anaphylaxis.

It is proposed that the attraction of eosinophils *in vitro* may be an exaggeration of *in vivo* events, and that selective eosinophil attraction may not be important in the occur-rence of eosinophilia *in vivo*. It is believed that eosinophilia in anaphylactic tissues results from selective retention of cells randomly infiltrating the site rather than selective attrac-tion to the site. Anaphylactic eosinophilia of the blood results from the rapid mobilisation of eosinophils stored in the tissues, and is controlled by substances released from the anaphylactic tissues, possibly enhanced by anaphylatoxin-associated substances generated in serum. Non-anaphylactic eosinophilia of the lymph-nodes, granulomatous lesions, blood and bone marrow appears to be a distinct phenomenon controlled by T lympho-cytes and possibly other lymphoid cells. The eosinophils are attracted to basophilic mono-nuclear cells of mesenteric milk spots and peritoneal exudates, with which they may form rosettes, and the infiltration into stimulated lymph-nodes probably reflects a similar attraction.

The concomitant activity of stimulated lymphocytes eliciting increased numbers of eosinophils in the blood and bone marrow, and the anaphylactic substances mediating rapid mobilisation of eosinophils in the blood and their retention in damaged tissues, ensures a rapid accumulation of eosinophils which probably have a beneficial function in anaphylactic or granulomatous lesions.

References

ARCHER, G. T. and HIRSCH, J. G.: Motion picture studies on degranulation of horse eosinophils during phagocytosis. J. exp. Med. *118:* 287 (1963).

ARNASON, B. G. and WAKSMAN, B. H.: The retest reaction in delayed sensibility. Lab Invest. *12:* 737 (1963).

BASTEN, A. and BEESON, P. B.: Mechanism of eosinophilia. II. Role of lymphocyte. J. exp. Med. *131:* 1288 (1970).

BASTEN, A.; BOYER, M. H., and BEESON, P. B.: Mechanism of eosinophilia. I. Factors affecting the eosinophil response of rats to *Trichinella spiralis*. J. exp. Med. *131:* 1271 (1970).

BOREL, J. F.; KELLER, H. U., and SORKIN, E.: Studies on chemotaxis. XI. Effect on neutrophils of lysosomal and other subcellular fractions from leukocytes. Int. Arch. Allergy *35:* 194 (1969).

BRAYTON, R. G.; STOKES, P. E.; SCHWARTZ, M. S., and LOURIA, D. B.: Effect of alcohol and various diseases on leukocyte mobilisation, phagocytosis and intracellular bacterial killing. New Engl. J. Med. *282:* 123 (1970).

COHEN, S.; VASSALLI, P.; BENACERRAF, B., and McCLUSKEY, R. T.: The distribution of antigenic and nonantigenic compounds within draining lymph nodes. Lab. Invest. *15:* 1143 (1966).

COHEN, S. and WARD, P. A.: *In vitro* and *in vivo* activity of a lymphocyte and immune complex-dependent chemotactic factor for eosinophils. J. exp. Med. *133:* 133 (1971).

COTTIER, H.; ODARTCHENKO, N.; KEISER, G.; HESS, M., and STONER, R. D.: Incorporation of tritiated nucleosides and amino acids into lymphoid and plasmocytoid cells during secondary response to tetanus toxoid in mice. Ann. N.Y. Acad. Sci. *113:* 612 (1964).

DAY, R. P.: Eosinophil cell separation from human peripheral blood. Immunology, Lond. *18:* 955 (1970).

DAY, R. P.: Basophil leucocyte separation from human peripheral blood. A technique for their isolation in high purity and high yield. Clin. Allergy *2:* 205 (1972).

KAY, A. B.: Studies on eosinophil leucocyte migration. II. Factors specifically chemotactic for eosinophils and neutrophils generated from guinea-pig serum by antigen-antibody complexes. Clin. exp. Immunol. *7:* 723 (1970).

KAY, A. B. and AUSTEN, K. F.: The IgE-mediated release of an eosinophil leukocyte chemotactic factor from human lung. J. Immunol. *107:* 899 (1971).

KAY, A. B.; STECHSCHULTE, J., and AUSTEN, K. F.: An eosinophil leukocyte chemotactic factor of anaphylaxis. J. exp. Med. *133:* 602 (1971).

KELLER, H. U. and SORKIN, E.: Studies on chemotaxis. XIII. Differences in the chemotactic response of neutrophil and eosinophil polymorphonuclear leucocytes. Int. Arch. Allergy *35:* 279 (1969).

KLEPSER, R. G. and NUNGESTER, W. J.: Effect of alcohol upon the chemotactic response of leukocytes. J. infect. Dis. *65:* 196 (1939).

KOSTAGE, S. T.; RIZZO, A. P., and COHEN, S. G.: Experimental eosinophilia. XI. Cell responses to particles of delineated size (32107). Proc. Soc. exp. Biol. Med. *125:* 413 (1967).

LACHMANN, P. J.; KAY, A. B., and THOMPSON, R. A.: The chemotactic activity for neutrophil and eosinophil leucocytes of the trimolecular complex of the fifth, sixth, and seventh components of human complement (C567) prepared in free solution by the 'reactive lysis' procedure. Immunology, Lond. *19:* 895 (1970).

LITT, M.: Studies in experimental eosinophilia. II. Induction of peritoneal eosinophilia by the transfer of tissues and tissue extracts. Blood *16:* 1330 (1960).

LITT, M.: Studies in experimental eosinophilia. III. The induction of peritoneal eosinophilia by the passive transfer of serum antibody. J. Immunol. *87:* 522 (1961).

LITT, M.: Eosinophils and antigen-antibody reactions. Ann. N.Y. Acad. Sci. *116:* 964 (1964).

LITT, M.: Studies in experimental eosinophilia. IX. Inhibition by puromycin of the eosinophil response which hemocyanin elicits in guinea pig lymph nodes. J. Immunol. *109:* 222 (1972).

OVARY, Z. and WARNER, N. L.: Electrophoretic and antigenic analysis of mouse and guinea-pig anaphylactic antibodies. J. Immunol. *108:* 1055 (1972).

PARISH, W. E.: Investigations on eosinophilia. The influence of histamine, antigen-antibody complexes containing γ1 or γ2 globulins, foreign bodies (phagocytosis) and disrupted mast cells. Brit. J. Derm. *82:* 42 (1970a).

PARISH, W. E.: Homologous serum passive cutaneous anaphylaxis in guinea pigs mediated by two γ1 or γ1-type heat-stable globulins and a non-γ1 heat-labile reagin. J. Immunol. *105:* 1296 (1970b).

PARISH, W. E.: Eosinophilia. III. The anaphylactic release from isolated human basophils of a substance that selectively attracts eosinophils. Clin. Allergy *2:* 381 (1972a).

PARISH, W. E.: Eosinophilia. II. Cutaneous eosinophilia in guinea-pigs mediated by passive anaphylaxis with IgG1 or reagin, and antigen-antibody complexes. Its relation to neutrophils and to mast cells. Immunology, Lond. *23:* 19 (1972b).

PARISH, W. E.: Eosinophilia. I. Eosinophilia in guinea-pigs mediated by passive anaphylaxis and by antigen-antibody complexes containing homologous IgG1a and IgG1b. Immunology, Lond. *22:* 1087 (1972c).

PARISH, W. E. and COOMBS, R. R. A.: Peripheral blood eosinophilia in guinea-pigs following implantation of anaphylactic guinea-pig and human lung. Brit. J. Haemat. *14:* 425 (1968).

RILEY, J. F.: The location of histamine in the body; in WOLSTENHOLME and O'CONNOR Ciba Foundation Symposium on Histamine, p. 398 (Churchill, London 1956).

RILEY, J. F.: The mast cells (Livingstone, Edinburgh 1959).

SAMTER, M.; KOFOED, M. A., and PIEPER, W.: A factor in the lungs of anaphylactically shocked guinea-pigs which can induce eosinophilia in normal animals. Blood *8:* 1078 (1953).

SORKIN, E.; BOREL, J. F., and STECHER, V. J.: Chemotaxis of mononuclear and polymorphonuclear phagocytes; in VAN FURTH Mononuclear phagocytes, p. 397 (Blackwell Scientific Publications, Oxford 1970).

SORKIN, E.; STECHER, V. J., and BOREL, J. F.: Chemotaxis of leucocytes and inflammation. Ser. Haemat. *3:* 131 (1970).

SPEIRS, R. S.: The action of antigen upon hypersensitive cells. Ann. N.Y. Acad. Sci. *113:* 819 (1964).

SPEIRS, R. S. and SPEIRS, E. E.: Cellular localization of radioactive antigen in immunized and nonimmunized mice. J. Immunol. *90:* 561 (1963).

SPRY, C. J. F.: Mechanism of eosinophilia. VI. Eosinophil mobilization. Cell Tissue Kinet. *4:* 365 (1971a).

SPRY, C. J. F.: Mechanism of eosinophilia. V. Kinetics of normal and accelerated eosino-poiesis. Cell Tissue Kinet. *4:* 351 (1971b).

WALLS, R. S.; BASTEN, A.; LEUCHARS, E., and DAVIES, A. J. S.: Mechanisms for eosino-philic and neutrophilic leucocytoses. Brit. med. J. *iii:* 157 (1971).

WALLS, R. S. and BEESON, P. B.: Mechanism of eosinophilia. VIII. Importance of local cellular reactions in stim ulating eosinophil production. Clin. exp. Immunol. *12:*111 (1972).

WARD, P. A.: Chemotaxis of human eosinophils. Amer. J. Path. *54:* 121 (1969).

WISSLER, J. H.; STECHER, V. J., and SORKIN, E.: Biochemistry and biology of a leucotactic binary serum peptide system related to anaphylatoxin. Int. Arch. Allergy *42:* 722 (1972).

Author's address: Dr. W. E. PARISH, Lister Institute of Preventive Medicine, *Elstree, Hertfordshire* (England)

Antibiotics and Chemotherapy, vol. 19, pp. 271–283
(Karger, Basel 1974)

Chemotaxis of Eosinophil Leucocytes in Relation to Immediate-Type Hypersensitivity and the Complement System[1]

A. B. KAY

Department of Respiratory Diseases, University of Edinburgh at the City Hospital, Edinburgh

Introduction

An increase in the number of eosinophils in the tissues and the circulation is a feature of various clinical conditions. These include diseases associated with high levels of IgE, such as extrinsic (or allergic) asthma, hay fever and parasitic infections [1–4], and disorders in which there is direct or circumstantial evidence of the presence of circulating, or fixed antigen antibody complexes capable of activating the complement system. The latter group of diseases includes polyarteritis nodosa with lung involvement, pulmonary aspergillosis and rheumatoid arthritis [5–8].

In experimental animals, an eosinophilia can be evoked following injection of specific antigen into a sensitised animal. Thus, an eosinophilia can occur following general or local anaphylaxis [9–11] or following the implantation of anaphylactic tissue into normal animals [12, 13]. Injections of antigen-antibody complexes into the peritoneal cavity of normal animals also results in a local eosinophilia [14].

In order to study the relationship between immediate-type (anaphylactic) hypersensitivity, the complement system and the accumulation of eosinophils, initial experiments were performed in guinea-pig skin [15]. It was shown that IgG_1 and IgG_2, as preformed complexes, prepared skin for a subsequent local eosinophilia 12 h after injection. However, if antibody was

1 Supported by the Medical Research Council, the Royal College of Physicians of London (T.K. Stubbins Fellow) and an anonymous gift to the Department of Respiratory Diseases.

first placed in the skin and after a variant latent period the animal was challenged by antigen and Evans blue dye intravenously (as in a usual passive cutaneous anaphylactic reaction) IgG_1, but not IgG_2, elicited a local eosinophil response 8–12 h after the initial blueing reaction. Intradermal histamine given at different skin sites in the same animals in doses which gave a comparable blueing reaction to the IgG_1-mediated response did not lead to an accumulation of eosinophils. Eosinophils also accumulated around the sites of injections of compound 48/80, an agent which depletes mast cell granules. Low mast cell counts also accompanied the infiltration of eosinophils observed with IgG_1-mediated passive cutaneous anaphylaxis (PCA) reactions. It was suggested at this time that the accumulation of eosinophils following PCA reactions was by a complement-independent mechanism subsequent to the release from mast cells of an agent other than histamine. The participation of the complement system was suggested in antigen-antibody complex mediated eosinophilia.

When animals were decomplemented with cobra venom factor this had no effect on eosinophil accumulation into the site of IgG_1-mediated PCA reactions in the guinea-pig [16]. Furthermore, eosinophil accumulation also followed PCA reactions in animals partially or totally deficient in the fourth component of complement. Surprisingly, intradermal injections of preformed antigen-antibody complexes prepared from guinea-pig IgG_1 or IgG_2 were also unaffected by decomplementation with purified cobra venom factor. The lesions produced with either immunoglobulin complex were similar in appearance in decomplemented and normal animals and were followed by comparable tissue accumulation of eosinophils. Since IgG_2 is tunable to sensitise skin for PCA reactions [17] or lung fragments for hisamine [18] and slow-reacting substance of anaphylaxis (SRS-A) release [19], but can fix complement by the classical pathway, it was expected that complement may play a part in the cellular infiltration following intradermal injections of IgG_2 complexes. These apparent discrepancies could be explained on a weight basis since IgG_1 is far more efficient in preparing tissue for an eosinophilia when administered in PCA reactions than when injected as preformed complexes. It was previously demonstrated that soluble complexes prepared from guinea-pig IgG_1 or IgG_2 both have the capacity to liberate histamine from perfused guinea-pig lung [20]. This suggests that IgG_2 may have similar properties to IgG_1, in terms of its effect on mast cells, when presented to tissue as a complex.

These observations *in vivo* prompted studies on the identification of chemotactic agents for eosinophils released during the anaphylactic reaction or following complement activation. A modification of the Millipore technique of Boyden was used for these experiments [21]. Eosinophils were obtained from the peritoneal cavity of the guinea-pig following multiple injections of horse serum. In the human studies peripheral blood leucocytes from patients with eosinophilia were used as a source of target cells.

An Eosinophil Chemotactic Factor of Anaphylaxis (ECF-A)

As an experimental model of anaphylaxis, sensitised guinea-pig lung was chosen since this has been employed by numerous workers as a model of immediate-type hypersensitivity and the organ can be perfused free of blood thereby providing a serum-free system. Similarly, human lung fragments taken at pneumonectomy can be thoroughly washed before and after sensitisation with serum from an allergic individual. When sensitised lung was challenged with specific antigen there appeared in the diffusate, along with histamine and SRS-A, an agent which selectively attracted eosinophils. This eosinophil chemotactic factor of anaphylaxis has been termed ECF-A [22].

Formation Mechanism

The release of ECF-A was accompanied by the release of histamine and SRS-A, but differed in its time course of release. In the experiment depicted in figure 1 guinea-pig lung was perfused free of blood by cannulating the pulmonary artery. Specific antigen was injected *via* the same route. Approximately 10 ml volumes of perfusate were collected every 15 min for 1 h. Virtually all the histamine and SRS-A was released during the first 15 min, whereas ECF-A continued to be released for the duration of the experiment. A similar time-course of release was observed with human lung fragments passively sensitised with serum from a ragweed-sensitive individual and challenged with ragweed antigen E [23]. Histamine, SRS-A and ECF-A were similar in terms of the time-course of passive sensitisation and the amount of antigen required for optimal release. ECF-A release was dependent on the presence of divalent cations and was strikingly enhanced by the presence of succinate or maleate; features characteristic of the release of histamine and SRS-A [24].

In the guinea-pig, decomplementation by the administration of purified cobra venom factor had no effect on the antigen-induced release of ECF-A

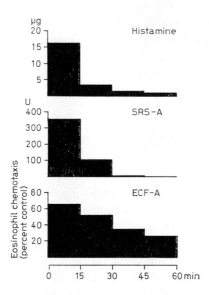

Fig. 1. Time-course of release of chemical mediators of anaphylaxis following antigen challenge of perfused, actively sensitised guinea-pig lung. The measurements represent the mean values from 16 animals. The experimental technique and the expression of results are described in ref. No. 22. SRS-A = slow-reacting substance of anaphylaxis, ECF-A = eosinophil chemotactic factor of anaphylaxis.

from actively or passively sensitised lung fragments. Since the lungs were then perfused free of blood and the fragments thoroughly washed this strongly suggests that ECF-A release is complement-independent [22].

There is evidence that increase in the level of intracellular 3'–5' cyclic adenosine monophosphate (cyclic AMP) inhibits the release of ECF-A since the antigen-induced release of ECF-A was inhibited from passively sensitised human lung fragments by either isoprenaline, in doses ranging from 10^{-5} to 10^{-8} M, or with dibutyryl cyclic AMP in a concentration of 5×10^{-4} M [25].

Antibody Requirement

ECF-A from the guinea-pig can be released from normal lung passively sensitised with IgG_1. The same immunoglobulin has been shown to sensitise guinea-pig lung slices for the antigen-induced release of histamine and SRS-A. The fractions of IgG_1 were prepared by ion exchange chromatography and lost no activity after heating at 56 °C for 4 h, a property characteristic of IgG_1 but not of an IgE-like immunoglobulin. Comparable amounts of IgG_2-containing fractions failed to sensitise lung fragments for the antigen-induced release of histamine, SRS-A or ECF-A. The capacity of guinea-pig IgE to sensitise tissue for ECF-A release is yet to be ascertained due to difficulties in raising antibodies of this class and in obtaining guinea-pig IgE free from IgG_1.

In the human, the release of ECF-A was shown to be mediated by IgE [23]. This was demonstrated by absorption of ragweed-sensitive serum with a rabbit antibody specific for IgE. Absorbed antibody could no longer passively sensitise lung for the release of ECF-A, histamine or SRS-A. ECF-A could also be released from normal lung by the reversed-type reaction employing a specific anti-IgE prepared in the rabbit.

The Identification of ECF-A as a Distinct Pharmacological Agent

Various chemical mediators of anaphylaxis were tested directly for their ability to evoke the migration of eosinophils [22]. It was found that histamine, bradykinin, serotonin and prostaglandins (PGE) PGE_1, PGE_2 and $PGF_{2\alpha}$ over a wide dose range, were not chemotactic for eosinophils *per se;* and, furthermore, when incubated with sensitised lung in the absence of antigen these agents did not secondarily affect the release of ECF-A. Although SRS-A is yet to be chemically characterised there were several reasons for considering ECF-A to be distinct from this agent. SRS-A survived boiling in alkaline solution for 20 min, whereas ECF-A activity was abolished by this procedure and, in addition, SRS-A and ECF-A in the guinea-pig could be separated by gel-filtration. Human ECF-A appears to be more closely associated with human SRS-A, but they could be separated functionally since the latter was partially destroyed by lyophilisation and boiling in acid solution, whereas ECF-A was inactivated by boiling for 10 min in alkaline solution. Further evidence that ECF-A is distinct from SRS-A and histamine was apparent from measurements of mediators in the perfusates described for figure 1. In 4 of the 16 lungs no SRS-A or histamine was detected in the 15- to 30-min perfusate, although this contained large quantities of ECF-A. Histamine and SRS-A were released during the first 15 min only.

Physico-Chemical Properties of ECF-A

Guinea-pig and human ECF-A were relatively resistant to heat, only about 25% of the activity being lost following boiling for 10 min or heating at 56 °C for 1 h. Both human and guinea-pig ECF-A activity was totally recovered following lyophilisation and multiple freezing and thawing. About 30% of their activity was destroyed by extraction in 80% ethanol, evaporation to dryness and re-suspension of the residue to the original volume. Some preliminary evidence suggested that ECF-A may be a small peptide [26]. It is fully recovered from a column of Sephadex G-25 and had an estimated molecular weight of between 500 and 1,000. The peak of ECF-A activity contained materials which had free amino groups (ninhydrin reaction) and peptide bonds (starch-iodide reaction). Recent experiments have shown, however, that guinea-pig ECF-A activity is *not* destroyed by relatively large doses of pronase, chymotrypsin and trypsin [27]. In these experiments advantage was taken of the relative heat stability of ECF-A. Partially purified guinea-pig ECF-A, prepared by G-25 Sephadex chromatography, and a control Tyrode's buffer with antigen subjected to the same procedure, were incubated with enzymes for 1 h at 30 °C and then heated at 56 °C for 30 min. Both chymotrypsin and trypsin completely lost activity by this heat treatment although pronase was unaffected. However, the heated enzymes were inactive in chemotaxis when tested alone, and when mixed with ECF-A did not affect its activity. The proteolytic activity of the enzymes was monitored by following the release of trichloroacetic acid insoluble material from casein using the same buffer and identical incubation conditions to those used for the samples.

Selectivity of ECF-A for Eosinophil Leucocytes

ECF-A selectively attracts eosinophils from a mixed leucocyte population. Guinea-pig neutrophils from a pure population (>96% purity) will migrate towards ECF-A, although when eosinophils are introduced into the suspension they are selectively attracted when they comprise 10% or more of the mixed cell population [28]. Human ECF-A also selectively attracts eosinophils but neutrophils and basophils will migrate when eosinophils are present in very small numbers. Thus, a broad specificity for ECF-A, as of other chemotactic factors can be shown when conditions for migration are optimal and when other cells which may be preferentially attracted are absent.

Table I. Tissue eosinophil counts at the sites of intrapulmonary injections of guinea-pig anaphylactic lung diffusates

Animal No.	Anaphylactic diffusate (right lung)	Control (left lung)
1	240	50
2	187	95
3	80	35
4	277	28
5	39	24
6	9	11
7	61	41
8	82	32
9	73	23
Mean	116	38

Volumes of 0.2 ml were injected into each lung, and tissue eosinophil counts were performed 12 h later using the technique previously described [39]. The anaphylactic diffusate contained 1.0 μg/ml of histamine, 60 U/ml of slow-reacting substance of anaphylaxis (SRS-A) and 50 μg/ml of ovalbumin. An equivalent amount of histamine and antigen was injected into the control lung.

Other Features of ECF-A

The source of lung ECF-A is not known; however, a comparable material has been identified from highly purified sensitised human peripheral blood basophils challenged with antigen [29]. There is evidence that ECF-A exists in a pre-formed state in highly sensitised human basophils [30] or guinea-pig lung [31], i.e. cells or tissue which will release relatively large amount of mediators following antigen challenge. In these situations slight mechanical manipulation, in the absence of antigen challenge, results in the elaboration of an eosinophil-attracting substance.

When a guinea-pig anaphylactic diffusate was injected into one lung of an animal and the appropriate control administered into the other lung, a 3-fold increase in local eosinophil accumulation could be demonstrated (table I) [32]. A similar local eosinophilia could be seen in the guinea-pig peritoneal cavity following injection of partially purified ECF-A (table II). A summary of the features of ECF-A is depicted in table III.

Table II. Eosinophilia following intraperitoneal injections of partially purified ECF-A

Animal No.	Treatment	Eosinophil count per millilitre of fluid recovered, $\times 10^3$
1		500
2	partially purified ECF-A	245
3	in Tyrode's solution	293
Mean		346
4		85
5	Tyrode's solution	140
6		120
Mean		115

ECF-A = eosinophil chemotactic factor of anaphylaxis.
Volumes of 5 ml were injected and eosinophil counts were performed 12 h later. Partially purified ECF-A, giving a chemotactic count of 45 cells/ml [22], was prepared by Sephadex G-25 chromatography and was free of histamine and SRS-A.

Table III. Some properties of human and guinea-pig ECF-A

Released by antigen challenge of actively or passively sensitised lung.
Passive sensitisation mediated by IgE in human lung and IgG_1 in the guinea-pig.
Release of ECF-A accompanied by histamine and SRS-A.
Release of ECF-A from perfused whole guinea-pig lung slower than histamine and SRS-A.
Optimal conditions for ECF-A release, in terms of antigen and antibody dose, similar to histamine and SRS-A.
Requirement for divalent cations.
Enhanced by succinate and maleate.
Complement not required.
Probably modulated by levels of intracellular cyclic AMP.
Distinct from histamine, bradykinin, serotonin, SRS-A, prostaglandins (PGE) PGE_1, PGE_2, $PGF_{2\alpha}$ and C5a.
Labile in alkaline solution.
Resistant to heat, lyophilisation and multiple freezing and thawing.
Molecular weight 500–1,000.

Complement-Derived Eosinophil Chemotactic Factors

Factors Generated from Serum by Antigen-Antibody Complexes

Although the *in vivo* significance of complement-dependent chemotactic factors is unknown, an eosinophil specific chemo-attractant can be generated from serum [21]. Preformed antigen-antibody complexes prepared either from guinea-pig IgG_2 or IgG_1 were equally capable of generating, from normal serum, heat-stable activities that were chemotactic for guinea-pig eosinophils and neutrophils. The generation of this activity was apparently dependent on the presence of an intact complement system. When serum, activated by complexes prepared from either of these sub-classes, was passed over a column of Sephadex G-100, two peaks of chemotactic activity could be demonstrated. The peak of activity for guinea-pig neutrophils had a molecular weight of approximately 75,000. The eosinophil chemotactic activity eluted with molecules having a molecular weight of between 15,000 and 20,000. The nature of the neutrophil chemotactic activity is unknown but a comparable activity has been generated from rat serum and is thought to be a cleavage product of the 5th component of complement [33]. The smaller fragment which was predominately chemotactic for eosinophils had a similar molecular size and sedimentation constant to that of C5a. It was subsequently confirmed that C5a prepared from highly purified C5 did, in fact, preferentially attract eosinophils from a mixed leucocyte population [28]. Although C5a, like ECF-A, is chemotactic for neutrophils, when eosinophils comprised approximately 10% or more of a mixed population they were preferentially attracted by this complement fragment.

Interaction of ECF-A with C5a

Since ECF-A and C5a are distinct both in their molecular weight and formation mechanism, it was of interest to determine the effect of combining the two agents in eosinophilotaxis [28]. When these agents were mixed together in the test compartment of the chemotactic chamber the resultant counts were three times or more greater than would have been expected by summation of counts when the agents were assayed alone. This suggested that ECF-A and C5a act synergistically in their ability to attract eosinophils. It is possible that eosinophils have more than one receptor for chemotaxis and that, if different types of receptors are stimulated at a low threshold, this produces an increased chemotactic response. These observations on synergism may be of significance in parasitic infestations, many of which are associated with a pronounced eosinophilia. Homocytotropic antibody and complement-fixing antibody can occur together in a variety of parasitic diseases, situations in which ECF-A and C5a might act together.

Other Complement-Derived Eosinophil Chemotactic Factors

A fragment from the third component of complement (C3a) [34] and the trimolecular complex of $C\overline{567}$ [35] have also been shown to attract eosinophils. When purified $C\overline{56}$ and $C\overline{7}$ were prepared as in the 'reactive-lysis' procedure their combination resulted in attraction of neutrophils, eosinophils [36] and basophils [37]. There was no evidence that $C\overline{567}$ was preferentially chemotactic for any of these cell types.

Cell-specific attraction for eosinophils or neutrophils has recently been demonstrated by activating serum from the hog, rat or guinea-pig with antigen-antibody complexes, yeast or dextran [38]. Following progressive purification of the activated serum, two peptides have been isolated which have been termed 'classical anaphylatoxin' and 'cocytotaxin'. By themselves these peptides have little chemotactic activity but on re-combination and depending on their molar ratio, cell-specific chemotactic activity for eosinophils or neutrophils can be demonstrated. These interesting observations suggest that low molecular weight anaphylatoxins are distinct from chemotactic agents although closely related in molecular size, and that the ratio in which they are combined determines the expression of various biological activities. Although the peptides are probably products of complement activation this has yet to be demonstrated conclusively.

Conclusion

Two chemotactic factors have been described which selectively attract eosinophil leucocytes. They differ both in molecular size and in formation mechanism. ECF-A is a product of the anaphylactic reaction and is distinct from previously described pharmacological mediators. The complement-derived factor, C5a, can show preferential attraction of eosinophils under certain experimental conditions. Marked synergism was observed between

ECF-A and C5a in their ability to attract eosinophil leucocytes. The elaboration of these agents in various hypersensitivity states may in part account for the infiltration of eosinophils.

References

1   JOHANSSON, S. G. O.: Raised levels of a new immunoglobulin class (IgND) in asthma. Lancet *ii:* 951 (1967).
2   BERG, T. and JOHANSSON, S. G. O.: IgE concentrations in children with atopic diseases. Int. Arch. Allergy *36:* 219 (1969).
3   ROWE, D. S. and WOOD, C. B. S.: The measurement of serum immunoglobulin E levels in healthy adults and children and in children with allergic asthma. Int. Arch. Allergy *39:* 1 (1970).
4   JOHANSSON, S. G. O.; MELLBIN, T., and VAHLQUIST, B.: Immunoglobulin levels in Ethiopian preschool children with special reference to high concentrations of immunoglobulin E (IgND). Lancet *i:* 1118 (1968).
5   ROSE, G. A. and SPENCER, H.: Polyarteritis nodosa. Quart. J. Med. *26:* 43 (1957).
6   CROFTON, J. W.; LIVINGSTONE, J. L.; OSWALD, N. C., and ROBERTS, A. R. M.: Pulmonary eosinophilia. Thorax *7:* 1 (1952).
7   PEPYS, J.; RIDDELL, R. W.; CITRON, K. W.; CLAYTON, J. M., and SHORT, E. I.: Clinical and immunologic significance of *Aspergillus fumigatus* in the sputum. Amer. Rev. Tuberc. *80:* 167 (1959).
8   FALCK, I. J. H. VON und SCHRODER, I.: Eosinophilie und Rheumatismus. Münch. med. Wschr. *105:* 574 (1963).
9   SAMTER, M.: The response of eosinophils in the guinea-pig to sensitization, anaphylaxis and various drugs. Blood *4:* 217 (1949).
10 PARISH, W. E.: Investigations on eosinophilia. The influence of histamine, antigen-antibody complexes containing γ_1 or γ_2 globulins, foreign bodies (phagocytosis) and disrupted mast cells. Brit. J. Derm. *82:* 42 (1970).
11 LITT, M.: Studies on experimental eosinophilia. VIII. Induction of eosinophilia by homologous $7S\gamma_1$ antibody and by extremely minute doses of antigen. Proc. 6th Congr. of the Int. Ass. of Allergology, Montreal. Excerpta med. int. Congr. Ser. *162:* 38 (1967).
12 SAMTER, M.; KOFOED, M. A., and PIEPER, W.: A factor in lungs of anaphylactically shocked guinea-pigs which can induce eosinophilia in normal animals. Blood *8:* 1078 (1953).
13 PARISH, W. E. and COOMBS R. R. A.: Peripheral blood eosinophilia in guinea-pigs following implantation of anaphylactic guinea-pig and human lung. Brit. J. Haemat. *14:* 425 (1968).
14 LITT, M.: Studies in experimental eosinophilia. III. The induction of peritoneal eosinophilia by the passive transfer of serum antibody. J. Immunol. *87:* 522 (1961).
15 KAY, A. B.: Studies on eosinophil leukocyte migration. I. Eosinophil and neutrophil accumulation following antigen-antibody reactions in guinea-pig skin. Clin. exp. Immunol. *6:* 75 (1970).

16   KAY, A. B. and AUSTEN K. F.: Antigen-antibody induced cutaneous eosinophilia in complement deficient guinea-pigs. Clin. exp. Immunol. *11:* 37 (1972).

17   OVARY, Z.; BENACERRAF, B., and BLOCH, K. J.: Properties of guinea-pig 7S antibodies. II. Identification of antibodies involved in passive cutaneous and systematic anaphylaxis. J. exp. Med. *117:* 951 (1963).

18   BAKER, A. R.; BLOCH, K. J., and AUSTEN, K. F.: *In vitro* passive sensitization of chopped guinea-pig lung by guinea-pig 7S antibodies. J. Immunol. *93:* 525 (1964).

19   STECHSCHULTE, D. J.; AUSTEN, K. F., and BLOCH, K. J.: Antibodies involved in antigen-induced release of slow reacting substance of anaphylaxis (SRS-A) in the guinea-pig and rat. J. exp. Med. *125:* 127 (1967).

20   BRODER, I.: Histamine release by soluble antigen-antibody complexes (SC) containing non-sensitizing antibody (abstract). Fed. Proc. *28:* 377 (1969).

21   KAY, A. B.: Studies on eosinophil leukocyte migration. II. Factors specifically chemotactic for eosinophils and neutrophils generated from guinea-pig serum by antigen-antibody complexes. Clin. exp. Immunol. *7:* 723 (1970).

22   KAY, A. B.; STECHSCHULTE, D. J., and AUSTEN, K. F.: An eosinophil leukocyte chemotactic factor of anaphylaxis. J. exp. Med. *133:* 602 (1971).

23   KAY, A. B. and AUSTEN, K. F.: The IgE-mediated release of an eosinophil leukocyte chemotactic factor from human lung. J. Immunol. *107:* 899 (1971).

24   AUSTEN, K. F. and BROCKLEHURST, W. E.: Anaphylaxis in chopped guinea-pig lung. II. Enhancement of the anaphylactic release of histamine and slow-reacting substance by certain dibasic aliphatic acids and inhibition by monobasic fatty acids. J. exp. Med. *113:* 541 (1961).

25   KAY, A. B.: Unpublished observation (1971).

26   KAY, A. B.; STECHSCHULTE, D. J.; KAPLAN, A. P., and AUSTEN, K. F.: The antigen-induced release of eosinophil leukocyte chemotactic factors from passively sensitized guinea-pig or human lung (abstract). Fed. Proc. *30:* 682 (1971).

27   BACH, M. K. and KAY, A. B.: Unpublished observations (1972).

28   KAY, A. B.; SHIN, H. S., and AUSTEN, K. F.: Selective attraction of eosinophils and synergism between eosinophil chemotactic factor of anaphylaxis (ECF-A) and a fragment cleaved from the fifth component of complement (C5a). Immunology *24:* 969 (1973).

29   PARISH, W. E.: Eosinophilia. III. The anaphylactic release from isolated human basophils of a substance that selectively attracts eosinophils. Clin. Allergy *2:* 381 (1972).

30   PARISH, W. E.: Personal communication (1973).

31   KAY, A. B.: Unpublished observations (1971).

32   KAY, A. B. and SAMTER, M.: Unpublished observations (1971).

33   WARD, P. A.: Chemotactic factors for neutrophils, eosinophils, mononuclear cells and lymphocytes; in AUSTEN and BECKER Biochemistry of the acute allergic reactions, p. 231 (Blackwell, Oxford 1971).

34   WARD, P. A.: Chemotactic factors for neutrophils, eosinophils, mononuclear cells and lymphocytes; in AUSTEN and BECKER Biochemistry of the acute allergic reactions, p. 230 (Blackwell, Oxford 1971).

35   WARD, P. A.: Chemotaxis of human eosinophils. Amer. J. Path. *54:* 121 (1969).

36   LACHMANN, P. J.; KAY, A. B., and THOMPSON, R. A.: The chemotactic activity for

neutrophil and eosinophil leukocytes of the trimolecular complex of the fifth, sixth and seventh components of human complement (C567) prepared in free solution by the 'reactive lysis' procedure. Immunology, Lond. *19:* 895 (1970).

37   KAY, A. B. and AUSTEN, K. F.: Chemotaxis of human basophil leukocytes. Clin. exp. Immunol. *11:* 557 (1972).

38   WISSLER, J. H.; STECHER, V. J., and SORKIN, E.: Biochemistry and biology of a leucotactic binary serum peptide system related to anaphylatoxin. Int. Arch. Allergy *42:* 722 (1972).

39   SAMTER, M.: Early eosinophilia induced in guinea-pigs by intrapulmonary injection of antigenic determinants and antigens. J. Allergy *45:* 234 (1970).

Author's address: Dr. A. B. KAY, Department of Respiratory Diseases, University of Edinburgh at the City Hospital, Greenbank Drive, *Edinburgh EH10 5SB* (Scotland)

Antibiotics and Chemotherapy, vol. 19, pp. 284–295
(Karger, Basel 1974)

Chemotaxis of Mononuclear Leukocytes

R. Snyderman[1] and C. E. Stahl

Division of Rheumatic and Genetic Diseases, Departments of Medicine and Immunology, Duke University Medical Center, Durham, N.C. and Division of Rheumatology, Durham Veterans Administration Hospital, Durham, N.C.

The role played by macrophages in inflammation has been the object of avid investigation for nearly a century, but the mechanism by which these cells accumulate at local inflammatory sites is still unclear. As phagocytes with intracellular digestive properties, macrophages promote wound-healing by eliminating tissue debris and fibrin deposits [5, 10]. During the course of immune-mediated inflammatory reactions, they bring about the degradation of the inciting antigen as well as the phagocytosis of neutrophil debris [5]. These functions are essential to host defense and require the accumulation of a sufficient number of cells at sites where antigen or tissue breakdown products are concentrated.

One mechanism which could account for the local accumulation of motile cells like macrophages is chemotaxis, the unidirectional migration of cells toward an increasing gradient of attractant substance. Since Boyden's development of a quantitative assay for leukocyte chemotaxis in vitro [4], a number of mononuclear cell types from a variety of species have been assayed for their chemotactic responsiveness to several humoral and cellular factors [3, 6, 9, 24, 28, 30, 32]. Only recently, concomitant with the development of a method for assaying peripheral blood monocytes, have human mononuclear cells been subject to a quantitative evaluation of their chemotactic function [18]. Using this method, we have defined two factors chemotactic for human monocytes [1, 18, 19] and evaluated the chemotactic responsiveness of monocytes from patients with various diseases [17]. In the following discussion, we will review our studies of human and non-human macrophage chemotaxis in vitro and speculate upon the relevance of our findings to macrophage function in vivo.

1   Howard Hughes Medical Investigator.

I. *Humoral Mediators of Macrophage Chemotaxis*

A. Chemotactic Factor Derived from Complement

In previous investigations we studied the role of the complement system as a mediator of polymorphonuclear leukocyte (PMN) chemotaxis and found that a fragmentation product of the fifth component of complement, C5a, was an important chemotactic factor *in vitro* and *in vivo* [15, 20–22, 25]. During the course of these studies, purified C5a was injected into the skin of rabbits [7], and guinea pigs [Snyderman and Shin, unpublished observation] and histological studies done at various times thereafter [7]. In addition to the massive accumulation of PMN at the site of C5a injection, we observed large numbers of macrophages. This prompted subsequent investigation to determine if C5a directly mediated the chemotaxis of macrophages as well as that of neutrophils.

Highly purified guinea pig C5 was cleaved with either the earlier acting components of complement (EAC$\overline{1}$, $\overline{4}$, $\overline{2}$, $\overline{3}$[2]) or trypsin. Cleavage of C5 by either of these means resulted in the production of chemotactic activity for guinea pig peritoneal macrophages [24]. In these experiments solutions containing $0.085\mu g$ C5a/ml (5.6×10^{-9} M) were strongly chemotactic for macrophages. In order to determine if the chemotactic activity for guinea pig macrophages and PMN resided on the same fragment of C5, reaction mixtures of activated C5 were fractionated by polyacrylamide gel electrophoresis and molecular sieve chromatography. Chemotactic activity for the two cell types could not be separated by these means, thus indicating that the activity resides on the same or similar 15,000 molecular weight fragment of C5 [24]. A purified acid-acting proteinase derived from homogenates of macrophages could also cleave purified C5 and so produce chemotactic activity for macrophages and PMN [23]. However, this proteinase cleaved C5 into a number of chemotactically active fragments. These studies suggest that the release of this proteinase by macrophages during phagocytosis could cleave C5 into a number of fragments chemotactic for additional macrophages, thus amplifying the inflammatory response.

The above data demonstrate that C5a derived from purified C5 is chemotactic for macrophages. We sought to determine if complement activated in the more physiologic milieu of whole serum would also result in generation of chemotactic activity for macrophages. Antigen-antibody

2   Signifies sensitized erythrocytes carrying the indicated components of complement in their activated form.

complexes, endotoxin and cobra venom factor were all used to activate serum complement. Each of these factors is known to produce C5a [21], and each generated chemotactic activity for macrophages when incubated with whole serum [6]. The amount of chemotactic activity generated was directly proportional to the amount of hemolytic complement activity consumed in the treated serum. Fractionation by molecular sieve chromatography of sera treated with endotoxin, antigen-antibody complexes and cobra venom factor demonstrated the presence of two peaks of chemotactic activity. One was a high molecular weight (ca. 90,000) heat-labile (56 °C for 30 min) factor present in normal as well as activated serum (to be discussed). The other was a low molecular weight (ca. 15,000) heat-stable (56 °C for 30 min) factor found only in activated serum and had an elution profile similar to C5a. This second peak of activity was completely destroyed by incubation with antibody to C5 or C5a, thus establishing the identity of the chemotactic factor generated in whole serum by complement activation as C5a [6].

An analogous set of experiments was performed to see if activation of human serum resulted in the production of chemotactic activity for human peripheral blood monocytes [18]. Human serum treated with antigen-antibody complexes, endotoxin or cobra venom factor contained chemotactic activity for human monocytes *in vitro*. The active material had a low molecular weight (ca. 15,000), was heat-stable (56 °C for 30 min) and was inhibited by antibody to human C5. Human C5a derived from activated serum was chemotactic for human monocytes at concentrations as low as $0.050 \mu g/ml$ (3.3×10^{-9} M) [18].

Clearly C5a is chemotactic for mononuclear leukocytes as well as PMN. Since PMN are not present in the mononuclear leukocyte preparations tested *in vitro*, it can be assumed that prior accumulation of neutrophils is not required for the chemotaxis of macrophages *in vivo*. This assumption is supported by the observation of SIMPSON and ROSS that macrophages accumulate normally at wound sites in neutropenic animals [16]. The role that C5a plays *in vivo* as a chemotactic factor for macrophages is still a matter of speculation. C5a may be responsible for macrophage accumulation at sites of immune complex induced inflammation. Macrophage accumulation at sites of nonspecific tissue trauma may also be mediated by C5 fragmentation products, since intracellular proteolytic enzymes released by damaged cells can cleave C5 to produce phlogistic peptides [23, 27, 29]. Similarly, virus-infected cells can release a factor which cleaves C5 to produce chemotactic activity for macrophages [26].

Table I. Comparison of chemotactic activity in human serum and plasma

Volume tested ml[2]	Chemotactic activity[1]	
	serum	plasma[3]
0.025	27.6 ± 8.8	33.3 ± 5.0
0.050	71.3 ± 12.9	62.2 ± 0.4
0.100	95.5 ± 15.1	86.4 ± 3.1
0.300	94.2 ± 5.3	84.8 ± 3.5
0.500	83.5 ± 4.4	83.5 ± 5.3
Medium alone	11.6 ± 2.6	

1 Chemotactic activity is expressed as the mean and standard error of migrating mononuclear leukocytes per oil immersion field ($\times 1,445$).
2 The indicated volume of serum or plasma was brought to 1.0 ml with gelatin veronal buffer (pH 7.4) then to 1.7 ml with Gey's balanced salt solution [20].
3 Contains 5 U heparin/ml.

B. Chemotactic Activity in Serum and Plasma

As previously mentioned, fresh untreated serum contains chemotactic activity for guinea pig macrophages. Since it has been reported that serum, but not plasma, is chemotactic for macrophages [2, 3], we studied fresh serum and plasma to determine if chemotactic activity for guinea pig peritoneal macrophages [6] and human peripheral blood monocytes is present. Substantial and similar amounts of chemotactic activity for macrophages [6] and human monocytes are found in both serum and plasma (table I). However, excessive amounts of heparin inhibit the chemotactic activity of both serum and plasma for guinea pig macrophages [6] and human monocytes (table II). These findings suggest that the amount of anticoagulant used in the formation of plasma could account for the low chemotactic activity of plasma reported by others [2, 3]. Characterization of the chemotactic activity in untreated guinea pig serum revealed it to be heat-labile (56 °C for 30 min) and to have a molecular weight of 90,000 [6].

The existence of chemotactic activity in normal serum and plasma suggests that similar activity may be circulating in whole blood, a puzzling phenomenon if chemotaxis is, in fact, the mechanism by which cells accumulate at local inflammation sites. Alternatively this activity may not be present in whole blood but may be produced by the conversion of whole blood to

Table II. Effect of heparin concentration on chemotactic activity of human serum and plasma

Heparin added U[2]	Chemotactic activity[1]	
	serum	plasma
None	59.7 ± 12.9	51.1 ± 0.4
10	40.4 ± 2.6	49.8 ± 6.9
50	27.2 ± 6.9	37.4 ± 2.6
100	27.2 ± 2.6	25.3 ± 3.4
500	10.5 ± 3.1	20.0 ± 0.4

1 Chemotactic activity is expressed as the mean and standard error of migrating mononuclear leukocytes per oil immersion field ($\times 1,445$). Background chemotactic activity of medium alone (11.6) was subtracted from all values.
2 The indicated amount of heparin contained in 0.1 ml gelatin veronal buffer was added to 0.05 ml serum or plasma (5 U heparin/ml) contained in 0.9 ml gelatin veronal buffer. All samples were brought to 1.7 ml with Gey's balanced salt solution [20].

serum or plasma. Such a conversion might activate the complement, kinin and/or clotting systems whose activated components possess a range of biologic activities, among them chemotactic activity [13]. We used various methods in an attempt to inhibit activation of these systems during the formation of plasma to determine how chemotactic activity found in plasma would be affected. Guinea pigs were bled directly into plastic tubes containing EDTA, citrate, heparin or hexadimethrene. Other blood samples were collected at 0 °C and formed elements removed immediately by centrifugation in the cold. Hexadimethrene, an inhibitor of the activation of Hageman factor [14], was also directly inhibitory to macrophage chemotaxis [SNYDERMAN and HAUSMAN, unpublished data], and experiments to determine if this agent inhibited the formation of chemotactic activity in plasma were, therefore, inconclusive. In all other cases the resultant plasma contained the same amount of chemotactic activity normally found in fresh serum [6]. Although these experiments suggest that the complement and clotting systems need not be activated to produce the chemotactic activity present in normal serum, partial activation may have occurred despite the use of various inhibitors. Moreover, activation of the kinin-forming system may be responsible for the generation of this activity since the conversion of prekallikrein to kallikrein does result in the formation of chemotactic activity for neutrophils [8] and human blood monocytes [SNYDERMAN et al., unpublished data].

In summary, the origin of the chemotactic activity in fresh serum and plasma remains undetermined. If this activity is present in whole blood, it could function to inhibit migration of monocytes out of the circulation until chemotactic activity at local inflammatory sites reaches a concentration above that found in whole blood. Thus, chemotactic activity in the circulation could have a modulating effect on the inflammatory response. If, on the other hand, this chemotactic activity is generated upon the conversion of whole blood to serum, it could play an important role in wound-healing by attracting macrophages to sites of coagulation at areas of tissue injury.

II. Chemotactic Activity Produced by Lymphocytes

Cellular immune reactions, like immune reactions in general, enable the host to recognize and eliminate non-self. The recognition phase is highly specific, and in cellular immunity is mediated by small lymphocytes. When small lymphocytes interact with non-self, they initiate an inflammatory reaction which is no longer specific in that not only the initiating antigen but other antigens may be destroyed. For example, the introduction of tumor cells into sites of ongoing cellular-immune inflammation can result in their death even though the inflammation was initiated in response to an unrelated antigen [33]. An important effector cell which mediates the destruction of foreign material in the cellular immune response is the macrophage. The observation that lymphocytes release biologically active effector molecules after interaction with antigen has greatly enhanced the understanding of how macrophages may accumulate at sites of cellular immune reactions [11]. One important observation was that guinea pig lymphocytes, when stimulated with specific antigen, release a factor which is chemotactic for homologous macrophages [30]. This factor was reported to be separable from migration inhibition factor [31].

Our laboratory has focused upon mechanisms of macrophage accumulation in human cellular immune inflammation [1, 17–19]. Human peripheral blood leukocytes have been isolated from normal individuals and from individuals of known sensitivity to tuberculoprotein (PPD). The incubation of normal leukocytes with nonspecific mitogen (phytohemagglutinin [PHA]) or of leukocytes from tuberculin-sensitive patients with specific antigen or

Table III. Chemotactic activity in supernatants of phytohemagglutinin (PHA) stimulated human leukocyte cultures for human mononuclear leukocytes[1]

Material tested[3]	Chemotactic activity[2]		
	experiment I	experiment II	experiment III
1. Supernatants of leukocytes incubated with PHA (0.9 μg/ml)	225 ± 19.0	76 ± 3.0	150 ± 13.8
2. Supernatants of leukocytes incubated with media alone	61 ± 4.7	16 ± 9.4	$-$[4]
3. PHA control (0.9 μg/ml)	41 ± 5.1[5]	10 ± 0.9[5]	39 ± 2.1[6]
4. Media alone	42	6 ± 1.8	19 ± 1.3

1 From reference 18.
2 Each value represents the mean and standard error of triplicate samples expressed as cells per oil field.
3 0.5 ml of the indicated material was brought to 1.0 ml with gelatin veronal buffer (pH 7.3) then to 1.7 ml with Gey's media.
4 Not done.
5 PHA in media not incubated with leukocytes.
6 PHA was added to the leukocyte cultures at time of final centrifugation.

nonspecific mitogen results in the production of chemotactic activity for autologous or homologous monocytes (table III, IV) [1, 18]. Since cultures of column-purified lymphocytes stimulated by a nonspecific mitogen produce as much chemotactic activity as similarly stimulated leukocyte cultures containing lymphocytes, macrophages and neutrophils, the chemotactic activity must be produced by lymphocytes [1]. Studies of the kinetics of lymphocyte-derived chemotactic factor (LDCF) production demonstrated that the factor is released prior to blastogenesis, and that indeed blastogenesis is not necessary for its production [1]. Release of chemotactic activity from lymphocytes can be detected within 6 h after stimulation with PHA or specific antigen *in vitro*. Human LDCF was characterized by molecular sieve chromatography and sucrose density gradient ultracentrifugation and found to have a molecular weight of approximately 12,000. In addition, LDCF is antigenically distinct from the C5a fragment of the complement system [1].

Table IV. Chemotactic activity in supernatants of tuberculoprotein (PPD) stimulated leukocyte cultures for human mononuclear leukocytes[1]

Material tested[3]	Chemotactic activity[2]	
	experiment I (PPD, 0.25 μg/ml)	experiment II (PPD, 1 μg/ml)
1. Supernatants of leukocytes incubated with PPD	120 ± 18.3	273 ± 6.0
2. Supernatants of leukocytes plus PPD added at harvest	22 ± 6.9	36 ± 2.6
3. Media alone	19 ± 3.0	41 ± 6.1

1 From reference 18.
2 Each value represents the mean and standard error of triplicate samples expressed as cells per oil field.
3 0.5 ml of the indicated material was brought to 1.0 ml with gelatin veronal buffer (pH 7.3), then to 1.7 ml with Gey's media.

Current studies investigating the production of lymphokines *in vivo* have demonstrated the appearance of macrophage chemotactic activity after the introduction of a specific antigen into the peritoneal cavities of sensitized guinea pigs. When animals were immunized with an antigen which induced primarily cellular immunity and were challenged with this antigen intraperitoneally, the majority of the chemotactic activity recovered was not C5a and had physical characteristics in common with LDCF [12].

The assay for chemotaxis of human monocytes has proved to be a valuable technique for at least two reasons. Firstly, it is a very sensitive indicator of the production of a human lymphokine. LDCF can be detected in unconcentrated leukocyte supernatants and in supernatants of leukocytes cultured in the absence of added plasma [18]. The absence of contaminating factors in added plasma will further allow the biochemical characterization of this important human lymphokine. Secondly, the monocyte chemotaxis assay has permitted evaluation of lymphokine production and monocyte chemotactic responsiveness in patients with various disease states to determine in which conditions defects in the effector limb of the cellular immune response exist. A previously unrecognized immune dysfunction, namely defective monocyte chemotactic responsiveness, has recently been defined [17]. This defect was found in a patient with chronic mucocutaneous candidiasis

Table V. The effect of transfer factor therapy on the chemotactic responsiveness of monocytes from a patient with chronic mucocutaneous candidiasis[1]

Date	Chemotactic index, %[2]	
	C5a	LDCF
06.25.71	11.1	—[3]
07.02.71	8.1	21.7
07.19.71[4]	2.0	19.4
08.09.71	33.0	26.6
10.27.71	64.4	38.7
01.26.72	48.6	—
04.19.72	76.4	79.6

LDCF = Lymphocyte-derived chemotactic factor.

1 From reference 17.

2 $\dfrac{\text{Mean chemotactic activity of patient's mononuclear leukocytes}}{\text{Mean chemotactic activity of normal mononuclear leukocytes}} \times 100.$

3 Not done.

4 Transfer factor therapy begun and continued for the duration of these experiments.

who had cutaneous anergy to a wide variety of antigens. Candida skin tests, in particular, failed completely to elicit an inflammatory reaction. This patient's monocytes *in vitro* were markedly unresponsive to preformed chemotactic factors. In initial studies the patient's mononuclear leukocyte chemotactic responsiveness *in vitro* to C5a was 2–11% of normal and to LDCF 19–22% of normal. After treatment with transfer factor, the patient's monocyte response to both C5a and LDCF improved to 75–80% of normal (table V). This improvement was initially accompanied by a definite though partial clinical remission as well as by the development of a positive skin test for candida antigen.

It is not yet clear how transfer factor affected the reversal of monocyte unresponsiveness but these studies indicate that cutaneous anergy can be caused by defective monocyte chemotaxis. In the future, therefore, complete evaluation of cellular immune function will require the quantitation of monocyte chemotactic function. To this end our laboratory, in collaboration with Dr. BETCHER and Dr. LEONARD at the National Institutes of Health and Dr. HAUSMAN and Dr. BROSMAN at the UCLA Medical Center, has begun a study of monocyte chemotactic responsiveness in patients with tumors. We hope to determine if subtle defects in monocyte function contribute to the tumor-bearing patient's inability to reject his own tumors. It is also apparent that similar studies can be performed in patients with increased susceptibility to infection and particularly those with cellular immune anergy.

Summary

Macrophage chemotaxis can be mediated by factors produced by both cellular and humoral immune reactions as well as by nonspecific tissue injury. C5a, a factor chemotactic for both PMN and mononuclear leukocytes *in vitro*, may mediate leukocyte accumulation at local inflammatory sites duiing humoral immune reactions. This factor is released when the complement system is activated by antigen-antibody complexes or endotoxin, or when C5 itself is cleaved directly by the action of proteolytic enzymes derived from lysed cells. Macrophage accumulation at sites of nonspecific tissue damage may be similarly mediated by the effects of cell lysis upon the complement system, and some evidence suggests that the clotting and kinin systems, which are activated at wound sites, could supply mediators of chemotaxis.

Work *in vitro* prompts the conclusion that macrophages at cellular immune sites have accumulated in response to a chemotactic lymphokine released by sensitized lymphocytes stimulated by specific antigen. The LDCF is distinct from C5a and from the chemotactic activity present in normal serum and plasma. The discovery of a new immune dysfunction, namely defective monocyte chemotaxis, emphasizes the importance of quantitating monocyte chemotactic responsiveness in evaluating patients with impaired resistance to infection or neoplasia. Assays *in vitro* of mononuclear leukocyte chemotactic responsiveness and chemotactic factor production may enable us to determine whether impaired resistance to infection or neoplasia results from abnormal chemotactic factor production or monocyte chemotactic responsiveness.

References

1 ALTMAN, L. C.; SNYDERMAN, R.; OPPENHEIM, J. J., and MERGENHAGEN, S. E.: A human mononuclear leukocyte chemotactic factor. Characterization, specificity, and kinetics of production by homologous leukocytes. J. Immunol. *110:* 801–810 (1973).

2 BOREL, J. F.: Studies on chemotaxis. Effect of subcellular leukocyte fractions on neutrophils and macrophages. Int. Arch. Allergy *39:* 247–271 (1970).

3 BOREL, J. F. and SORKIN, E.: Differences between plasma and serum mediated chemotaxis of leukocytes. Experientia *25:* 1333–1335 (1969).

4 BOYDEN, S.: The chemotactic effect of mixtures of antibody and antigen on polymorphonuclear leukocytes. J. exp. Med. *115:* 453–466 (1962).

5 COHN, Z. A.: The structure and function of monocytes and macrophages. Adv. Immunol. *9:* 163–214 (1968).

6   HAUSMAN, M. S.; SNYDERMAN, R., and MERGENHAGEN, S. E.: Humoral mediators
    of chemotaxis of mononuclear leukocytes. J. infect. Dis. *125:* 595–602 (1972).

7   JENSEN, J.; SNYDERMAN, R., and MERGENHAGEN, S. E.: Chemotactic activity. A prop-
    erty of guinea pig C'5 anaphylatoxin. 3rd Int. Congr. of Allergy and Anaphylaxis,
    Basel 1969, pp. 265–278 (Karger, Basel 1969).

8   KAPLAN, A. P.; KAY, A. B., and AUSTEN, K. F.: A prealbumin activator of pre-
    kallikrein. III. Appearance of chemotactic activity for human neutrophils by the
    conversion of human prekallikrein to kallikrein. J. exp. Med. *135:* 81–97 (1972).

9   KELLER, H. U. and SORKIN, E.: Studies on chemotaxis. VI. Specific chemotaxis in
    rabbit polymorphonuclear leucocytes and mononuclear cells. Int. Arch. Allergy *31:*
    575–586 (1967).

10 ODLAND, G. and ROSS, R.: Human wound repair. I. Epidermal regeneration. J. Cell
    Biol. *39:* 135–151 (1968).

11 PICK, E. and TURK, J. L.: The biological activities of soluble lymphocyte products.
    Clin. exp. Immunol. *10:* 1–23 (1972).

12 POSTLETHWAITE, A. and SNYDERMAN, R.: Mononuclear leukocyte chemotactic
    activity *in vivo* in delayed hypersensitivity. Fed. Proc. *31:* 988 (1973).

13 RATNOFF, O.: Some relationships among hemostasis, fibrinolytic phenomena, im-
    munity, and the inflammatory response. Adv. Immunol. *10:* 146–227 (1969).

14 RATNOFF, O. D. and MILES, A. A.: The induction of permeability-increasing
    activity in human plasma by activated Hageman factor. Brit. J. exp. Path. *45:* 328–345
    (1964).

15 SHIN, H. S.; SNYDERMAN, R.; FRIEDMAN, E.; MELLORS, A., and MAYER, M. M.:
    Chemotactic and anaphylatoxic fragment cleaved from the fifth component of
    complement. Science *162:* 361–363 (1968).

16 SIMPSON, D. M. and ROSS, R.: Wound healing in neutropenic guinea pigs treated
    with anti-neutrophil serum. Fed. Proc. *30:* 569 (1971).

17 SNYDERMAN, R.; ALTMAN, L. C.; FRANKEL, A., and BLAESE, R. M.: Defective
    mononuclear leukocyte chemotaxis. A previously unrecognized immune dysfunction.
    Studies in a patient with chronic mucocutaneous candidiasis. Ann. intern. Med. *78:*
    509–513 (1973).

18 SNYDERMAN, R.; ALTMAN, L. C.; HAUSMAN, M. S., and MERGENHAGEN, S. E.:
    Human mononuclear leukocyte chemotaxis. A quantitative assay for humoral and
    cellular chemotactic factors. J. Immunol. *108:* 857–860 (1972).

19 SNYDERMAN, R.; ALTMAN, L. C., and MERGENHAGEN, S. E.: Human mononuclear
    leukocyte chemotaxis. Definition of two chemotactic factors and a previously un-
    recognized immune dysfunction. Proceedings on 'non-specific' mediators of in-
    flammation (in press).

20 SNYDERMAN, R.; GEWURZ, H., and MERGENHAGEN, S. E.: Interactions of the comple-
    ment system with endotoxic lipopolysaccharide. Generation of a factor chemotactic
    for polymorphonuclear leukocytes. J. exp. Med. *128:* 259–275 (1968).

21 SNYDERMAN, R.; PHILLIPS, J. K., and MERGENHAGEN, S. E.: Polymorphonuclear
    leukocyte chemotactic activity in rabbit serum and guinea pig serum treated with
    immune complexes. Evidence for C5a as the major chemotactic factor. Infect. Im-
    mun. *1:* 521–525 (1970).

22 SNYDERMAN, R.; PHILLIPS, J. K., and MERGENHAGEN, S. E.: Biological activity of

complement *in vivo*. Role of C5 in the accumulation of polymorphonuclear leuko-cytes in inflammatory exudates. J. exp. Med. *134:* 1131–1143 (1971).

23 SNYDERMAN, R.; SHIN, H. S., and DANNENBERG, A. M.: Macrophage proteinase and inflammation. The production of chemotactic activity from the fifth component of complement. J. Immunol. *109:* 896–898 (1972).

24 SNYDERMAN, R.; SHIN, H. S., and HAUSMAN, M. H.: A chemotactic factor for mononuclear leukocytes. Proc. Soc. exp. Biol. Med. *138:* 387–390 (1971).

25 SNYDERMAN, R.; SHIN, H. S.; PHILLIPS, J. K.; GEWURZ, H., and MERGENHAGEN, S. E.: A neutrophil chemotactic factor derived from C'5 upon interaction of guinea pig serum with endotoxin. J. Immunol. *103:* 413–422 (1969).

26 SNYDERMAN, R.; WOHLENBURG, C., and NOTKINS, A. L.: Inflammation and virus infection. Chemotactic activity resulting from the interaction of antiviral antibody and complement with cells infected with herpes simplex virus. J. infect. Dis. *126:* 207–209 (1972).

27 TAUBMAN, S. B.; GOLDSCHMIDT, P. R., and LEPOW, I. H.: Effects of lysosomal enzymes from human leukocytes on human complement components. Fed. Proc. *29:* 434 (1970).

28 WARD, P. A.: Chemotaxis of mononuclear cells. J. exp. Med. *128:* 1201–1221 (1968).

29 WARD, P. A. and HILL, J. H.: C5 chemotactic fragments produced by an enzyme in lysosomal granules of neutrophils. J. Immunol. *104:* 535–543 (1970).

30 WARD, P. A.; REMOLD, H. G., and DAVID, J. R.: Leukotactic factor produced by sensitized lymphocytes. Science *163:* 1079–1081 (1969).

31 WARD, P. A.; REMOLD, H. G., and DAVID, J. R.: The production by antigen-stimulated lymphocytes of a leukotactic factor distinct from migration inhibitory factor. Cell. Immunol. *1:* 162–174 (1970).

32 WILKINSON, P. C.; BOREL, J. F.; STECHER-LEVIN, V. J., and SORKIN, E.: Macrophage and neutrophil specific chemotactic factors in serum. Nature, Lond. *222:* 244–247 (1969).

33 ZBAR, B.; WEPSIC, H. T.; RAPP, H. J.; STEWART, L. C., and BORSOS, T.: Tumor-graft rejection in syngeneic guinea pigs. Evidence for a two-step mechanism. J. nat. Cancer Inst. *44:* 473–481 (1970).

Authors' address: Dr. R. SNYDERMAN and C. E. STAHL, Division of Rheumatic and Genetic Diseases, Departments of Medicine and Immunology, Duke University Medical Center, *Durham, NC 27710* (USA)

Antibiotics and Chemotherapy, vol. 19, pp. 296–332
(Karger, Basel 1974)

The Nature of a Mediator of
Leucocyte Chemotaxis in Inflammation

H. Hayashi, M. Yoshinaga and S. Yamamoto

Department of Pathology, Kumamoto University Medical School, Kumamoto

I. Introduction

In spite of the importance of leucocyte emigration in inflammation, the problem of its mediation had remained confused for many years; there had been a voluminous literature concerned with endogenous substances which might be responsible for emigration of leucocytes from inflamed vessels, but most of this work is now of historical interest only. The reason for its irrelevance seemed to lie to a large extent in the unsatisfactory methods used to demonstrate stimulation of leucocyte movement [1]. Following the introduction of the Boyden technique for measuring leucotaxis [2] numerous investigations aimed at determining the naturally occurring chemotactic host factor for leucocytes; thus, different types of chemotactic factors have been proposed from many laboratories. However, present knowledge on the natural mediators for leucocyte emigration in inflammation still seems incomplete.

One line of investigation aimed at elucidating the role of the complement system in chemotaxis [3–7]. These observations were largely based on *in vitro* experiments, but in some instances attempts were made to isolate chemotactic factors from inflammatory tissue [8–11].

A second line of investigation concerned with leucocyte chemotaxis has been concentrated on the discovery and characterization of chemotactic factors present in inflammatory tissue [12, 13]. While this type of approach is essential to clarify the mediation of inflammatory leucotaxis, there remains the difficult problem of how to obtain sufficient chemotactic factor from tissues.

The third line of work relates to the study of bacterial metabolites or products [14–17]. The observations were usually made *in vitro* and seemed to be helpful in explaining leucotaxis in the infected but not in the non-infected tissue.

Research in this laboratory has largely concentrated on the second approach mentioned above. According to our view, any chemotactic factor claimed to be involved in inflammatory leucotaxis should satisfy the following criteria [12, 13].

1. The chemotactic factor is locally available to induce the inflammatory leucotaxis.

2. The amount (or activity) of the chemotactic factor parallels the time-course of the leucotaxis.

3. The chemotactic factor, when injected in concentrations reasonably comparable to those detected in the inflamed site, can produce morphologic changes similar to the inflammatory reaction.

4. The action of chemotactic factor is specific, being active in causing leucocyte emigration but not in inducing vascular permeability change and hemorrhagic change.

5. The chemotactic factor is inhibited by the specific antagonistic substance locally available.

6. Inflammatory leucotaxis is suppressed by a specific antagonistic substance.

7. Depletion of chemotactic factor or its precursor can cause a decrease in the leucotaxis.

8. The precursor of chemotactic factor or the enzyme associated with the chemotactic generation is locally available.

II. Leucoegresin, a Chemotactic Factor for Neutrophilic Polymorphonuclear (PMN) Leucocytes from Inflammatory Tissue

A. Time-Course of Leucocyte Emigration in Inflammatory Process

As is well-known, increased vascular permeability and leucocyte emigration are consistent and significant events in inflammation. In general, the vascular permeability change precedes PMN leucocyte emigration; and those emigrated PMN leucocytes often become replaced by mononuclear cells (i.e., monocytes, macrophages and their derivatives) as well as by lymphocytes, though there are some modifications according to the nature of inflammatory stimuli. Up till quite recently it was assumed that leucocyte emigration

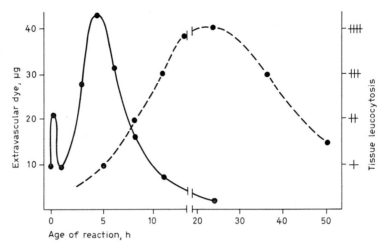

Fig. 1. Different time-course in increased vascular permeability and polymorpho-nuclear (PMN) leucocyte emigration in cutaneous Arthus reaction in the rabbit. Inflam-mation was induced by intradermal injection of 2.5 mg bovine serum albumin (BSA) into sensitized rabbit. Vascular permeability change was shown by the amount of pont-amine blue extracted from the skin site, and PMN infiltration was graded histologically. Similar results were observed in the burned skin lesions of rabbits and rats. Vascular permeability change was followed by PMN emigration. ——— = Extravascular dye; ---- = tissue leucocytosis.

occurred at the same time and from the same blood vessels as did leakage of plasma protein. It was shown, however, that after injury, leucocyte emigration did not commence until much later than increased vascular permeability. The time-course of these inflammatory events was distinctive, suggesting different chemical mediators responsible for each event.

As a satisfactory model of inflammation, the Arthus skin lesions were induced by intradermal injection of bovine serum albumin (BSA, 2.5 mg) on the back of sensitized rabbits [18]. Macroscopically, the reactions such as diffuse edema and erythema commenced in about 90 min were maximal in about 12 h and then declined in intensity. Moderate to severe hemorrhage occurred in the lesion center. The vascular permeability change was shown by the amounts of pontamine sky blue extracted from inflamed site, and PMN leucocyte emigration was graded histologically. Similarly, burned skin lesions of rabbits and rats induced by heating at 56 °C for 20 sec were studied as the second model of inflammation [19].

As shown in figure 1, the increased vascular permeability was constantly followed by extensive emigration of PMN leucocytes [13]. Vascular permeability change appeared to be biphasic; immediate and delayed [19]. The immediate vascular response was clearly mediated by an antihistamine-susceptible substance (possibly histamine). By contrast, the delayed vascular response was apparently associated with an antihistamine-unsusceptible substance; and the possible permeability factor was extracted in the pseudo-globulin fraction of the inflamed site during the period of delayed vascular response, and then purified by chromatography using DEAE-cellulose, hydroxylapatite and DEAE-Sephadex G-50 [20–22]. The substance was a heat-stable peptide with a long-lasting effect and its molecular weight was approximately 7,000; it was quite different from the plasmakinins. The peptidic permeability factor was termed *vasoexin* [12, 23], because a similar permeability factor was isolated from various sources of different types of inflammation in animal and man.

By the mechanism of increased vascular permeability, various types of plasma protein were exuded into the extravascular spaces. As described below, immunoglobulin G (IgG) among plasma proteins exuded was locally converted to a chemotactic factor for neutrophilic PMN leucocytes.

B. Isolation and Purification of Leucoegresin from Inflammatory Tissue

The skin bearing Arthus lesions at 4.5, 12, 24 and 48 h after antigen (BSA) injection was excised, frozen and cut into slices about 50 μm thick with a freezing microtome. After dehydration with cold acetone, the powdered skin was extracted with M/15 phosphate buffer (pH 7.4) at 4 °C for 4 h. From the extracts, three protein fractions containing euglobulin, pseudo-globulin and albumin were prepared with ammonium sulfate and then desalted through a column of Sephadex G-50 [12, 24, 25].

The capacity to promote leucocyte emigration was measured *in vitro* by a modification of BOYDEN's method [2], with millipore filters (pore size 0.65 μm) [25, 26]. Modified stainless steel chambers consisting of two 1-ml compartments were used. The test samples, in M/15 phosphate buffer (pH 7.4) were placed in the lower compartments. PMN leucocytes (1.5×10^6/ml), were collected from the peritoneal exudates produced by injection of gly-cogen. They were suspended in Gey's buffered solution containing 2% human serum albumin (pH 7.4) and placed in the upper compartment.

The chemotactic activity of the extracts was predominantly concentrated in the pseudoglobulin fraction. The specific activities (PMN count/E $_{280nm}$) of pseudoglobulin fractions from Arthus lesions at 4.5, 12, 24 and 48 h

Table I. Procedures for isolation and purification of leucoegresin

	Inflamed skin	
	50-μm slices powdered with cold acetone	
	extracted with M/15 phosphate buffer	
	(pH 7.4) at 4°C for 4 h	
	Extract (SPA = 4)	
	fractionated with ammonium sulfate	
Albumin fraction	Pseudoglobulin fraction	Euglobulin fraction
	desalted on Sephadex G-50 (SPA = 24)	
	eluted in M/15 phosphate buffer (pH 7.4)	
	on DEAE-Sephadex A-50 (SPA = 240)	
	eluted in the first peak of linear gradient elution	
	with NaCl on CM-Sephadex C-50[1]	
	(SPA = 1,450)	
	eluted in M/10 glycine-HCl buffer (pH 3.0)	
	on anti-IgG affinity column (SPA = 1,500)	
	Leucoegresin purified	

1 If necessary, this step was repeated until the elution profile became homogeneous.
SPA = specific activity, numbers of polymorphonuclear leucocytes (PMN) migrated/
E 280 nm.

were respectively 5, 24, 14 and 10 [24]. The degrees of tissue leucocytosis observed in lesions at the different times corresponded roughly to the activities of the fractions, suggesting that the chemotactic factor in the pseudoglobulin fraction may be a mediator of inflammatory leucotaxis.

As summarized in table I, for purification the active pseudoglobulin fraction was eluted on DEAE-Sephadex A-50, CM-Sephadex C-50 [25], and antirabbit IgG affinity column [27] in that order. IgG affinity column was prepared with goat antirabbit IgG γ-globulin which was conjugated with Sepharose 4B by means of cyanogen bromide.

The chemotactic factor was also isolated from burned skin lesions and purified in the same way as described above; it was indistinguishable from the chemotactic factor obtained from Arthus skin lesions as noted below. The substance was therefore named *leucoegresin* [24]. The amounts of leucoegresin isolated were approximately 250 μg per Arthus site (12 h old) showing maximal PMN emigration, and 50 μg per burned site (4 h old) with maximal leucotaxis. Such difference in amounts of leucoegresin was reasonable, because PMN infiltration in the Arthus reaction was more intensive than that in the burned lesion.

C. Physico-Chemical and Immunological Assay of Leucoegresin

The chemotactic factor obtained was further eluted on Sephadex G-200; the elution profile of the protein showed a symmetric homogeneous pattern, and the chemotactic activity of the protein paralleled the absorbancy at 280 nm. The molecular weight of the chemotactic factor was approximately 140,000 when measured by gel filtration on Sephadex G-200. Its homogeneity was further confirmed by boundary electrophoresis and ultracentrifugation; and no heterogeneity was detected on Schlieren profiles during the electrophoresis between pH 2 and 9, and also no heterogeneity on the Schlieren profiles during sedimentation at 59,780 r.p.m. up to 120 min. Its isoelectric point was pH 5.0 and sedimentation coefficient (Sw20) was 6.58 S [25].

The chemotactic factor was nondiffusible and relatively thermostable. It showed an ultraviolet absorption maximally at 280 nm, a positive reaction with Folin-phenol reagent and with biuret, indicating that this substance was a protein. No nucleic acid and lipid were detectable.

In further experiments, antisera against rabbit serum, skin extract of Arthus lesions (12 h old) and purified leucoegresin were prepared in goats. From these antisera, γ-globulin fractions were separated by fractionation with sodium sulfate. Leucoegresin was examined with antileucoegresin or antiskin extract in agar immunoelectrophoresis and immunodiffusion. Antileucoegresin was also tested with Arthus skin extracts. In all cases, only a single precipitin line was revealed, and it was confirmed that the chemotactic factor obtained did not contain any immunologically detectable impurity.

D. Physico-Chemical Comparison of Leucoegresin and Other Chemotactic Factors

Leucoegresin seemed difficult to relate with complement or its products. The chemotactic activity of leucoegresin was not influenced by n-CBZ-α-glutamyl L-tyrosine (M/100 and M/50), which inactivated the chemotactic complex of complement, i.e. C567 [28]. The molecular weight of the trimolecular complex was more than 300,000, but that of leucoegresin was about 140,000. The molecular weight of previously described complement-derived chemotactic factors was clearly smaller than that of leucoegresin. It was of interest to note that the chemotactic factor from chick embryos carrying a viral infection had a molecular size similar to that of leucoegresin [10]. The comparison in the molecular weight and

Table II. Comparison of the molecular weight of chemotactic factors for PMN leucocytes

Chemotactic factors	Molecular weight	Remarks	Reference No.
Exogenous			
Diplococcus factor	<3,600	metabolite	15
Mycobacterial factor	<3,000	bacterial cell constituent	17
Staphylococcal factor	>20,000	culture filtrate	16
Endogenous			
Leucoegresin	140,000	IgG-derived, from inflammatory tissue	23,24
C_{567} complex	>300,000	activation of fresh serum	3
C_3 fragment	6,000	by plasmin	4
C_3 fragment	7,000	by C_3 convertase	30
C_5 fragment	8,500	by trypsin	5
C_5 fragment	70,000	from inflammatory lesion	9
Unknown	5,000–35,000	normal serum component	31
Unknown	150,000	chick embryo with viral infection	10

some remarks of previously reported chemotactic factors are summarized in table II. Other chemotactic factors were not referred to in the table because their molecular weights have not yet been measured. The molecular weights of chemotactic factors derived from C_5 were reported to range from 8,000 to 23,000 and even up to 70,000 [9, 29].

III. Biological Action of Leucoegresin

A. Macro- and Microscopic Effect of Leucoegresin

Biological action of leucoegresin was assayed in rabbit, guinea pig, rat and mouse [13, 25]. Rabbit leucoegresin (20 and 50 μg in M/15 phosphate buffer, pH 7.4) was intradermally injected into the left flanks of these animals; the same volumes of permeability factor (vasoexin) into the right flanks. At various intervals from 0 to 60 min after intradermal injection, the animals

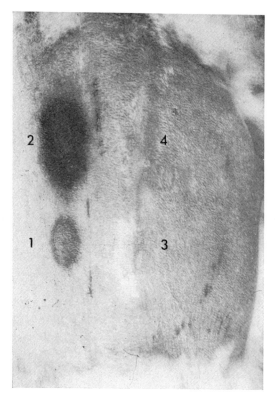

Fig. 2. Macroscopic reaction after intradermal injections of vasoexin and leucoegresin in rabbit. Skin sites (1, 2) injected with vasoexin (20 and 50 μg) showed marked edematous change accompanied by pontamine blue leakage. Skin sites (3, 4) injected with leucoegresin (20 and 50 μg) showed no inflammatory response. The photograph was taken 2 h after intradermal injections.

were given pontamine sky blue intravenously and then the blueing of the injected skin sites were observed for 3–4 h.

As illustrated in figure 2, the skin sites injected with vasoexin showed marked edema which was accompanied by strong dye leakage; the intensity of edema and blueing was proportional to the concentrations of vasoexin injected. Microscopically, such skin sites showed edematous change with slight deposition of eosinophilic amorphous substance in the intercellular spaces and slight eosinophilic swelling of collagen fibers, but PMN leucocyte emigration was only slight during the period of observation.

In contrast, the skin sites injected with leucoegresin showed no macro-

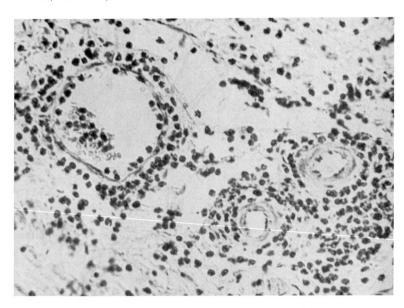

Fig. 3. Pronounced PMN emigration and infiltration 3 h after intradermal leucoegresin (50 μg). The cells, emigrated through the venular wall, gathered around the
arterioles and some entered the lymphatic vessel. No cells other than PMN leucocytes
were observed. H and E. ×215.

scopic change during the period of observation; the substance did not
increase vascular permeability to circulating pontamine blue. As shown in
figure 3, intradermal leucoegresin induced pronounced PMN emigration
followed by widespread infiltration in a diffuse and focal (perivascular)
form throughout the skin. PMN leucocytes were found sticking to the
endothelium of the venules within 15–30 min of intradermal injection and
emigrating through the venular walls after 30–60 min. PMN emigration
became maximal at 3–4 h after injection. No or little emigration of cells
other than neutrophilic PMN leucocyte was revealed, indicating the specific
action of leucoegresin for the cells. Since leucoegresin was isolated in amounts
of 250 μg/12-hour-old Arthus site, this amount was administered intradermally. The macroscopic response was not observed in the skin sites
injected, but the microscopic response was characterized by clearly more
intense infiltration of PMN leucocytes, resembling Arthus leucotaxis. The
amount of leucoegresin was about 50 μg/4-hour-old burned site; the intensity
of PMN infiltration induced by this amount was similar to that in thermal
leucotaxis.

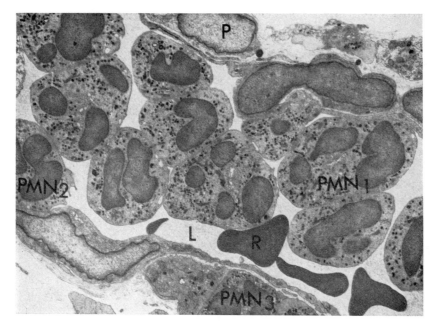

Fig. 4. Characteristic accumulation of a number of neutrophilic PMN leucocytes in the lumen (L) of the venule in the cremaster muscle of the rat. The cells adhered loosely to each other. PMN_1 and PMN_2 adhered to the endothelial surface. PMN_3 had already passed through the venular wall. R = red blood cell; P = periendothelial cell. The photograph was taken 2 h after injection of leucoegresin. × 3,300.

The site of PMN emigration was found only at the site of the venules, but not of the capillaries when assayed on the blood vessels traced by carbon particles given intravenously in the panniculus carnosus muscle of rabbit, which was removed from the skin [32]. No cells other than neutrophilic PMN leucocytes were found in the preparations.

The effect of leucoegresin was also tested in the skin of guinea pig, rat and mouse. Intradermal injection of this chemotactic factor (50 μg) induced considerable PMN emigration followed by a widespread infiltration, though the intensity was found to be less marked. No macroscopic response was found in the skin sites of these animals. These observations indicated that rabbit leucoegresin was active not only for PMN leucocytes of rabbit but also for those of guinea pig, rat and mouse. By cinemicrophotography [33], a detailed sequence of neutrophilic PMN emigration was recorded when leucoegresin from different sources was applied to the mesentery of the guinea-pig. PMN emigration occurred only at the site of the venules.

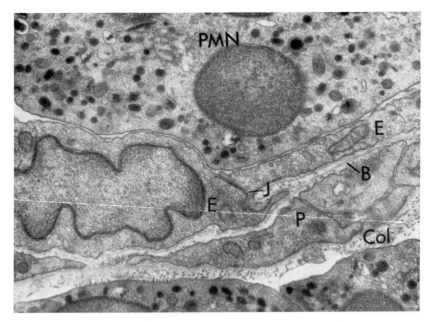

Fig. 5. Adhering of PMN leucocyte to the endothelial cells (E); it became flattened against the endothelial cells and increased the area of contact, leaving a narrow space (100–400 Å). The cell started to extend a pseudopod into the intercellular junction (J). Demonstrable structural change was not found in the endothelial cells. B = basement membrane; P = periendothelial cell; Col = collagen fibers. The photograph was taken 2 h after leucoegresin. × 18,000.

B. Electron-Microscopic Study of Leucoegresin Effects

The morphological sequence of emigration of PMN leucocytes by leucoegresin (50–70 μg) was studied on the cremaster muscles of rat and rabbit by means of electron microscopy [34].

The emigration of the cells was essentially found at the site of the venules 30–50 μm in diameter. As can be seen in figure 4, PMN leucocytes were characteristically accumulated in the lumina of the venules as the first step of emigration, and then adhered to the endothelial surfaces. As shown in figure 5, the cell frequently became flattened against the vessel wall and increased the contact surface, leaving a narrow space (100–400 Å in distance) showing no particular electron density. The sticking cell developed a small pseudopod from the flattened cytoplasm, and inserted it into the junction of two adjacent endothelial cells. As illustrated in figure 6, as the cell started

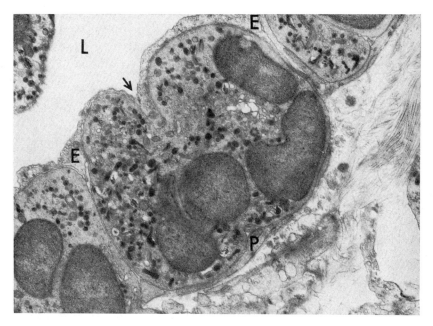

Fig. 6. The PMN leucocyte was perfectly enveloped by marginal fold-like cytoplasmic processes from two adjacent endothelial cells (E) by which the cell was fixed on
the endothelial lining against the blood stream. A hollow in the cell at the luminal side
(as shown by an arrow) suggested the existence of the mechanical force of these cytoplasmic processes. PMN emigration appeared to be prevented temporarily by the periendothelial cell (P). L = lumen. × 9,100.

its passage through the intercellular junction by protruding cytoplasm, the
marginal fold-like processes were formed from the endothelial cytoplasm,
and the cell was enveloped partially or completely. The appearance of such
characteristic cytoplasmic processes was noted only in those endothelial
cells with PMN leucocytes adhering to them, suggesting reasonable fixation
of the cell to the endothelial surface and successful prevention of pushing
by the bloodstream. While the PMN leucocyte was completely surrounded
by such cytoplasmic processes at the luminal side and intercepted from the
lumen, the cell itself penetrated by ameboid motion into the intercellular
junction, and then reached the space (periendothelial sheath) lying between
the endothelial cells and periendothelial cells. As illustrated in figure 7, as
the PMN leucocyte moved, the intercellular junction was promptly re-
formed by the cytoplasmic processes described above; and no free endothelial
gap communicating from the lumen to the periendothelial sheath throughout

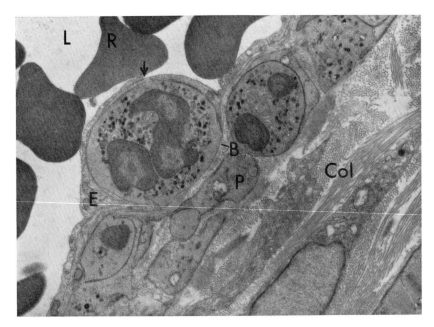

Fig. 7. As the PMN leucocyte passed through the intercellular junction, cytoplasmic processes from the adjacent endothelial cells (E) were piled up and formed a tight junction (as indicated by arrow). Basement membrane (B) was extended by an emigrating PMN leucocyte and became irregular. L = vessel lumen; R = red blood cell; P = periendothelial cell; Col = collagen fibers. The photograph was taken 2 h after leucoegresin injection. × 5,500.

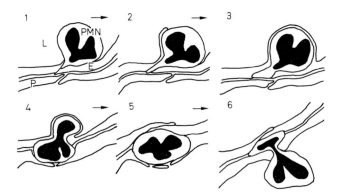

Fig. 8. Schematic drawing of morphological sequence of PMN emigration through the venular wall. Note the embracement of the cell by endothelial (E) processes and the prompt reformation of tight junction. P = periendothelial cell; L = lumen.

the course of emigration was observed. Finally, the PMN leucocyte passed through the space between the neighboring periendothelial cells, and emigrated into the adjacent connective tissue. Emigration of other blood cells, such as monocyte, lymphocyte, red blood cell and platelet, was not found in the cremaster muscle preparations. Each stage of emigration of PMN leucocyte is summarized in the schematic drawing in figure 8.

The basement membrane of the venules showed a continuous structure with smooth appearance before emigration of the PMN leucocyte, but the structure was clearly changed as the cell penetrated; this layer was strongly extended by the cell and became very irregular. The thin portion of the layer was finally dissociated and permitted the passage of the cell. However, shortly after the cell passed through, the dissociated portion appeared to return to the normal structure of the basement membrane. Electron-microscopic features in the emigration of PMN leucocyte by leucoegresin strikingly resembled those observed in experimentally induced pancreatitis in dogs [35].

As described above, emigration of PMN leucocytes by leucoegresin was characterized by the active involvement of the endothelial cells as shown in the form of reasonable envelopment of the cells by the marginal fold-like cytoplasmic processes and of prompt reformation of the intercellular junction. By contrast, extravascular leaking of plasma protein by histamine, plasmakinin and vasoexin was characterized by the structural change in the intercellular junction in the form of a widely opened gap [32, 34]. The formation of such an open gap was not observed at all stages of PMN emigration. Such differences in the electron-microscopic pictures seemed to explain the biological difference between leucoegresin and permeability factors such as vasoexin.

IV. Immunological Properties of Leucoegresin

A. Demonstration of Common Antigenicity between Leucoegresin and IgG

The immunological properties of leucoegresin were further analyzed in the following ways. Besides antibodies against rabbit serum, extracts of Arthus skin lesions and purified leucoegresin noted above, antisera against purified rabbit IgG, its Fc fragment, Fab fragment and light chain fragment were respectively prepared in goats, and γ-globulin fractions were separated from these antisera by fractionation with sodium sulfate.

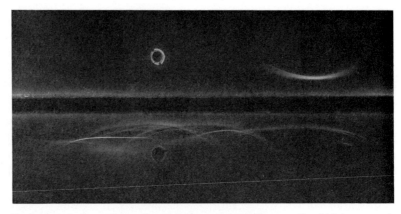

Fig. 9. Agar immunoelectrophoresis of leucoegresin. Upper well: rabbit leucoegresin (4 mg/ml); lower well: rabbit serum. Trough: goat antirabbit serum antibody.

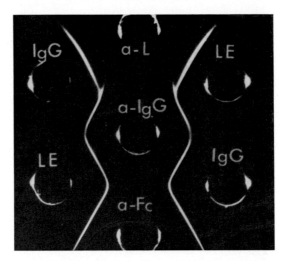

Fig. 10. Immunodiffusion of leucoegresin. Upper right: IgG = rabbit IgG (5 mg/ml). Bottom right: LE = leucoegresin (4 mg/ml). Top center: a-L = goat antirabbit light chain antibody. Middle center: a-IgG = goat antirabbit IgG. Bottom center: a-Fc = goat antirabbit Fc fragment antibody. Upper left: LE = leucoegresin. Bottom left: IgG = rabbit IgG.

As illustrated in figures 9 and 10, rabbit leucoegresin gave a single precipitin line with antirabbit serum, anti-IgG, anti-Fc or anti-light-chain antibody. These precipitin lines joined each other and a line formed between rabbit IgG and these antisera. These observations suggested that rabbit leucoegresin shared at least some of the antigenicity with rabbit IgG [26].

Table III. Inhibitory effect of anti-IgG, anti-Fc or anti-light chain antibody on leucoegresin chemotactic activity *in vitro*

Samples in lower compartment, 1 ml	Number of PMN migrated (per 5 fields)
Leucoegresin[1] plus buffer[2]	211
Buffer	26
Leucoegresin plus anti-IgG[3] plus buffer	4
Anti-IgG plus buffer	48
Leucoegresin plus anti-Fc[4] plus buffer	12
Anti-Fc plus buffer	22
Leucoegresin plus anti-light chain[5]	2
Anti-light chain plus buffer	32
Anti-IgG plus IgG[6]	15
Anti-Fc plus IgG	24
Anti-light chain plus IgG	20

1 100 μg (in phosphate buffer, M/15, pH 7.4).
2 Phosphate buffer, M/15, pH 7.4.
3 540 μg (in phosphate buffer, M/15, pH 7.4).
4 530 μg (in phosphate buffer, M/15, pH 7.4).
5 500 μg (in phosphate buffer, M/15, pH 7.4).
6 100 μg (in phosphate buffer, M/15, pH 7.4), separated from normal rabbit serum.

In spite of evidence described above, there remained the question of whether the chemotactic activity of the leucoegresin under observation was directly associated with leucoegresin or with contaminant in the sample. As summarized in table III, when leucoegresin was respectively absorbed by antileucoegresin, anti-IgG, anti-Fc or anti-light-chain antibody at their optimal conditions, the chemotactic activity completely disappeared from the fluid phase [26]. These experimental results confirmed that rabbit leucoegresin had a common antigenicity with rabbit IgG, and its activity was associated with leucoegresin itself, but not with any contaminant, if present.

B. Immunoadsorption of Chemotactic Activity of Inflammatory Extracts by Antileucoegresin

In a further experiment it was demonstrated that antileucoegresin gave a single precipitin line in immunoelectrophoresis with crude extracts of

Table IV. Immunoabsorption of chemotactic activity in the extract of inflamed skin by anti-leucoegresin antibody

Samples in lower compartment, 1 ml	Number of PMN migrated (per 5 fields)
LE[1] plus buffer[2]	225
LE plus anti-LE[3]	8
Buffer plus anti-LE	11
Extract[4] plus buffer	205
Extract plus normal goat IgG[5]	212
Extract plus anti-LE	70

1 100 μg in phosphate buffer (M/15, pH 7.4).
2 Phosphate buffer (M/15, pH 7.4).
3 Goat anti-LE antibody, 550 μg in phosphate buffer (M/15, pH 7.4).
4 Extract from 12-hour-old skin lesion of Arthus reaction, 20.0 at 280 nm.
5 500 μg in phosphate buffer (M/15, pH 7.4).
LE = leucoegresin.

12-hour-old Arthus skin lesions, and that the chemotactic activity of the extracts was significantly absorbed by adding antileucoegresin in amounts capable of inducing an optimal precipitation (table IV). The same experimental results were also obtained with the extracts of burned skin lesions. These observations indicated that most of the chemotactic activity of the extracts of inflamed skin was due to leucoegresin itself. It was therefore obvious that this chemotactic factor played an essential part in an inflammatory leucotaxis. As cited above, the extraction of skin poweder was performed with M/15 phosphate buffer (pH 7.4) at 4°C. This procedure has proven to be a mild but very effective one for extraction of different types of natural mediators, including leucoegresin and proteases.

V. Production of Leucoegresin by Inflammatory Neutral SH-Dependent Protease

A. Isolation and Purification of the Protease from Inflammatory Tissue and Its Inflammation-Inducing Effect

The neutral SH-dependent proteases were isolated from the skin site of Arthus reaction and thermal injury in rabbits and then purified as follows:

from the extracts of powdered inflamed skin, euglobulin fraction was prepared with ammonium sulfate, and then desalted through a Sephadex G-50 column [18, 36]. The protein fraction was successfully eluted on DEAE-Sephadex A-50 and then on GE-cellulose by successive elution with phosphate buffer of the following molarity and pH: 1/100, pH 8.0; 1/50, pH 7.4; 1/20, pH 7.4; 1/10, pH 7.4; and finally with 1.0 M sodium chloride [37]. The molecular size of the SH-dependent protease I (eluted in M/50 phosphate buffer, pH 7.4 on GE-cellulose) was approximately 200,000; that of the SH-dependent protease II (eluted in M/10 phosphate buffer, pH 7.4 on GE-cellulose) was approximately 14,000 [38]. The SH-dependent protease I was found to convert to the protease II when tested by chromatography on GE-cellulose column [38].

These proteases were similarly inactivated by *p*-chloromercuribenzoate or molecular oxygen, but readily reactivated by an excess of cysteine or reduced glutathione. The optimum pH of these proteases, was 7.1 when tested against casein, and between 6.0 and 7.0 when tested against hemoglobin. The arbitrary unit of proteolytic activity was defined as the amount of enzyme which would cause an increase of 0.001 U of extinction per minute of digestion [38]. Accordingly, one proteolytic unit represented 0.06 absorption units at 276 nm. Both enzyme preparations contained no acid proteases such as cathepsins D and E and both were thermolabile.

The SH-dependent protease has been shown to satisfy many of the criteria necessary for an inflammatory agent [12, 13]: (1) the protease was locally available to induce the inflammatory reactions; (2) protease activities paralleled the time-course of the inflammatory reactions; (3) the protease could produce morphologic changes similar to those of the reactions when injected in low concentrations comparable to those detected in sites of inflammation; (4) the protease was inhibited by the specific antagonistic substance locally available, and (5) the inflammatory reactions were suppressed by specifically antagonistic substances.

Two types of inhibitors for the SH-dependent protease have been isolated from inflamed site or serum, and were highly purified by chromatography. The inhibitor from inflamed skin was a heat-stable peptide with a molecular weight of approximately 12,000 [39, 40], and the amount of the protease inhibitor increased throughout the course of inflammatory reactions, e.g. Arthus reactions and burn in rabbits, but its increase became particularly marked as the reactions began to subside [36]. The protease inhibitor of rabbit serum was a glycoprotein with a molecular weight of approximately 72,000 [41], and was exuded by vascular permeability change into the adjacent connective tissue. A similar inhibitor was also separated from the serum of rabbit, cow, guinea pig and man.

Intradermal injection of the SH-dependent proteases I and II (0.35, 0.85 and 1.65 U) in small amounts produced a rapid development of local cutaneous lesions strikingly similar to those of the Arthus reaction in both their gross and microscopic features; the intensity of the inflammatory response induced was clearly dependent on the amounts of the enzyme injected. Maximal increase in vascular permeability (to circulating dye) was found in about 2 h, and pronounced PMN emigration in about 6 h [12, 42]. The inflammation-inducing effects of the protease II were more pronounced than those of the protease I, and the inflammatory response was similarly suppressed when the proteases were injected together with the inhibitor described above [43].

B. *In vivo* Production of Leucoegresin by the Protease

In a first experimental step, the permeability factor was isolated from the protease-induced skin lesions after 2 h, and purified by chromatography; the permeability factor obtained was indistinguishable from vasoexin in both its physico-chemical and biological properties. These observations suggested that local production of vasoexin was associated with action of the neutral SH-dependent protease injected. By the action of this permeability factor, various types of plasma protein were exuded into the adjacent connective tissue.

In further experiment, the chemotactic factor for PMN leucocytes was isolated from the same protease-induced skin lesions after 6 h, showing a pronounced PMN leucocyte infiltration [12], and then purified according to the procedures summarized in table I. The chemotactic substance obtained was indistinguishable from leucoegresin in the physico-chemical, immunological and biological properties. Protease II was found more active in producing the chemotactic factor than protease I, when both the proteases were separately injected intradermally at the same level of activity. From a single protease II-induced skin lesion, about 210 μg of leucoegresin were isolated (about 250 μg of leucoegresin was isolated from a *single* Arthus site at 12 h). These observations lead us to postulate that local production of leucoegresin was probably associated with the combined action of the SH-dependent proteases injected, especially of protease II, and the precursor of leucoegresin which is probably amongst the plasma proteins exuded by the action of permeability factors described above.

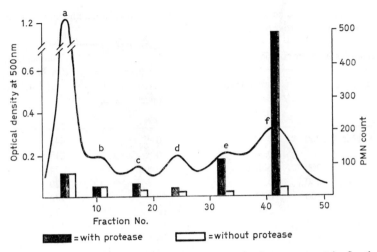

Fig. 11. *In vitro* generation of chemotactic activity in serum protein fraction by inflammatory SH-dependent protease. The mixture of each protein fraction (0.5 ml) and the protease (0.5 ml, 0.66 U/ml) was incubated at 37 °C for 60 min and then tested for chemotactic activity. Protein concentrations at 280 nm of each fraction were as follows: a = albumin, 1.40; b = α_1-α_2-globulin, 0.28; c = α_2-β_1:-globulin, 0.24; d = β_1:-γ_1:-globulin 0.30; e = γ_1:-γ_2-globulin, 0.28; f = γ_2-globulin, 0.54. Shaded columns: chemotactic potency before treatment with the protease activated by 10^{-3} M cysteine. Hatched columns: chemotactic potency after treatment with the activated protease.

C. Production of Leucoegresin by Protease Action on Serum IgG *in vitro*

Serum proteins were tested as possible precursors of leucoegresin. By preparative zone electrophoresis on Pevikon C-870, 6 protein fractions were separated from normal rabbit serum. They contained mainly albumin, α_1-α_2, α_2-β_1, β_1-γ_1, γ_1-γ_2, and γ_2-globulin fractions respectively. After dialysis against M/15 phosphate buffer (pH 7.4), equal volumes (0.5 ml) of each protein fraction and the SH-dependent protease II (0.33 U) were incubated for 1 h at 37 °C and tested for chemotactic activity [44].

As shown in figure 11, the γ_2-globulin fraction itself had no chemotactic activity, but became strongly chemotactic in the presence of SH-dependent protease activated by 10^{-3} M cysteine. However, none of the other protein fractions showed chemotactic effects under the same conditions. The protease itself was ineffective for PMN leucocytes. Accordingly, the γ_2-globulin is a *cytotaxigen* [45], and the chemotactic factor produced from it is

Table V. In vitro generation of chemotactic activity from rabbit and human myeloma IgG by inflammatory SH-dependent protease

Samples in lower compartment, 1 ml				Number of PMN (rabbit) migrated (per 5 fields)
IgG, μg		inflammatory protease proteolytic U	cysteine, final concentration, M	
Rabbit	200	0.33	10^{-3}	449
	100	0.33	10^{-3}	215
	200	buffer	10^{-3}	24
Human	200	0.33	10^{-3}	320
	100	0.33	10^{-3}	182
	200	buffer	10^{-3}	22
Myeloma	200	0.33	10^{-3}	471
	100	0.33	10^{-3}	249
	200	buffer	10^{-3}	20
Buffer		0.33	10^{-3}	25
Buffer		buffer	buffer	14

The mixture of each sample (0.5 ml) was previously incubated at 37°C for 60 min and placed in the lower compartment for assay of chemotactic activity.

a *cytotaxin* [45]. These observations indicated that the precursor of the chemotactic factor was contained in the γ_2-globulin fraction of rabbit serum.

On the basis of the observations indicating that leucoegresin showed common antigenicity with rabbit IgG which was the main component of the γ_2-globulin fraction of rabbit serum, further experiments were undertaken. As summarized in table V, rabbit IgG itself had no chemotactic activity for PMN leucocytes, but became strongly chemotactic when treated with the SH-dependent protease (0.33 U) activated in the presence of cysteine [44].

The same type of experiment was also performed with human IgG and human myeloma protein; these purified IgG samples, as summarized in table V, were ineffective for PMN leucocytes before treatment with SH-dependent protease, but became strongly chemotactic after treatment with SH-dependent protease. The molecular size of the chemotactic factor produced was approximately 140,000 when measured by gel filtration, suggesting a minor structural change of the IgG molecule. Its molecular size was indistinguishable from that of leucoegresin and of untreated IgG. It was thus suggested that serum IgG of rabbit and man was a possible precursor of leucoegresin for PMN leucocytes [44].

In a series of experiments, the activity of SH-dependent protease added was almost comparable to that of SH-dependent protease detected in an early stage of Arthus lesions showing an initial emigration of PMN leucocytes. Since serum IgG was considered to be exuded into the inflamed site in a satisfactory amount, it seemed reasonable that the production of leucoegresin was associated with the local activity of SH-dependent protease.

Generation of chemotactic activity from IgG by SH-dependent protease was further attempted for each subclass of serum IgG of rabbit and man. Electrophoretic migration of rabbit IgG, eluted on M/200 phosphate buffer (pH 8.0) was slow, whereas rabbit IgG eluted by M/100 phosphate buffer (pH 8.0) was fast-moving. Human IgG subclasses, i.e. IgG_1, IgG_2, IgG_3 and IgG_4, were also tested. All IgG samples had no or little chemotactic potency before treatment with the active SH-dependent protease II, but they all became strongly chemotactic after treatment with the protease, though some differences in their chemotactic effects were observed between the treated IgG subclasses [46]. The molecular size of chemotactic factors produced was again about 140,000 when measured by gel filtration. It is thus possible that the inflammatory SH-dependent protease might produce chemotactic factors bearing some relation to a particular structural specificity of the IgG molecule.

D. Production of Chemotactic Factor by Papain Action on Serum IgG in vitro

The same type of experiment as mentioned above was carried out with papain (0.5 μg) with the same proteolytic activity (0.33 U) as the inflammatory protease used in the previous experiments. Besides IgG subclasses of rabbit and man, mouse myeloma IgG subclasses, referred to as IgG_1, IgG_{2a} and IgG_{2b}, were also tested.

As summarized in table VI, fast-moving rabbit IgG became strongly chemotactic only after treatment with papain, while slow rabbit IgG was very slightly chemotactic before treatment with papain, but the chemotactic potency was not influenced by the enzymatic treatment [47]. The enzyme itself did not show any chemotactic effect. Similar chemotactic generation was also revealed for human IgG_2 and IgG_4 and mouse IgG_1, but not for human IgG_1 and IgG_3 and mouse IgG_{2a} and IgG_{2b}. These observations strongly suggested that the production of a chemotactic factor by papain was associated with a structure specificity of the IgG molecule [46].

Table VI. In vitro generation of chemotactic activity by papain digestion of IgG subclasses of rabbit, mouse and man

Samples in lower comparement, 1 ml				Number of PMN migrated (per 5 fields)
	μg	type	papain, μg[1]	
Rabbit fast IgG	200		0.5	398
	200		buffer[2]	25
Rabbit slow IgG	200		0.5	70
	200		buffer	98
Human IgG$_1$	200	$\varkappa$	0.5	37
	100		0.5	45
	200		buffer	22
	200	λ	0.5	28
	200		buffer	25
Human IgG$_2$	200	$\varkappa$	0.5	444
	100		0.5	222
	200		buffer	25
	100	λ	0.5	214
	100		buffer	24
Human IgG$_3$	200	λ	0.5	13
	200		buffer	28
	100	$\varkappa$	0.5	28
	100		buffer	25
Human IgG$_4$	100	$\varkappa$	0.5	256
	100		buffer	5
Mouse IgG$_1$	100		0.5	216
	100		buffer	7
Mouse IgG$_{2a}$	100		0.5	72
	100		buffer	6
Mouse IgG$_{2b}$	100		0.5	35
	100		buffer	5
Buffer			0.5	15
Buffer			buffer	8

The mixture of each sample (0.5 ml) was incubated at 37°C for 60 min and placed in the lower compartment for assay of chemotactic activity for PMN leucocyte of rabbit.
1 Caseinolytic activity of papain added was 0.33 U.
2 M/15 phosphate buffer (pH 7.4).

The molecular size of the chemotactic factor produced was also about 140,000.

As is widely accepted, the antigenic and structural differences between the 4 subclasses of human IgG have been considered to be associated with those of the heavy chains of the molecule [48]; they were divided into groups according to their susceptibility to papain: IgG_1 and IgG_3 were papain-sensitive, whereas IgG_2 and IgG_4 were papain-resistant [49]. Mouse IgG_1 was found to be more resistant to papain than IgG_{2a} and IgG_{2b} [50]; fast rabbit IgG was more resistant to papain than slow IgG [51]. Accordingly, our observations suggested that generation of chemotactic activity from IgG molecules by their treatment with papain was associated with the suscepti-bility to the enzyme, namely the specificity of heavy chain structure of the IgG molecule.

The experimental results with the inflammatory protease or papain suggested that the generation of a chemotactic IgG derivative was associated with some minor structural change in the IgG molecule. This assumption is supported as follows: (1) its molecular weight was similar to that of leuco-egresin or untreated IgG, and (2) the Fab and Fc fragments produced by treatment with papain had no chemotactic potency (the inflammatory protease did not produce such fragments).

Such a successful generation of chemotactic IgG molecules was con-sidered to occur only under the mild treatment with a small amount of the inflammatory protease or papain. In these experiments, IgG molecules were similarly digested with the lower concentration (0.33 U) of these enzymes. It seems reasonable that similar results were obtained with water-insoluble papains (free of cysteine), because papain in this form failed to release Fab and Fc fragments. Since short-term enzymatic digestion of IgG with papain or with water-insoluble papain produced some intermediate products with similar molecular size as IgG [52], it seemed possible that the intermediate products showed chemotactic potency. The intermediate products may be cleaved enzymatically in the successive steps and the chemotactic effect then be abolished.

VI. Structural Change during Generation of Chemotactic IgG by Inflammatory Neutral SH-Dependent Protease

A. Release of Dialysable Peptides by the Protease from IgG

Since it was confirmed that generation of chemotactic IgG derivative by the SH-dependent protease II was associated with some minor structural change of the molecule, further experiments were performed with anti-BSA IgG; the assay of antibody activity during generation of chemotactic activity

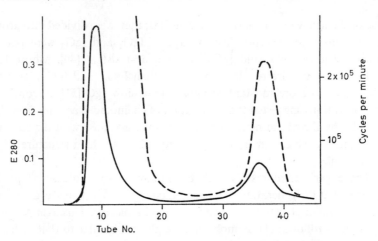

Fig. 12. Release of dialysable peptides from ^{125}I-labeled rabbit IgG at 1 h of digestion with inflammatory SH-dependent protease II. The mixture of IgG and the protease was chromatographed on a Sephadex G-75 column. The second peak contained dialysable peptides and the protease. The first peak showed chemotactic potency. ——— = E280; ------ = cycles per minute.

seemed a useful indicator of structural changes in the IgG molecule. The IgG fraction containing anti-BSA was separated in the cold from fresh sera of BSA-sensitized rabbits, and then eluted on an immunoadsorbent column which was prepared with Sepharose 4B conjugated with BSA. Anti-BSA antibody itself showed no chemotactic effect for PMN leucocytes, but it became strongly chemotactic after treatment with the SH-dependent protease (0.33 U). The generated chemotactic activity was quite similar to that of protease-treated IgG from normal rabbit serum.

As illustrated in figure 12, the generation of chemotactic activity by the enzyme was accompanied by release of dialysable peptides from the antibody IgG molecule without concomitant release of fragments like Fab or Fc. Release of dialysable peptides became evident after 1-hour assay with IgG labeled with ^{125}I. Prolongation of the digestion period resulted in a gradually decreased chemotactic activity which abolished after a 24-hour digestion period. Such decrease in the chemotactic activity was undoubtedly associated with progressive digestion of the chemotactic factor, as revealed by increased formation of dialysable peptides. It was thus confirmed that for the structural change to occur in the IgG molecule required for generation of chemotactic activity only a mild digestion with small amount of the inflammatory protese was needed.

In contrast to papain, the inflammatory protease was characterized by its failure to produce fragments like Fab and Fc even after prolonged digestion of IgG (24 h); only production of dialysable peptides was observed. Recent work [54] has demonstrated the production of a high molecular weight fragment similar to leucoegresin and of peptides from IgG by a neutral lysosomal enzyme from the spleen and bone marrow, resembling the inflammatory SH-dependent protease. Other work [55] has also suggested increased amounts of various types of IgG products in the lymph of severely burned patients. These observations postulated the significance of enzymatic digestion of IgG molecules in inflammation.

B. Possible Site of Structural Change in IgG during Generation of Chemotactic Activity

The structural changes in anti-BSA IgG under the influence of SH-dependent protease were further investigated [53]. As illustrated in figure 13, the amounts of precipitates, produced by protease-pretreated anti-BSA IgG and BSA, were closely similar to those obtained by nonenzyme-treated anti-BSA IgG and BSA. Thus, antigen-binding activity of anti-BSA IgG was not affected even by prolonged digestion (10 and 24 h) with the SH-dependent protease.

In a further experiment, the potency of protease-treated anti-BSA IgG to induce passive cutaneous anaphylaxis (PCA) and reversed PCA reaction in the guinea pig was assayed. As summarized in table VII, the potency of the antibody to induce PCA was not affected by SH-dependent protease in 1 and 3 h but was clearly diminished when the digestion was prolonged (10 and 24 h). Negative results were observed with leucoegresin with no antibody potency for BSA. A similar tendency was obtained in the reversed PCA reaction using anti-IgG, anti-Fc and anti-light chain; positive results were also observed with leucoegresin which had a common antigenicity with IgG. Thus, the SH-dependent protease progressively affected the structure of the Fc portion of the antibody IgG molecule, because the Fc site had been widely accepted as responsible for the skin sensitization. Autoradiographic studies using leucoegresin labeled with ^{131}I or ^{125}I has suggested that this substance may bind with PMN leucocytes at the Fc site for its chemotactic action [56]. An electron-microscopic study using leucoegresin labeled with ferritin has also demonstrated that the material may bind to the membrane of PMN leucocytes [56].

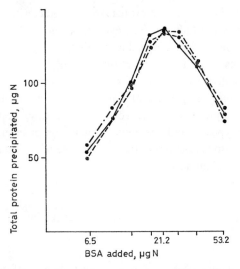

Fig. 13. Quantitative precipitin reaction with anti-BSA IgG treated with inflammatory SH-dependent protease and then incubated with BSA. Rabbit anti-BSA IgG (7.5 mg) was treated with the SH-dependent protease (3.0 U) for 3 or 24 h. ——— = Treated anti-BSA IgG for 3 h; ·—·—· = treated anti-BSA IgG for 24 h; ------ = untreated anti-BSA IgG.

Table VII. Induction of passive cutaneous anaphylaxis (PCA) in guinea-pigs by anti-BSA IgG treated with inflammatory SH-dependent protease

Amount of injected materials µg N	Mean diameter of reaction induced by				
	IgG untreated mm	IgG after 1-hour digestion mm	IgG after 10-hour digestion mm	IgG after 24-hour digestion mm	Leuco-egresin mm
0.5	22	20	17	5	0
0.24	16	16	13	0	0
0.12	15	15	5	0	0
0.06	13	12	0	0	0
0.03	10	11	0	0	0
0.015	5	4	0	0	0

The samples (0.1 ml) were injected intradermally, and 3 h later each animal was given intravenously 0.25 ml of a mixture of equal parts of BSA solution (80 µg N) and dye solution. The reactions were measured after 45 min.

Further analysis of structural changes in IgG was performed by determination of the complement-fixing ability of the protease-treated anti-BSA IgG. The complement-fixing ability of digested anti-BSA IgG was not affected when the digestion of the antibody sample lasted only 1 and 3 h, but it was clearly diminished when the digestion was prolonged (10 and 24 h). These results were also convenient to explain also the PCA data described above. Taken together, these observations suggest that the structural site of IgG degraded by the SH-dependent protease was not the antigen binding (Fab) site.

VII. Leucoegresin as an Inflammatory Chemotactic Factor

Leucoegresin was confirmed to satisfy many of the criteria proposed for a natural mediator of leucotaxis in inflammation.

1. Leucoegresin was locally available to initiate and maintain the tissue leucotaxis. The amount of purified leucoegresin, which was isolated from a *single* inflamed site, seemed to reproduce to a large extent the original leucocyte infiltration.

2. The activity of leucoegresin paralleled the time-course of leucocyte emigration. The chemotactic activity of extracts of inflammatory tissue was significantly decreased by adsorption with antileucoegresin.

3. Leucoegresin elicited a long-lasting effect which would account for the prolonged duration of leucotyce infiltration.

4. The action of leucoegresin was specific, in that it caused neutrophilic PMN leucocytes to adhere to the endothelium, emigration and chemotaxis, but it was not capable of inducing vascular permeability change and hemorrhages. Leucoegresin was active not only for rabbit, guinea pig, rat and mouse PMN leucocytes. The dissociation of the two phenomena of increased vascular permeability and leucocyte emigration seems to be explained by the discovery of the two different natural mediators, namely vasoexin and leucoegresin.

5. Leucoegresin could produce characteristic morphologic changes similar to those seen in the original inflammatory site, as confirmed by microscopic and electron-microscopic observation. The site of emigration of PMN leucocytes was the venules, but not the capillaries. Characteristic involvement of the endothelial cells of the venules was needed.

6. Leucoegresin was produced *in vivo* and *in vitro* from serum IgG by the neutral SH-dependent protease from inflammatory tissue. Both the SH-

dependent protease and IgG were locally available. The SH-dependent protease seemed to be released from histiocytic cells at an early stage of inflammation [57, 58] and later on from emigrated PMN leucocytes.

Whether a specific antagonistic substance for leucoegresin exists in the inflammatory tissue remains to be ascertained. Experiments on the isolation of an SH-independent protease [59] and an anti-SH-dependent protease [39] suggested the possibility of agents capable of inhibiting leucoegresin action in an inflammatory tissue.

VIII. Possible Chemotactic Factors for Lymphocytes and Mononuclear Cells

A. Production of Serum IgM-Derived Chemotactic Factor for Lymphocytes by Neutral SH-Dependent Protease of PMN Leucocytes in vitro

It has been shown recently that rabbit PMN leucocytes when incubated with antigen-antibody complexes in vitro [60, 61] release two types of neutral proteases. These proteases were isolated from the lysosomal fractions of the cells and then purified by chromatography using DEAE-Sephadex A-50 and CM-Sephadex C-50 by increasing the salt gradient; the neutral SH-dependent protease was eluted in M/50 phosphate buffer (pH 6.8) on CM-Sephadex, and the neutral SH-independent protease eluted in 1.0 M sodium chloride on CM-Sephadex [60]. The neutral SH-dependent protease essentially seemed to be identical with the inflammatory SH-dependent protease in its physico-chemical and biological properties.

The problem of whether these neutral proteases from PMN leucocytes could produce a chemotactic factor for lymphocytes from serum immuno-globulins was studied in a similar way as that described above [62]. Lympho-cytes (4×10^9 cells/ml) were collected from the thoracic duct lymph of rats, and suspended in Eagle's MEM containing 2% human serum albumin (pH 7.4). Most of the cells collected were small lymphocytes. Millipore filters of 3-μm pore size were used.

Rabbit IgM had no chemotactic activity for lymphocytes, but it became strongly chemotactic when treated with activated SH-dependent protease for 3 h. The intensity of generated chemotactic activity was dependent on the activity of the enzyme (0.08 and 0.33 U) added. Chemotactic activity generated decreased in parallel with a prolongation of the enzymatic treatment (10 and 24 h). Rabbit IgM was a more potent precursor than rabbit IgG. This may be due to the structure specificity of the IgM molecule or to its higher affinity for small lymphocytes.

The protease-IgM mixture containing the lymphocyte chemotactic factor was chromatographed on Sephadex G-200; chemotactic activity was found predominantly in the chromatographic fraction with a molecular size of the protease (molecular weight 14,000).

The problem remains as to whether this type of chemotactic factor for lymphocytes can be isolated from an inflammatory tissue. Unlike PMN leucocytes or monocytes, the lymphocytes do not pass through the inter- cellular junctions but cross the vascular wall by entering the endothelial cells and traversing their cytoplasm. It has yet to be demonstrated whether the desribed chemotactic factor for lymphocytes can, when injected, induce similar events.

In contrast, no chemotactic generation of IgM for lymphocytes was revealed after treatment with the SH-independent protease at the same level of activity (0.33 U, 3 h), suggesting different enzymatic characteristics for these proteases.

B. Production of Serum IgG-Derived Chemotactic Factor for Mononuclear Cells by Neutral SH-Independent Protease from PMN Leucocytes *in vitro*

Mononuclear cells were collected from oil-induced peritoneal exudates of guinea pigs and suspended in Eagle's MEM containing 2% human serum albumin (pH 7.4). The *in vitro* assay for chemotaxis of mononuclear cells seemed more difficult than that of PMN leucocytes or lymphocytes. This was possibly in part due to the tendency of mononuclear cells to adhere to and spread on the millipore membranes and in part to the slower movement of mononuclear cells.

Rabbit IgG was not chemotactic for these cells, but it became chemo- tactic when treated with the SH-independent protease (0.33 U) for 6 h. In contrast, the SH-dependent protease did not generate chemotactic activity during incubation with IgG molecule when tested at the same level of activity (0.33 U). These preliminary observations [63] suggested that the production of different types of chemotactic factors might be associated with the mode of the cleavage of immunoglobulin molecules by the different proteases.

Recently, a chemotactic factor for mononuclear cells was extracted from a 24-hour-old skin site of a reaction in guinea pigs which was induced

Table VIII. In vitro chemotactic generation from immunoglobulin M by neutral SH-dependent protease from PMN leucocytes

Samples	Number of lymphocytes (rat) migrated
Rabbit IgM[1] plus SH-protease[2]	234
Buffer plus SH-protease	41
Rabbit IgM plus buffer	41
Buffer alone	46

1 IgM, 100 μg.
2 SH-protease, 0.33 U.
The mixture of IgM (0.5 ml) and protease (0.5 ml) or buffer was previously incubated at 37°C for 3 h, and placed in the lower compartment for assay of chemotactic activity.

by intradermally injected tuberculin purified protein derivative (PPD) [64]. This chemotactic factor was largely concentrated in the pseudoglobulin fraction of the skin extracts. In what way these two chemotactic factors are related is unknown. On the other hand, a permeability factor was also extracted from lymph node cells [65] and skin sites of tuberculin reaction [66]; it was termed lymph node permeability factor (LNPF). The material was probably a mixture of several biologically active factors, because intra-dermal injection of this material induced vascular permeability change, emigration of PMN leucocytes and mononuclear cells, and deposition of fibrinoid substance. The separation of chemotactic factor(s) for mono-nuclear cells from LNPF samples should be attempted.

Conclusion

1. *Leucoegresin*, a chemotactic factor for neutrophilic PMN leucocytes of rabbits, guinea pigs, rats and mice was isolated from inflammatory tissue. It was found in the pseudoglobulin fraction of the extract and purified by chromatography.

Leucoegresin behaved as a homogeneous substance on ultracentrifugation and in moving-boundry electrophoresis. Immunological analysis of the chemotactic factor revealed the absence of impurity. The chemotactic factor was a protein free of nucleic acid, and its molecular weight was approximately

140,000 when assayed by gel filtration. The sedimentation coefficient of this substance was 6.58 S and its isoelectric point was pH 5.0. The substance was relatively heat-stable.

2. In contrast to the permeability factor (vasoexin) which was separated from the same pseudoglobulin fraction of inflammatory tissue, leucoegresin induced pronounced PMN emigration, but did not increase vascular permeability; PMN emigration occurred characteristically only at the site of the venules.

3. Leucoegresin satisfied several criteria to make it acceptable as a natural mediator of inflammatory leucotaxis, i.e. local availability, parallelity between activity (amount) and time-course of the reaction, and morphologic resemblance to the reaction. This chemotactic factor was locally produced by the action of inflammatory neutral SH-dependent protease.

4. Leucoegresin shared common antigenic sites with serum IgG, and was produced *in vitro* from IgG by the inflammatory protease; IgG (fast and slow) of rabbit, and 4 IgG subclasses of man became chemotactic by the enzymatic treatment.

5. Generation of chemotactic activity was accompanied by release of inactive dialysable peptides from the IgG molecule without concomitant release of Fab or Fc fragments. The minor structural change induced by the enzyme in the IgG molecule leading to its chemotactic potency seemed to occur exclusively at the Fc portion.

6. Formation of activity by papain action on IgG was found only for papain-resistant IgG; it included fast rabbit IgG, IgG_2 and IgG_4 of man and mouse IgG_1. In contrast, papain-sensitive IgG, such as slow rabbit IgG, IgG_1 and IgG_3 of man and mouse IgG_{2a} and IgG_{2b} were not cytotaxigenic. This served to emphasize the distinct enzymatic characteristics of the inflammatory protease.

7. Only short-term digestion of IgG with small amounts of the enzymes was needed for generation of chemotactic activity.

8. A possible chemotactic factor for lymphocytes was produced *in vitro* from immunoglobulins, especially IgM, by action of the neutral SH-dependent protease from PMN leucocytes. This enzyme seems identical with the inflammatory protease.

9. The neutral SH-independent protease from PMN leucocytes converted IgG into a chemotactic factor for mononuclear cells.

10. It is suggested that immunoglobulin molecules, besides their function as antibodies, may play an immunologically nonspecific part as precursors of chemotactic factors in inflammation.

11. Our data suggest that following the increase in vascular permeability, the neutral SH-dependent protease which was activated in and released from the mesenchymal connective cells by inflammatory stimulation, e.g. antigen-antibody reaction, first converted the exuded IgG molecules to a chemotactically active molecule i.e. leucoegresin. The latter might then also cause the emigration of PMN leucocytes into the inflammatory lesion. Later on, the neutral SH-dependent protease, which was activated in and released from emigrated PMN leucocytes by antigen-antibody complexes, might produce from the exuded immunoglobulin (probably IgM) molecules a chemotactic factor which might induce the accumulation of lymphocytes into the inflamed site. The emigration of mononuclear cells might be associated with the chemotactic factor produced from IgG by the SH-independent protease released from PMN leucocytes.

12. The neutral SH-dependent protease, which was termed an inflammatory protease, was inactivated by two types of inhibitors locally available. It is suggested that the inactivation of the enzyme by these inhibitors might be associated with decreased production of leucoegresin and lymphocyte-chemotactic factor.

Acknowledgements

The authors are very grateful to Dr. M. KOONO and Dr. T. KOUNO in this laboratory for purification of proteases and for assay of proteolytic activity, and to Dr. K. KAGEYAMA, Keio University School of Medicine, Tokyo, for cinemicro-photographic observation on effects of leucoegresin. Our thanks are also due to Dr. K. YOSHIDA, Dr. A. TASHIRO, Dr. T. OGATA, Dr. S. KIYOTA, Dr. NISHIURA and Dr. Y. HIGUCHI for their generous collaboration at different steps of the investigation.

The authors wish to acknowledge the permission of the Publishers of *Immunology* to reproduce tables III, V, VI, and figures 2, 9, 11 and 13, and of *Kumamoto Medical Journal* to reproduce figures 4–8.

References

1 HARRIS, H.: Mobilization of defensive cells in inflammatory tissue. Bact. Rev. *24:* 3–15 (1960).
2 BOYDEN, S.: The chemotactic effect of mixtures of antibody and antigen on polymorphonuclear leucocytes. J. exp. Med. *115:* 454–466 (1962).
3 WARD, P. A.; COCHRANE, C. G., and MÜLLER-EBERHARD, H. J.: The role of serum complement in chemotaxis of leucocytes *in vitro*. J. exp. Med. *122:* 327–346 (1965).

4   WARD, P. A.: A plasmin split fragment of C'3 as a new chemotactic factor. J. exp. Med. *126:* 189–206 (1967).

5   WARD, P. A. and NEWMAN, L. J.: A neutrophile chemotactic factor from human C'5. J. Immunol. *102:* 93–99 (1969).

6   SNYDERMAN, R.; SHIN, H. S.; PHILLIPS, J. K.; GEWURZ, H., and MERGENHAGEN, S. E.: A neutrophile chemotactic factor derived from C'5 upon interaction of guinea-pig serum with endotoxin. J. Immunol. *103:* 413–422 (1969).

7   KELLER, H. U. and SORKIN, E.: Studies on chemotaxis. I. On the chemotactic and complement-fixing activity of γ-globulins. Immunology, Lond. *9:* 241–247 (1965).

8   WARD, P. A.: Complement derived leucotactic factors in pathological fluids. J. exp. Med. *134:* 109–113 (1971).

9   WARD, P. A. and HILL, J. H.: Biologic role of complement products. Complement derived leucotactic activity extractable from lesions of immunologic vasculitis. J. Immunol. *108:* 1137–1145 (1972).

10 WARD, P. A.; COHEN, S., and FLANAGAN, T. D.: Leucotactic factors elaborated by virus infected tissues. J. exp. Med. *135:* 1095–1103 (1972).

11 BERENBERG, J. L.; WARD, P. A., and SONENSHINE, D. E.: Tick-bite injury. Mediation by a complement-derived chemotactic factor. J. Immunol. *109:* 451–456 (1972).

12 HAYASHI, H.; KOONO, M.; YOSHINAGA, M., and MUTO, M.: The role of an SH-dependent protease and its inhibitor in the Arthus-type hypersensitivity reaction, with special reference to chemical mediation of increased vascular permeability and leucocyte emigration; in BERTELLI and HOUCK Inflammation biochemistry and drug interaction, pp. 34–52 (Excerpta Medica, Amsterdam 1969).

13 HAYASHI, H.: Mechanism of development and inhibition of inflammatory process (in Japanese). Trans. Soc. Path., Japan *56:* 37–63 (1967).

14 KELLER, H. U. and SORKIN, E.: Studies on chemotaxis. V. On the chemotactic effect of bacteria. Int. Arch. Allergy *31:* 505–517 (1967).

15 WARD, P. A.; LEPOW, I. H., and NEWMAN, L. J.: Bacterial factors chemotactic for polymorphonuclear leucocytes. Amer. J. Path. *52:* 725–736 (1968).

16 WALKER, W. S.; BARLET, R. C., and KURTZ, H. M.: Isolation and partial characterization of a staphylococcal leucocyte cytotaxin. J. Bact. *97:* 1005–1008 (1969).

17 AOKI, T.: Leucocyte chemotaxis factors (in Japanese). Nihon-Rinsho *27:* 171–184 (1969).

18 HAYASHI, H.; MIYOSHI, H.; NITTA, R., and UDAKA, K.: Proteolytic mechanism in recurrence of Arthus-type inflammation by thiol compounds. Brit. J. exp. Path. *43:* 564–573 (1962).

19 HAYASHI, H.; YOSHINAGA, M.; KOONO, M.; MIYOSHI, H., and MATSUMURA, M.: Endogenous permeability factors and their inhibitors affecting vascular permeability in cutaneous Arthus reactions and thermal injury. Brit. J. exp. Path. *45:* 419–435 (1964).

20 YOSHINAGA, M.; TASAKI, I., and HAYASHI, H.: Purification of permeability factors mediating increased vascular response in cutaneous Arthus-type inflammation. Biochim. biophys. Acta *127:* 172–178 (1966).

21 TASAKI, I.; YOSHINAGA, M., and HAYASHI, H.: Further study of permeability factors mediating increased vascular response in cutaneous Arthus-type inflammation. Biochim. biophys. Acta *130:* 260–262 (1966).

22 HAYASHI, H.; TASAKI, I., and YOSHINAGA, M.: Permeability factors of rabbit skin associated with delayed vascular response to thermal injury. Nature, Lond. *215:* 759–760 (1967).

23 HAYASHI, H.: A permeability factor in the vascular Arthus reaction; in SCHILD 3rd Pharmacological Congr. Immunopharmacology, pp. 91–96 (Pergamon Press, London 1968).

24 YOSHIDA, K.; YOSHINAGA, M., and HAYASHI, H.: Leucoegresin. A factor associated with leucocyte emigration in Arthus lesions. Nature, Lond. *218:* 977–978 (1968).

25 YOSHINAGA, M.; YOSHIDA, K.; TASHIRO, A., and HAYASHI, H.: The natural mediator for PMN emigration in inflammation. I. Purification and characterization of leucoegresin from Arthus skin site. Immunology, Lond. *21:* 281–298 (1971).

26 YAMAMOTO, S.; YOSHINAGA, M., and HAYASHI, H.: The natural mediator for PMN emigration in inflammation. II. Common antigenicity of leucoegresin with immunoglobulin G. Immunology, Lond. *20:* 803–808 (1971).

27 YOSHINAGA, M.; YAMAMOTO, S., and HAYASHI, H.: Immunologically nonspecific role of immunoglobulins; in KROYANAGI, OHTAKA and MATSUHASHI Immunology series, vol. 7, pp. 31–52 (in Japanese) (Igaku-Shoin, Tokyo 1972).

28 WARD, P. A.; COCHRANE, C. G., and MÜLLER-EBERHARD, H. J.: Further studies on the chemotactic factor of complement and its formation *in vivo.* Immunology, Lond. *11:* 141–153 (1966).

29 WARD, P. A. and HILL, J. H.: C′5 chemotactic fragments produced by an enzyme in lysosomal granules of neutrophils. J. Immunol. *104:* 535–543 (1970).

30 BOKISH, V. A.; MÜLLER-EBERHARD, H. J., and COCHRANE, C. G.: Isolation of fragment of human complement containing anaphylatoxin and chemotactic activity and description of an anaphylatoxin inactivation of human serum. J. exp. Med. *129:* 1109–1130 (1969).

31 SORKIN, E.; BOREL, J. F., and STECHER, V. J.: Chemotaxis of mononuclear and polymorphonuclear phagocytes; in VAN FURTH Mononuclear phagocytes, pp. 397–421 (1970).

32 TASAKI, I.: Mechanisms in the delayed and prolonged vascular permeability changes in inflammation. II. Topography of the leaking vessels after thermal injury and the site of action of burns permeability factors. Kumamoto med. J. *21:* 13–27 (1968).

33 KAGEYAMA, K.; YAMAGUCHI, H.; FUKUDA, J.; TAKEUCHI, H., and TORIGATA, T.: Action mode of chemical mediators (in Japanese). Jap. J. Allergy *17:* 432–434 (1968).

34 OGATA, T.: The role of inflammatory chemotactic factor (leucoegresin) and permeability factor (vasoexin) in acute inflammation. An electron microscopic observation of biologic action of these natural mediators. Kumamoto med. J. *24:* 103–123 (1971).

35 WILLIAMSON, J. R. and GRISHAM, J. W.: Electron microscopy of leucocytic margination and emigration in acute inflammation in dog pancreas. Amer. J. Path. *39:* 239–256 (1961).

36 HAYASHI, H.; UDAKA, K.; MIYOSHI, H., and KUDO, S.: Further study of correlative behavior between specific protease and its inhibitor in cutaneous Arthus reactions. Lab. Invest. *14:* 665–673 (1965).

37   KOONO, M.; MUTO, M., and HAYASHI, H.: Proteases associated with Arthus skin lesions: their purification and biological significance. Tohoku J. exp. Med. *94:* 231–235 (1968).

38   KOONO, M. and HAYASHI, H.: Purification of inflammatory SH-dependent proteases I and II from Arthus skin lesion and their significance in inflammatory changes (in Japanese). Jap. J. biochem. Soc. *41:* 436 (1969).

39   UDAKA, K. and HAYASHI, H.: Further purification of a protease inhibitor from rabbit skin with healing inflammation. Biochim. biophys. Acta *97:* 251–261 (1965).

40   UDAKA, K. and HAYASHI, H.: Molecular weight determination of a protease inhibitor from rabbit skin with healing inflammation. Biochim. biophys. Acta *104:* 600–603 (1965).

41   TOKAJI, G.: The chemical pathology of thermal injury, with special reference to burn SH-dependent protease and its serum inhibitor. Kumamoto med. J. *24:* 68–86 (1971).

42   MUTO, M.: Studies on the role of SH-dependent proteases in Arthus-type hypersensitivity reaction. II. Production of leucoegresin by Arthus proteases (in Japanese). J. Kumamoto med. Soc. *43:* 690–706 (1969).

43   UDAKA, K.: Studies on the role of sulfhydryl groups in the biochemical mechanisms of allergic inflammation. V. Antiprotease in the cutaneous lesions of Arthus-reaction and its significance. Kumamoto med. J. *16:* 70–81 (1963).

44   YOSHINAGA, M.; YAMAMOTO, S.; MAEDA, S., and HAYASHI, H.: The natural mediator for PMN emigration in inflammation. III. *In vitro* production of a chemotactic factor by inflammatory SH-dependent protease from serum immunoglobulin G. Immunology, Lond. *20:* 809–815 (1971).

45   KELLER, H. U. and SORKIN, E.: Chemotaxis of leucocytes. Experientia *24:* 641–652 (1968).

46   NISHIURA, M.; YAMAMOTO, S., and HAYASHI, H.: Conversion of various IgG to chemotactic factor by inflammatory SH-dependent protease (in Japanese). Trans. Soc. Path. Jap. (in press).

47   YOSHINAGA, M.; YAMAMOTO, S.; KIYOTA, S., and HAYASHI, H.: The natural mediator for PMN emigration in inflammation. IV. *In vitro* production of a chemotactic factor by papain from immunoglobulin G. Immunology, Lond. *22:* 393–399 (1972).

48   TERRY, W. D. and FAHEY, J. L.: Subclasses of human γ-globulin based on differences in the heavy polypeptide chains. Science *146:* 400–401 (1964).

49   JEFFERIS, R.; WESTON, P. D.; STANWORTH, D. R., and CLAMP, J. R.: Relationship between the papain sensitivity of human γG immunoglobulins and their heavy chain subclass. Nature, Lond. *219:* 646–649 (1968).

50   GORINI, G.; MEDGYESI, G. A., and DORIA, G.: Heterogeneity of mouse myeloma γG globulins as revealed by enzymatic proteolysis. J. Immunol. *103:* 1132–1142 (1969).

51   GOODMAN, J. W.: Heterogeneity of rabbit γ-globulin with respect to cleavage by papain. Biochemistry *4:* 2350–2357 (1965).

52   CEBRA, J. J.; GIVOL, D.; SILMAN, H. I., and KATCHALSKI, E.: A two-stage cleavage of rabbit γ-globulin by a water insoluble papain preparation followed by cystein. J. biol. Chem. *236:* 1720–1725 (1961).

53   YAMAMOTO, S.; NISHIURA, M., and HAYASHI, H.: The natural mediator for PMN emigration in inflammation. V. The site of structural change in the chemotactic

generation of immunoglobulin G by inflammatory SH-dependent protease. Immunology, Lond. *24:* 791–801 (1973).

54 LoSpalluto, J.; Fehr, K., and Ziff, M.: Degradation of immunoglobulins by intracellular proteases in the range of neutral pH. J. Immunol. *105:* 886–897 (1970).

55 Goldberg, C. B. and Whitehouse, E., jr.: F(ab')²-like fragments from severely burned patients provide a new serum immunoglobulin component. Nature, Lond. *228:* 160–162 (1970).

56 Yamamoto, S.; Nishiura, M., and Ishimaru, Y.: Role of immunoglobulins in chemotaxis of inflammatory cells (in Japanese). Infect. Inflamm. Immun. *1:* 69–80 (1972).

57 Hayashi, H.: Antigen-antibody activation of macrophage enzymes; in Austen and Becker Biochemistry of the acute allergic reactions, pp. 141–152 (Blackwell Scientific Publications, Oxford 1968).

58 Udaka, K. and Udaka, K.: The possible role of an specific SH-dependent protease in the cutaneous Arthus reaction. Fed. Proc. *28:* 616 (1969).

59 Koono, M.: An endogenous mechanism of increased vascular permeability in allergic inflammation. Anti-permeability factor newly isolated from Arthus lesion and its biologic significance. Kumamoto med. J. *19:* 79–89 (1966).

60 Kouno, T.: The role of neutral SH-dependent protease of PMN leucocyte lysosome in inflammation. Kumamoto med. J. *24:* 135–150 (1971).

61 Kouno, T.; Katsuya, H.; Maeda, S., and Hayashi, H.: The isolation of lysosomal protease of PMN leucocyte and its biological significance in inflammation (in Japanese). Jap. J. Allergy *21:* 156 (1972).

62 Higuchi, Y.; Kouno, T., and Hayashi, H.: Generation of lymphocyte chemotactic factor from IgM fraction (in Japanese). Jap. J. Allergy *22:* 150 (1973).

63 Maeda, S.; Katsuya, H., and Hayashi, H.: Pathological chemistry of delayed hypersensitivity reaction. I. Macrophage chemotactic factor (in Japanese). Jap. J. Allergy *19:* 110–111 (1970).

64 Kambara, T.; Katsuya, H., and Maeda, S.: Experimental study on the mechanisms of delayed hypersensitivity reaction, with special reference to emigration and function of mononuclear cells. Acta path. jap. *22:* 465–476 (1972).

65 Willoughby, D. A.; Boughton, B.; Spector, W. G., and Schild, H. O.: A vascular permeability factor extracted from normal and sensitized guinea pig lymph node cells. Life Sci. *5:* 347 (1962).

66 Willoughby, D. A.; Boughton, B.; Spector, W. G., and Boughton, B.: A lymph node permeability factor in the tuberculin reaction. J. Path. Bact. *87:* 353–363 (1964).

Author's address: Dr. H. Hayashi, Department of Pathology, Kumamoto University Medical School, *Kumamoto 860* (Japan)

Antibiotics and Chemotherapy, vol. 19, pp. 333–337
(Karger, Basel 1974)

The Regulation of Leukotactic Mediators[1]

P. A. WARD

Department of Pathology, University of Connecticut Health Center, Farmington,
Conn.

Due to the presence of inhibitors or inactivators in human serum or
plasma, most inflammatory mediators are under strict control. Examples of
these regulatory mechanisms include the natural inactivator of the kinins and
the anaphylatoxins [1]. Human serum contains a carboxypeptidase B-type
enzyme that irreversibly inactivates by cleavage of the C terminal arginine
the kallikrein-produced kinins as well as the anaphylatoxin-active peptide
C3a [2]. The presence of this inactivator in serum explains why neither kinin
activity nor anaphylatoxin activity can be demonstrated in human serum
after appropriate activation of kallikrein or the C3 convertase. However, if
the carboxypeptidase is blocked by the presence of high concentrations of
ε-amino caproic acid, substantial amounts of C3a anaphylatoxin can be
demonstrated in serum following appropriate activation procedures [2].

The Chemotactic Factor Inactivator

Recently in collaboration with Dr. J. L. BERENBERG and Dr. G. TILL it
has been found that human serum contains a factor that blocks the biological
activity associated with several chemotactic factors (table I). The chemotactic
factor inactivator (CFI) is present in normal human serum in low concentrations. Its activity can be demonstrated by fractionation and concentration
of the pseudoglobulin portion of serum after ammonium sulfate precipitation
of the γ-globulin [3]. The mechanism by which CFI acts is not known.
However, two points can be made. First, the interaction of CFI with the

1   These studies were supported in part by NIH grants AI-09651 and AI-11526.

chemotactic factor appears to be irreversible. This observation was made possible by the fact that CFI is heat-labile while the bacterial chemotactic factor is heat-stable. If the mixture of CFI and bacterial chemotactic factor is first incubated with 37 °C and then the temperature raised to 56 °C so as to inactivate CFI, the chemotactic activity does not reappear, indicating that the CFI has reacted in an irreversible manner. The second point is that the interaction between CFI and chemotactic factor does not seem to involve irreversible binding between the two reactants [3]. Both of these observations would be consonant with the interpretation that CFI enzymatically inactivates chemotactic factors, but direct evidence for this hypothesis is lacking.

The relationship between the CFI and the anaphylatoxin inactivator is not clearly defined, although there is some evidence that would indicate the inactivators are dissimilar. This is based on the differences in physical-chemical features of the two inactivators as well as by the observations that CFI appears relatively insusceptible to inhibitors of the anaphylatoxin inactivator, including the chelators ethylenediaminetetraacetate sodium and o-phenanthrolene [4]. The two inactivators are both resistant to inhibition by diiosopropylfluorophosphate [1, 4]. The question of the functional relationship between the two inactivators is emphasized by the observations that the anaphylatoxin inactivator can destroy both the anaphylatoxin as well as the chemotactic activity associated with the C3a fragment of C3. This would suggest that the C terminal arginine is critical for both activities. However, whether CFI can destroy both the anaphylatoxin activity of C3a as well as the chemotactic activity (the latter being proven) is not yet established. The lack of understanding of the structures of the chemotactic factors has hampered the biochemical explanation for the basis of CFI activity.

Abnormal Levels of CFI in Human Sera

As indicated above, CFI exists in normal human sera in low amounts. This accounts for the fact that human serum can be readily chemotactically activated by incubation with complement triggers such as immune complexes, bacterial endotoxin or the yeast particle zymosan. The fact that anaphylatoxin can not be generated (or expressed) under such conditions indicates either that the anaphylatoxin inactivator is present in substantially higher amounts than CFI, or that the chemotactic products generated in serum are resistant to the action of either the anaphylatoxin inactivator or CFI, or

Table I. Chemotactic factors inactivated by the chemotactic factor inactivator

1. Activity present in human serum activated by incubation with immune complexes
2. $C\overline{567}$
3. C3 fragment
4. C5 fragment
5. Bacterial factor from *Escherichia coli*

Table II. Abnormalities in chemotactic factor inactivator (CFI) levels in human sera[1]

I. Elevated levels: Hodgkin's disease
II. Depressed levels: α_1-antitrypsin deficiency

1 See text for details.

that the chemotactic assay is substantially more sensitive than the anaphylatoxin assay (which involves contraction of guinea pig ileum that is blocked by antihistamines).

Two general categories of abnormal levels of CFI have been found: elevated levels and depressed levels (table II). The former has been found in nearly half of all sera from Hodgkin's patients examined [5]. In some cases the CFI level appears to be at least 5-fold above the level observed in normal serum. In most respects the CFI in the pathological serum resembles that present in normal human serum on the basis of electrophoretic and ultracentrifugal characteristics. However, one qualitative difference has been found. This relates to the finding that CFI in the Hodgkin's serum can inactivate the chemotactic activity in the bacterial chemotactic factor and that activity associated with the C5, but not the C3, fragment. In contrast, CFI isolated from normal human serum can inactivate all three of the chemotactic factors.

In the presence of elevated CFI levels in Hodgkin's sera, the amount of chemotactic activity generated in the same sera after addition of zymosan is substantially below amounts generated in similarly treated normal human sera. Further, the addition of preformed chemotactic factors to Hodgkin's (but not in the case of normal human) sera results in substantial inactivation of the chemotactic activity. These observations suggest that because of elevated levels of CFI in Hodgkin's sera there is a significant abnormality in the apparatus that regulates inflammatory (chemotactic) mediators.

The second category of abnormalities in CFI activity in human serum relates to significantly diminished levels of CFI found in sera from patients demonstrating severe deficiency of α_1-antitrypsin. These patients concomitantly demonstrate progressive pulmonary emphysema. Studies of companion sera indicate that CFI deficiency is not present in sera from patients with severe chronic bronchitis and/or advanced pulmonary emphysema, as long as substantially normal levels of α_1-antitrypsin are present [6]. Thus, there is a general correlation between the levels of α_1-antitrypsin and CFI. Most of the current evidence suggests that these two proteins are dissimilar on the basis of physical-chemical comparisons. This would further suggest that the two proteins are under the same or similar genetic control.

When sera deficient in CFI are fractionated according to some of the procedures outlined above, little, if any, inactivator activity for chemotactic factors has been found. Correspondingly, if the same sera are chemotactically activated with zymosan, supernormal amounts of chemotactic activity are generated [6]. Thus, there appears to be a clear regulatory defect in the control of inflammatory (chemotactic) mediators in these sera. How this relates to the progressive pulmonary emphysema seen in these patients is unknown, although this disease may reflect long-standing, unregulated (excessive) inflammatory reactions in lungs. Whether conditions other than the α_1-antitrypsin deficiency are associated with diminished levels of CFI in serum is unknown at the present time.

The CFI appears to be an important regulator of inflammatory responses in humans. Its function is almost surely to maintain a homeostatic balance when the inflammatory response has been incited by the generation of chemotactic mediators. Excessive or depressed levels of CFI would be expected to be translated, respectively, into inadequate or exaggerated inflammatory responses. There are some clinical data that support this hypothesis. There are well documented examples of insufficient inflammatory responses in Hodgkin's patients, involving both acute (neutrophils) and chronic (delayed-type hypersensitivity) inflammatory cells. Excessive levels of CFI may result in a relative insufficiency of chemotactic mediators and, thus, a diminished inflammatory response. In the case of deficiency of CFI there are no data bearing on the relative excess (or inadequacy) of the inflammatory response. It would seem that CFI represents an important, naturally occurring anti-inflammatory substance in serum. A better understanding of this factor may yield novel approaches to therapeutic intervention in the face of inflammatory diseases.

Summary

The CFI is a natural regulator of leukotactic mediators and exists in small quantities in normal human serum. CFI irreversibly inactivates the neutrophil chemotactic activity associated with zymosan-activited human serum, with $C\overline{567}$, the fragments of C3 and C5 and the bacterial chemotactic factor. CFI is present in considerably elevated amounts in the sera of patients with Hodgkins disease. These sera are defective in the generation of chemotactic activity and show the ability to block the activity of preformed chemotactic factors when added to the sera. CFI appears to be present in abnormally low levels in sera of patients with severe deficiency of α_1-antitrypsin. When activated with zymosan, these sera generate supernormal amounts of chemotactic activity. These findings emphasize the biological significance of the CFI.

References

1   BOKISH, V. A. and MÜLLER-EBERHARD, H. J.: Anaphylatoxin inactivator of human plasma. Its isolation and characterization as a carboxypeptidase. J. clin. Invest. 49: 2427 (1970).

2   VALLOTA, E. H. and MÜLLER-EBERHARD, H. J.: Formation of C3a and C5a anaphylatoxins in whole human serum after inhibition of the anaphylatoxin inactivator. J. exp. Med. 137: 1109 (1973).

3   BERENBERG, J. L. and WARD, P. A.: The chemotactic factor inactivator in normal human serum. J. clin. Invest. (in press).

4   TILL, G. O. and WARD, P. A.: The chemotactic factor inactivator. J. Immunol. (in press).

5   BERENBERG, J. L. and WARD, P. A.: Serum inhibitors of chemotactic factors (abstract). J. clin. Invest. 51: 9 (1972).

6   WARD, P. A. and TALAMO, R. C.: Deficiency of the chemotactic factor inactivator in human sera with α_1-antitrypsin deficiency. J. clin. Invest. 52: 516 (1973).

Author's address: PETER A. WARD, Department of Pathology, University of Connecticut Health Center, Farmington, CT 06032 (USA)

Antibiotics and Chemotherapy, vol. 19, pp. 338–349
(Karger, Basel 1974)

Chemotaxis and Random Mobility

Clinical and Biologic Differentiation[1]

M. E. MILLER

Charles R. Drew Postgraduate Medical School, and Martin Luther King, jr. General
Hospital, Los Angeles, Calif.

Recent clinical observations on disorders of neutrophil mobility have
increased our understanding of the mechanisms of cell movement and their
relationships to normal inflammatory function. In this chapter, we summarize the current knowledge of the clinical disorders involving leukocyte
movement, and review *in vitro* data which demonstrates that: (1) at least
two types of neutrophil movement are of significance in human disease-
chemotaxis and random mobility, and (2) measurement of random mobility
by the capillary tube method measures an entirely different set of parameters
than as measured by the method of chamber migration.

I. Historical Aspects

In the first disorder involving a primary abnormality of neutrophil
movement, the 'lazy leukocyte syndrome' [1], two unrelated patients (a 4-
year-old boy and a 3-year-old girl) were studied. Each child had been troubled
with recurrent episodes of low-grade gingivitis, stomatitis, otitis media and
recurrent fevers. Of particular note was the disproportionately low-grade
clinical symptomatology of these children. In other words, fever and the
usual signs of toxicity were mild, or absent, during their episodes of infection.
Upon laboratory examination it was noted that each child had a marked,
severe neutropenia, with peripheral blood polymorphonuclear leukocyte
(PMN) counts in the range of 2–500 absolute PMN/cc. Examination of
bone marrow specimens from each child showed an entirely normal matura-

1   Supported by USPHS grant 1-RO-1-AIHD-10689.

tion series, with morphologically normal appearing neutrophils in all stages of development, thus offering no apparent explanation for the peripheral neutropenia.

Evaluation of the ability of the patients to mobilize this apparently normal store of neutrophils to the peripheral blood stream showed three significant abnormalities: (1) stimulation of 'marginal pool', or extra-marrow neutrophils by administration of epinephrine yielded no effective increase in circulating PMN; (2) stimulation of marrow pool neutrophils by administration of endotoxin yielded no effective increase in circulating PMN, and (3) *in vivo* inflammatory response as measured by the inflammatory cycle method of REBUCK and CROWLEY [2] yielded no PMN. Thus, the paradox of adequate neutrophil stores in the bone marrow, but peripheral neutropenia and poor mobilization of these marrow neutrophils by epinephrine, endotoxin or inflammatory stimulation had to be explained.

A number of possible explanations were considered: (1) the patients' plasmas might lack some factor(s) essential for the recruitment of PMN from the marrow; (2) the patients' plasmas might contain an inhibitory factor(s) against leukotaxis; (3) a primary deficiency of the serum complement system resulting in an inability to generate sufficient amounts of chemotactic factor(s) might be present, and (4) a primary cellular defect of neutrophil function might be present.

Through studies under appropriate *in vitro* and *in vivo* conditions, the first three possibilities were clearly ruled out. Upon study of PMN from the patients *in vitro*, however, a primary defect of cell movement involving *both* chemotaxis and random tube migration was demonstrated. By contrast, *in vitro* phagocytic and bactericidal activities of the PMN were normal. The defect, thus, seemed quite specific, and an apparent 'cause and effect' relationship seemed established – i.e. the defect in cell movement was reflected in the abnormal epinephrine, endotoxin and skin window responses, and the patients' symptoms were presumably related to their inability to mobilize PMN to the site of infection.

That these relationships might not be that simple soon became apparent with the recognition of a second primary defect of PMN movement, in this case a *familial* chemotactic defect [3]. Two families were studied, in one family a brother and sister being involved and in the other an 11-year-old girl. All three patients had an unusual clinical course characterized by congenital ichthyosis and recurrent trichophyton rubrum infections. Neutrophil counts were relatively normal as were bone marrow examinations. PMN from the patients were normal in phagocytic and bactericidal activities but, as in the 'lazy leukocyte syndrome', were deficient in chemotactic activity. *Unlike* the 'lazy leukocyte syndrome', however, the measurement of random mobility was normal. Further, stimulation of neutrophils by epinephrine, endotoxin or Rebuck window was normal. The familial nature of the defect was revealed by finding the same abnormalities of chemotaxis with normal random mobility in PMN from the fathers.

Through the characterization of these two clinical entities, it was therefore apparent that the relationships between *in vitro* assays of neutrophil movement and *in vivo* correlates, such as neutrophil counts, and mobilization of PMN were more complex than had originally been thought, and that random mobility and chemotaxis might involve separate mechanisms or functions of PMN. In the following sections, current data is summarized which supports both of these contentions.

II. In vitro *Observations on Random Mobility and Chemotaxis*

Two types of neutrophil movement, *directed* migration (or chemotaxis) and *non-directed* migration (or random mobility) have long been recognized. There has, however, been much controversy as to whether these two types of cell movement involve the same or different mechanisms of cell function [4]. Much of the confusion on this point derives from the fact that the several methods used by investigators to measure 'random mobility' have not been measuring the same phenomena. The data now summarized shows that random migration, as measured by the method of capillary tube movement, involves distinctly different mechanisms of cell activity than random mobility as measured by migration through a micropore filter. Following this, clinical data will be summarized which demonstrates the *in vivo* significance of these studies.

In these experiments human, peripheral blood PMN have been studied in three different assay systems.

1. *Chemotaxis* has been studied by a previously described modification of the Boyden assay [1]. Washed suspensions of 5×10^5 PMN in medium 1066 with 10% heat-inactivated fetal calf serum were placed in the upper portion of a Sykes-Moore tissue culture chamber and separated from a source of chemotactically active material by a 3-μm pore size filter. The chemotactically active materials were generated by the addition of pre-formed antigen-antibody complexes to fresh, normal human serum. Follow-

ing a 3-hour incubation period at 37°C, the filters were removed, stained, and the average number of PMN per high power field reaching the lowermost portion of the filter determined.

2. In the second assay, *random migration* by the *filter* method was studied by a modification of the method of CARRUTHERS [5]. The assays were performed in the same Sykes-Moore chambers as above, but the PMN were prepared in concentrations of 10^5 PMN/cc and placed over an 8-μm instead of a 3-μm filter. Concentrations of buffer or plasma or serum were identical on both sides of the filter. In other words, unlike the chemotactic assay, no gradient was present.

3. *Random migration* in *capillary tubes* was measured by a previously described modification of the method of Ketchel and Favour [1, 6]. Washed, *plasma-free* suspensions of 5×10^5 PMN/cc in 0.1% human albumin were placed in *siliconized*, microhematocrit tubes placed upon a microscope stage which was turned to the vertical position, thus placing the tubes in an upright position, like sedimentation rate preparations. The distance migrated upwards by the leading cells in the field after a 3-hour incubation period at room temperature was then measured with an ocular micrometer. The optics of the system were improved remarkably by placing the tubes in a chamber constructed out of capillary tubes sandwiched between two microscope slides. The chamber was filled with immersion oil and the random migration tubes placed in the chamber. This modification permitted accurate observation of the cells within the tubes.

Comparative effects of a variety of additive materials upon each of the three assays were then determined. The basic experimental protocol was as follows: normal PMN were incubated at 37°C for 30 min at constant rotation in the presence of a particular additive. The cells were then washed, resuspended as described and tested in each of the three assays. Incubations in the following agents were compared.

  1. Dinitrophenol in concentrations of 10^{-3}, 10^{-5} and 10^{-7} M.
  2. Potassium cyanide in concentrations of 10^{-3}, 10^{-5} and 10^{-7} M.
  3. Sodium fluoride in concentrations of 2×10^{-3}, 2×10^{-5} and 2×10^{-7} M.
  4. Iodoacetate in concentrations of 10^{-3}, 10^{-5} and 10^{-7} M.
  5. Mitomycin C at a concentration of 25 μg/ml.
  6. Sodium EDTA at a concentration of 1 mg/ml.
  7. Deoxyglucose in concentrations of 6×10^{-2}, 6×10^{-3} and 6×10^{-4} M.
  8. Chemotactic factors generated as described for the chemotactic assay.

Table I. Comparative effects upon chemotaxis, random chamber migration and random tube migration

	Chemotaxis	Chamber migration	Tube migration
Dinitrophenol			
10⁻³ M	NE	NE	NE
10⁻⁵ M	NE	NE	NE
10⁻⁷ M	NE	NE	NE
Potassium cyanide			
10⁻³ M	NE	NE	NE
10⁻⁵ M	NE	NE	NE
10⁻⁷ M	NE	NE	NE
Sodium fluoride			
2 × 10⁻⁷ M	D	D	NE
2 × 10⁻⁵ M	+D	+D	NE
2 × 10⁻³ M	+ +D	+ +D	NE
Iodoacetate			
10⁻⁷ M	D	D	NE
10⁻⁵ M	+D	+D	NE
10⁻³ M	+ +D	+ +D	NE
Mitomycin C, 25 μg/ml	D	D	NE
Sodium EDTA, 1 mg/ml	D	D	NE
Deoxyglucose			
6 × 10⁻⁴ M	D	D	NE
6 × 10⁻³ M	+D	+D	NE
6 × 10⁻² M	+ +D	+ +D	NE
Chemotactic factors	D	D	NE

NE = no effect; D = decreased by at least 2 SD from the mean value for that particular assay.

The results of these experiments are summarized in table I. A significant decrease following incubation in a particular additive is recorded only when the decrease exceeds by at least 2 SD the mean value for that particular assay. As can be seen, incubations with inhibitors of glycolysis such as sodium fluoride or iodoacetate effected comparable decreases upon chemotaxis and chamber migration, but had no effect upon tube migration. By contrast, incubations with inhibitors of oxidative metabolism such as dinitrophenol or potassium cyanide had no effect upon any of the three assays. PMN which had been incubated in the presence of mitomycin C, sodium EDTA, deoxyglucose or chemotactically active materials all showed markedly diminished chemotaxis and chamber migration but retained normal tube migration.

Thus, incubation of PMN in a wide variety of agents resulted in similar decreases of chemotaxis and random filter migration but had no effect upon random tube migration. One explanation might be that the tube migration assay is simply less sensitive than the filter assays. This, however, does not appear to be the case. Effects of temperature, for example, are identical for each assay, as are the effects of sodium arsenite. The latter uniformly decreases all three activities at comparable concentrations. Further, there are *in vitro* and *in vivo* situations in which random tube mobility is decreased but chemotaxis is normal.

The *in vivo* situations will be summarized in the following section on clinical disorders of PMN movement.

In *in vitro* studies, normal human PMN which have been passaged through a Dow hollow fiber artificial kidney membrane (cellulose acetate hollow fibers 200–250 μm internal diameter, approximately 20 cm long and 1 m² nominal area potted in silicone rubber) show modest decreases in chemotactic activity but are rendered entirely devoid of random tube mobility [7]. That this decrease is related to properties of the membrane, rather than to mechanical factors of the procedure, is suggested by the fact that identical treatment of cells utilizing the anisotropic polysulfone XM-50 renal dialysis membrane (Amicon) has no effect upon random mobility. It should, further, be noted that patients who have undergone renal dialysis in which the Dow membrane has been employed have considerably greater degrees of postdialysis neutropenia than those who have been dialyzed with the Amicon membrane.

The data summarized demonstrates that random tube migration and chemotaxis involve distinct mechanisms of cell activity. What is not presently clear, and forms one of the more compelling research questions in this field, however, is the issue of what process(es) is (are) actually being measured in the tube migration system. That is, is this truly 'random' migration, or is it glass adhesiveness, or both? KETCHEL and FAVOUR [8] and BRYANT et al. [6] proposed that glass adhesiveness was involved and that this was mediated in large part by plasma factors. Our data, however, does not support this. The tube assays were all performed in siliconized tubes in the absence of plasma. Further, EDELSON et al. have recently described two patients with deficient tube migration but normal adhesiveness of their PMN to nylon columns [9].

III. *Chemotaxis and Random Mobility – Clinical Correlates*

Clinical entities involving isolated deficiencies of random mobility, isolated deficiencies of chemotaxis and combinations involving deficiencies of both have now been identified. Four basic types of defects of abnormal cell movement are recognized: (1) intrinsic defects of the leukocyte involving only deficiencies of movement; (2) intrinsic defects of the leukocyte involving combined defects of movement and at least one other function such as phagocytosis, bactericidal activity or adhesiveness; (3) disorders of neutrophil movement related to abnormalities of the serum complement system, and (4) disorders involving the presence of humoral inhibitors of leukotaxis which mediate their effects either directly upon the leukocyte, or indirectly through activity against the serum complement system. We deal here specifically with categories 1 and 2. The disorders involving humoral factors are discussed elsewhere [10–12].

Table II lists the abnormalities of PMN movement involving intrinsic leukocyte defects of movement alone [1, 3, 13–19]. No humoral abnormalities have been observed among these defects, nor are there present any known defects of other neutrophil functions. As can be seen, isolated disorders of chemotaxis or random mobility, or combined defects of both have been reported. With these abnormalities, as with those of combined functional neutrophil defects of movement and at least one other PMN function (table III) [9, 20–22], many problems remain in the firm establishment of clinical significance, and the manner in which a particular *in vitro* defect(s) results in clinical increased susceptibility to infections.

IV. In vitro *and* in vivo *Correlates*

As emphasized, the relationships between *in vitro* abnormalities of PMN movement and *in vivo* observations such as neutrophil counts, responses to epinephrine and endotoxin, and the inflammatory cycle ('skin window') of REBUCK and CROWLEY are not nearly as simple as appeared in the description of the 'lazy leukocyte syndrome'. This is demonstrated in table IV, which lists the abnormalities in table II for which data exists concerning neutrophil counts, epinephrine and endotoxin and skin window responses. As can be seen, no single parameter correlates entirely with *in vitro* abnormalities. Examples now exist of normal, or nearly normal skin windows in patients with primary defects of chemotaxis, primary defects of random

Table II. Intrinsic defects of polymorphonuclear leukocytes (PMN) involving only movement

Defect and text reference	Chemotaxis	Random mobility
1. Chediak-Higashi syndrome [13]	diminished	normal
2. Lazy leukocyte syndrome [1]	diminished	diminished
3. Normal human neonate [14]	diminished	diminished
4. Diabetes mellitus		
Adult [15]	diminished	normal
Juvenile [16]	diminished	normal
5. Familial chemotactic defect [3]	diminished	normal
6. Rheumatoid arthritis [17]	diminished	–
7. Chronic mucocutaneous candidiasis with decreased cellular immunity [18]	diminished	normal
8. Defective mononuclear chemotaxis and chronic mucocutaneous candidiasis [19]	diminished (mononuclear)	–
9. Postrenal dialysis with Dow cellular membrane [7]	slightly diminished	totally absent

Table III. Intrinsic disorders of PMN movement associated with at least one other abnormality of PMN function

Text reference	Disorder
1. EDELSON *et al.* [9]	PMN deficient in chemotaxis, random tube mobility, nylon column adherence and bactericidal activity against gram-negative bacteria
2. EDELSON *et al.* [9]	PMN deficient in chemotaxis, random mobility and bactericidal activity against gram-negative bacteria
3. EDELSON *et al.* [9]	PMN deficient in chemotaxis, random mobility, NBT reduction and bactericidal activities against *S. aureus* and gram-negative bacteria
4. STEERMAN *et al.* [21]	intrinsic deficiencies of PMN and sex-linked hypogammaglobulinemia
5. HIGGINS *et al.* [22]	leukocytosis and hypergammaglobulinemia with recurrent infections; PMN deficient in random mobility, chemotaxis and glass bead adherence
6. MILLER and DOOLEY [20]	2 siblings with pyoderma gangrenosum; PMN deficient in phagocytosis and random mobility; chemotaxis normal

Table IV. In vitro and *in vivo* correlates of PMN movement defects

Defect	PMN count	Rebuck	Chemotaxis	Random mobility	Epineph-rine and endotoxin
1. Chediak-Higashi [13]	diminished	diminished or normal	diminished	normal	–
2. Lazy leuko-cyte syndrome	diminished	diminished	diminished	diminished	diminished
3. Neonate [14]	moderately diminished	moderately diminished	diminished	moderately diminished	–
4. Diabetes [15, 16]	normal	moderately diminished	diminished	normal	normal
5. Familial chemotactic defect [3]	normal	normal	diminished	normal	normal
6. Rheumatoid arthritis [17]	variable	–	diminished	–	–
7. Candidiasis with de-creased cellular immunity		moderately diminished	diminished	normal	normal
8. Dialysis with Dow membrane [7]	diminished	–	moderately diminished	totally absent	–

mobility and even in patients with primary abnormalities of both. It is also seen that neutrophil counts among these patients are variable. Random tube migration appears to correlate more strongly with neutrophil counts than chemotaxis.

The relationships between neutrophil response in skin window in-flammation and response to challenge with endotoxin are also inconsistent. For example, in the patient recently described by CLARK *et al.* with defective neutrophil chemotaxis [18], a 4-fold rise of peripheral granulocyte count was noted upon administration of endotoxin, while the skin window showed only a modest decrease in cell migration, particularly during the early poly-morphonuclear phase.

In the intrinsic defects of neutrophils involving combined functional defects, yet another consideration is introduced. In these disorders (table III), deficiencies of apparently unrelated neutrophil activities have been observed in combination. In the three patients described by EDELSON *et al.* [9], for example, the first had PMN deficient in chemotaxis, random tube mobility, adherence to nylon columns and bactericidal activity against gram-negative bacteria, the second had PMN deficient in chemotaxis and random mobility, bactericidal activity against gram-negative bacteria but normal nylon column adherence, and the third had PMN deficient in chemotaxis and random mobility, inability to reduce nitro-blue tetrazolium dye and deficient killing of both *Staphylococcus aureus* and gram-negative bacteria. We have recently described two brothers with pyoderma gangrenosum, whose PMN showed a combined defect of phagocytosis and random tube migration, with normal chemotaxis [20]. These two functions were compared in a series of *in vitro* experiments utilizing the same panel of inhibitors employed in the already discussed comparative studies of chemotaxis and random mobility. The results showed that phagocytosis is inhibited by these agents in an identical fashion to chemotaxis, but as before, random mobility (tube) was unaffected. In other words, the combined PMN deficiencies of phagocytosis and random mobility are very difficult to explain on the basis of a common PMN mechanism.

How, then, can we explain these combined defects of PMN function? While not yet definitive, the suggestion made by EDELSON *et al.* of early cell-membrane associated phenomena being involved seems quite resaonable [9]. To this end, in preliminary experiments in our laboratory, the treatment of normal PMN with trypsin under varying conditions has produced a variety of combined defects of leukocyte function. While much more work needs to be done in this area, the basic importance of differentiation of these combined defects of neutrophil activity from those of more specific nature involving movement, ingestion and killing lies in the recognition that the former group may involve membrane-associated deficiencies while, in the latter, specific metabolic pathways and/or receptor-type structures may be involved.

Summary

As should be apparent from the review in this chapter, disorders of leukocyte movement have become more widely recognized in recent years, and have revealed an increasingly complex set of humoral and cellular interactions. Rather than becoming discouraged at the prospects of working out these complex interactions, however, it should be remembered that this field has many parallels to other areas of the inflammatory-immune response, the disorders of which at one point also seemed hopelessly complex. With the development of more sophisticated methods of study, the exciting challenges presented to the understanding of these disorders should be successfully attacked and point to specific therapy in patients afflicted with one of these disorders.

References

1 MILLER, M. E.; OSKI, F. A., and HARRIS, M. B.: Lazy-leucocyte syndrome. A new disorder of neutrophil function. Lancet *i:* 665–669 (1971).

2 REBUCK, J. W. and CROWLEY, J. H.: A method of studying leukocytic functions *in vivo.* Ann. N.Y. Acad. Sci. *59:* 757–805 (1955).

3 MILLER, M. E.; NORMAN, M. E.; KOBLENZER, P. J., and SCHONAUER, T. J.: A new familial defect of neutrophil movement. J. Lab. clin. Med. *82:* 1–8 (1973).

4 ZIGMOND, S. H. and HIRSCH, J. G.: Leukocyte locomotion and chemotaxis. New methods for evaluation and demonstration of a cell-derived chemotactic factor. J. exp. Med. *137:* 387–410 (1973).

5 CARRUTHERS, B. M.: Leukocyte motility. I. Method of study, normal variation, effect of physical alterations in environment, and effect of iodoacetate. Canad. J. Physiol. Pharmacol. *44:* 475–485 (1966).

6 BRYANT, R. E.; DESPREZ, R. M.; VANWAY, M. H., and ROGERS, D. E.: Studies of human leukocyte motility. I. Effects of alterations in pH, electrolyte concentration, and phagocytosis on leukocyte migration, adhesiveness, and aggregation. J. exp. Med. *124:* 483–499 (1966).

7 HENDERSON, L. W.; MILLER, M. E.; HAMILTON, R. W., and NORMAN, M. E.: Dialysis leukopenia, polymorph random mobility and the control of peripheral white blood cell levels. A preliminary observation (submitted for publication).

8 KETCHEL, M. M. and FAVOUR, C. B.: The acceleration and inhibition of migration of human leucocytes in vitro by plasma protein fractions. J. exp. Med. *101:* 647–663 (1955).

9 EDELSON, P. J.; STITES, D. P.; GOLD, S., and FUDENBERG, H. H.: Disorders of neutrophil function. Defects in the early stages of the phagocytic process. Clin. exp. Immunol. *13:* 21–28 (1973).

10 WARD, P. A.: Insubstantial leukotaxis. J. Lab. clin. Med. *79:* 873–877 (1972).

11 MILLER, M. E.: Cell movement and host defenses. Ann. intern. Med. *78:* 601–603 (1973).

12 MILLER, M. E.: Leukocyte movement. *In vitro* and *in vivo* correlates. J. Pediat. (in press).

13 CLARK, R. A. and KIMBALL, H. R.: Defective granulocyte chemotaxis in the Chediak-Higashi syndrome. J. clin. Invest. *50:* 2645–2652 (1971).

14 MILLER, M. E.: Deficiency of chemotactic function in the human neonate. A previously unrecognized defect of the inflammatory response. Pediat. Res. *3:* 497–498 (1969).

15 MOWAT, A. G. and BAUM, J.: Chemotaxis of polymorphonuclear leukocytes from patients with diabetes mellitus. New Engl. J. Med. *284:* 621–627 (1971).

16 MILLER, M. E. and BAKER, L.: Leukocyte functions in juvenile diabetes mellitus. Humoral and cellular aspects. J. Pediat. *81:* 979–983 (1972).

17 MOWAT, A. G. and BAUM, J.: Chemotaxis of polymorphonuclear leukocytes from patients with rheumatoid arthritis. J. clin. Invest. *50:* 2541–2549 (1972).

18 CLARK, R. A.; ROOT, R. K.; KIMBALL, H. R., and KIRKPATRICK, C. H.: Defective neutrophil chemotaxis and cellular immunity in a child with recurrent infections. Ann. intern. Med. *78:* 515–519 (1973).

19 SNYDERMAN, R.; ALTMAN, L. C.; FRANKEL, A., and BLAESE, R. M.: Defective mononuclear leukocyte chemotaxis. A previously unrecognized immune dysfunction studied in a patient with chronic mucocutaneous candidiasis. Ann. intern. Med. *78:* 509–513 (1973).

20 MILLER, M. E. and DOOLEY, R.: Deficient random mobility, normal chemotaxis and impaired phagocytosis in two sibs with pyoderma gangrenosum (abstract). Pediat. Res. *7:* 137 (1973).

21 STEERMAN, R. L.; SNYDERMAN, R.; LEIKEN, S. L., and COLTEN, H. R.: Intrinsic defect of the polymorphonuclear leucocyte resulting in impaired chemotaxis and phagocytosis. Clin. exp. Immunol. *9:* 939–946 (1971).

22 HIGGINS, G. R.; SWANSON, V., and YAMAZAKI, J.: Granulocytasthenia. A unique leukocyte dysfunction associated with decreased resistance to infection. Clin. Res. *18:* 209 (1970).

Author's address: MICHAEL E. MILLER, MD. Department of Pediatrics, Martin Luther King, jr. General Hospital, 1620 E. 119th Street, *Los Angeles, CA 90059* (USA)

Antibiotics and Chemotherapy, vol. 19, pp. 350–361
(Karger, Basel 1974)

Chemotaxis of Human Polymorphonuclears *in vitro*

Critical Study of Clinical Interpretations

P. C. Frei, M. H. Baisero and Michèle Ochsner

Département d'Immunologie et d'Allergie, Service de Médecine, Hôpital Cantonal Universitaire, Lausanne

Introduction

Defective chemotaxis of polymorphonuclear leukocytes (PMN) was recently described in human disease by several authors [4, 5, 7, 8, 10, 13, 15–19, 21, 22, 24] and editorials have been published on this subject [14, 23]. Many of these papers report isolated cases in children [4, 5, 8, 10, 13, 16, 21, 22, 24] but the defect has also been observed in adults suffering from cirrhosis of the liver [7], infections [17], diabetes [18] and rheumatoid arthritis [19]. In three cases [1, 7, 15], plasma chemotactic factors were found to be deficient, an observation that was ascribed to the C3 component of complement in one case [1] and to C5 in another [15]. In two cases [21, 24], a serum inhibitor was described. The impairment in chemotactic response was shown to be limited to an intrinsic defect of PMN in three infected children with [4] and three without [5, 10, 22] Chediak-Higashi syndrome, or in two neutropenic children [16]. One group has demonstrated the same deficiency in two infected children and one adult [8]. In the groups of adults, described by Mowat and Baum [17–19], the PMN response was, on an average, lower than in control groups, but the plasma chemotactic factors were not tested.

All these experiments were performed with the help of Boyden's filter method. Polycarbonate (Nuclepore) filters were used in one instance [8], whereas Millipore filters were used by all other workers. Unfortunately, none of these authors mentioned the lowest normal values below which chemotaxis can be considered as pathological. Reproducibility is only given by Mowat and Baum [17–19], who calculated the variance and mentioned an accuracy of 7%.

Therefore, we reinvestigated the methodological problems inherent in the microfilter procedure until reproducible results in man were obtained and lowest normal values could be established. Apart from some minor modifications [20], the principal change consisted of a new way of counting the number of the migrating PMN [9].

The relations between recurrent infections and the deficiencies of chemotaxis are not very clear in the cited papers. They generally do not indicate whether the patients were repeatedly investigated and whether they were reversibly or irreversibly deficient. However, MOWAT and BAUM [17] pointed out that 5 of their 15 infected patients could be retested after the infection had subsided and PMN migration was then found to be normal. The question arises as to whether the intrinsic PMN deficiency may be the cause or a mere consequence of the clinical situation. To test this latter hypothesis chemotaxis was measured, in the present work, during and after benign acute infections in 21 otherwise healthy individuals. Since a relationship was found in these patients between the impairment of PMN migration and the increase in nonsegmented PMN, the study was further extended to patients with a 'shift to the left' due to other reasons, namely acute hemorrhage and diseases of the blood.

Material and Methods

Patients: (1) 51 normal individuals; (2) 21 usually healthy individuals suffering from a benign acute bacterial infection at the moment of the experiment; (3) 6 usually healthy individuals suffering from an acute hemorrhage, and (4) 14 patients with leukemia, Hodgkin's disease, carcinoma or systemic lupus erythematodes. Patients of group 2 and 3 were recontrolled after recovery. All subjects were adult.

Migration. The technique, which has been described in detail elsewhere [9, 20], will be given here only briefly. The chamber of CORNELY [6] was used with slight modifications [20]: Its dimensions were recalculated in order to obtain an optimal number of PMN on the filter. The upper compartment was made of perspex tubing, on the end of which a 13-mm millipore filter was fixed with MF cement (Millipore XX 70,000,000). The lower or distal polystyrene compartment was chosen with a flat thin bottom so that the readings of cells could be performed as described below.

The upper compartment was filled with 10^6 leukocytes suspended in 0.5 ml of buffer [20] with 2% human serum albumin (HSA). The lower compartment was filled with 1.5 ml of human plasma diluted 1:5 in buffer and previously activated by a 30-min incubation with 0.8 mg human aggregated γ-globulin (AGG) [12]. Some chambers were filled with nonactivated plasma for the simultaneous assessment of nonstimulated migration. Cells were allowed to migrate for $2^1/_2$ h at 37°C.

Readings. After incubation, both compartments were separated. Upper compartment: the filter was fixed and stained as usual and the number of leukocytes still adhering to its distal surface was counted with a microscope equipped with a reticle. Ten squares of 0.1 mm² were counted and the number of cells on the entire distal surface was calculated.

The lower compartment was homogenized on a Vortex mixer and briefly centrifuged: the PMN detached from the filter were thus equally distributed on the bottom They were directly counted without fixing or staining, with a reversed microscope equipped with a reticle. Four squares of 0.1 mm² were counted and the total number of cells which had fallen into the lower compartment was calculated.

The addition of the number of cells on the distal surface of the filter to the number of cells which had fallen into the lower compartment gave the total number of migrating cells.

Experimental pattern. In order to distinguish intrinsic deficiencies of PMN from deficiencies of plasma 'cytotaxins', the following scheme was used.

1. Patient's PMN attracted by autologous plasma.
2. Patient's PMN attracted by plasma from a normal individual.
3. PMN from a normal individual attracted by patient's plasma.
4. PMN from a normal individual attracted by autologous plasma (control).

Each of these experimental conditions was further divided into two groups, depending on whether the plasma was, or was not, activated by addition of AGG. Finally, each of these 8 groups consisted of 3–4 identical chambers, so that 24–32 chambers were needed to test one patient.

Results

In 51 normal adults of both sexes, the average total number of PMN migrating when chemotactic factors were generated by AGG was 178,500 (standard error of mean [SEM] = 13,800) per chamber. This figure represents 17.8% of the incubated leukocytes. Since the extreme values in this population were 38,500 and 452,000, it was found reasonable to consider any value lower than 30,000 as abnormal.

The average total number of PMN migrating without activation of the plasma by AGG was 23,000 (SEM = 2,100). The ratio between the two components of the total number of migrating cells (number of cells in the lower compartment per number of cells on distal surface of filter) was calculated in this control group and in patients. It was found to vary greatly between individuals with values being between 19.0 and 0.7. In some experiments, the layer of PMN lying on the bottom of the lower chamber was fixed, stained and examined: at least 99% of these cells were segmented PMN.

Table I shows the results obtained in 21 adults with different infections. Chemotaxis was impaired in nearly all of them. Since the deficiency was observed only when patient's cells were attracted by autologous or control

Table I. Impaired chemotaxis of PMN in adults during acute infection

Case No.	Age years	Diagnosis	Total number of migrating PMN (in thousands)					
			patient's PMN plus patient's plasma		patient's PMN plus normal plasma		normal PMN plus patient's plasma	
			non-activated	activated by AGG	non-activated	activated by AGG	non-activated	activated by AGG
1	84	bronchopneumonia	4.2	5.7	3.6	5.4	11.1	66.0
2	66	bronchopneumonia	3.6	32.7	9.4	21.4	4.9	68.5
3	35	bronchopneumonia	9.8	10.6	ND	ND	29.0	73.7
4	71	bronchopneumonia	1.2	4.1	14.7	23.9	67.2	341.8
5	52	bronchopneumonia	13.7	20.9	ND	ND	37.6	182.7
6	67	bronchopneumonia	3.5	6.8	8.0	16.2	12.2	104.2
7	52	bronchopneumonia	1.7	5.0	1.9	4.6	94.1	197.5
8	72	bronchopneumonia	2.5	2.6	1.0	2.2	10.6	197.4
9	26	bronchopneumonia	18.6	26.7	3.3	26.8	2.1	66.3
10	38	bronchopneumonia	1.0	1.2	1.0	1.3	ND	ND
11	20	septicemia (streptococcus)	3.7	4.5	1.8	2.6	18.3	180.2
12	48	septicemia (staphylo-)	1.0	1.3	1.0	1.0	20.8	55.7
13	69	septicemia (*Escherichia coli*)	0.9	1.4	1.3	2.9	14.7	168.9
14	73	septicemia (*Escherichia coli*)	1.5	2.7	0.2	4.0	107.8	119.7

Table I (continued)

Case No.	Age years	Diagnosis	Total number of migrating PMN (in thousands)					
			patient's PMN plus patient's plasma		patient's PMN plus normal plasma		normal PMN plus patient's plasma	
			non-activated	activated by AGG	non-activated	activated by AGG	non-activated	activated by AGG
15	28	urinary tract infection	2.4	6.7	4.0	13.2	4.3	82.9
16	26	urinary tract infection	6.6	21.4	0.6	16.6	41.2	134.5
17	37	pyelonephritis	2.4	3.0	1.5	1.9	7.3	109.7
18	80	perinephric abscess	8.5	14.8	1.4	17.3	12.0	93.5
19	48	pyoderma	1.2	1.3	0.8	1.2	76.8	229.9
20	29	meningitis (meningococcus)	2.7	18.3	2.8	16.6	25.7	107.4
21	52	acute bronchitis	0.9	4.2	1.2	4.5	15.0	187.5

AGG = aggregated γ-globulin. ND = not done.
Figures express the average total number of migrated cells per chamber on adding the number of cells counted on the distal surface of the filter to the number of cells counted in the bottom of the lower compartment. Values are decreased (lower than 30,000) in all cases, in experiments in which patient's PMN are attracted either by autologous or control plasma. Values are normal in experiments where control PMN are attracted by patient's plasma and by control plasma (this last variant was omitted from the table).

Table II. Normalization of PMN chemotaxis after termination of infection

Diagnosis	Number of days after acute phase	Total number of migrating PMN (in thousands); patient's PMN and patient's plasma		Proportion (in percent) of non-segmented PMN
		non-activated	activated by AGG	
Bronchopneumonia	0	1.4	4.3	75
	49	87.7	347.0	6
Bronchopneumonia	0	3.6	32.7	70
	196	15.7	229.0	8
Bronchopneumonia	0	9.8	10.6	38
	2	63.3	67.3	ND
	45	20.6	253.5	7
Bronchopneumonia	0	2.8	4.6	33
	3	12.6	18.7	23
	7	123.6	213.3	6
Bronchopneumonia	0	3.5	6.8	21
	24	6.2	41.4	ND
	224	79.7	342.3	8
Bronchopneumonia	0	18.6	26.7	36
	5	30.7	224.1	11
Pyelonephritis	0	2.4	3.0	22
	5	5.0	16.2	21
	11	42.6	103.2	10
Meningitis	0	2.7	18.3	37
	11	13.1	159.2	3

ND = not done.

Chemotaxis of PMN measured during and after acute infections in 8 adults. Figures express the average total number of migrated cells per chamber. Only results of chambers with patient's cells and autologous plasma are given here. Chambers with patient's cells and control plasma (decreased), control cells and patient's plasma (normal), control cells and control plasma (normal) have been omitted from the table. Figures of migration were compared with the ratio of nonsegmented PMN to the total number of cells used in the experiment.

Table III. Inverse relationship between chemotaxis and proportion of nonsegmented PMN in hemorrhages

Diagnosis	Number of days after (+) or before (−) hemorrhage	Total number of migrating PMN (in thousands)				Proportion (in percent) of nonsegmented PMN
		patient's PMN plus patient's plasma		patient's PMN plus normal plasma		
		non-activated	activated by AGG	non-activated	activated by AGG	
Acute gastrointestinal hemorrhage	0	2.6	21.3	1.1	9.7	30
	+1	2.8	64.8	ND	ND	8
	+15	30.6	78.2	1.5	147.8	9
Acute gastrointestinal hemorrhage	0	2.2	15.5	1.4	6.7	21
	+14	84.1	207.7	23.1	197.6	5
Traumatic hemorrhage	0	ND	6.1	ND	2.0	43
	+15	55.6	190.9	35.2	113.9	12
Normal menstruation	0	3.5	4.7	1.4	3.8	50
	+18	24.9	452.1	73.3	253.1	7
Normal menstruation	−15	20.3	107.0	9.2	91.0	9
	0	2.0	4.9	1.6	1.3	23
	+15	90.4	220.1	28.8	81.9	10
Normal menstruation	−20	84.7	366.0	47.5	278.9	5
	0	0.9	1.1	1.0	1.2	20
	+19	107.9	298.9	21.6	180.1	5

ND = not done.

Chemotaxis of PMN measured during and after acute hemorrhages in 6 adults. Figures express the average total number of migrated cells per chamber. Values lower than 30,000 after activation by AGG are considered as abnormal. Results of chambers with control PMN plus patient's plasma, and with control PMN plus control plasma were omitted (normal results). The percent of nonsegmented PMN is given in the last column: the shift to the left seems to correlate with the impairment of migration.

Table IV. Inverse relationship between chemotaxis and proportion of nonsegmented PMN in malignant diseases

Diagnosis	Total number of migrating PMN (in thousands)				Proportion (in percent) of nonsegmented PMN	
	patient's PMN plus patient's plasma		patient's PMN plus normal plasma			
	nonactivated	activated by AGG	nonactivated	activated by AGG		
Acute myeloblastic leukemia	0.6	1.1	0.6	1.7	98	PMN defect
Acute myeloblastic leukemia	1.0	1.3	1.0	1.5	59	PMN defect
Acute myeloblastic leukemia	26.9	73.6	11.9	60.7	17	normal
Hodgkin's disease	0.8	0.9	0.8	1.0	22	PMN defect
Hodgkin's disease	10.6	20.7	4.9	18.0	26	PMN defect
Hodgkin's disease	0.5	0.6	0.6	1.3	47	PMN defect
Hodgkin's disease	3.1	333.0	1.4	137.7	10	normal
Hodgkin's disease	33.0	246.9	4.9	84.8	15	normal
Carcinoma of the protstate	1.2	1.4	1.1	1.3	41	PMN defect
Carcinoma of the lung	0.7	1.1	0.9	1.3	21	PMN defect
Carcinoma of the bile ducts	113.9	192.6	8.9	123.0	4	normal
Carcinoma of the lung	9.9	70.5	3.2	55.2	10	normal
Systemic lupus (plus septicemia)	1.5	2.7	0.2	4.0	41	PMN defect
Systemic lupus	8.5	76.6	1.0	54.8	7	normal

Chemotaxis of PMN measured in 4 groups of diseases. Cases with and without increase in nonsegmented PMN are present in each group. The intrinsic defect of PMN migration was shown in patients with increase in nonsegmented PMN, independently of the underlying diagnosis.

plasma, but not when control cells were attracted by patient's plasma, it can be concluded that the impairment was limited to an intrinsic function of the PMN. In most of the cases, the values were much lower than the lowest normal limit of 30,000. The values obtained without activation by AGG were also lower with patient's cells than with control cells. In 8 cases of infection (table II), the experiment could be repeated later. In all cases the number of migrating PMN, with and without activation, returned to normal when the infection was eliminated. At the moment of the first determination of chemotaxis, all patients presented a 'shift to the left': the proportion of non-segmented PMN ('bands') among the incubated cells was abnormally high (between 21 and 75%). At the moment of the second and third determination, no patient revealed a 'shift to the left': the proportion of nonsegmented PMN was normal (between 3 and 11%).

The same inverse relationship between chemotaxis and the proportion of nonsegmented PMN was observed in 6 patients with a transitory 'shift to the left' due to a hemorrhage (table III). The migration of the PMN from these patients was transitorily impaired. Plasma chemotactic factors were normal, since migration of normal cells with patient's plasma (omitted in table III) was normal in all cases. When chemotaxis was abnormal, the proportion of 'bands' was between 20 and 43%. When chemotaxis returned to normal values, the proportion of nonsegmented PMN was found to be between 12 and 5%.

Table IV shows the results obtained in 14 cases of leukemia, Hodgkin's disease, carcinoma and systemic lupus erythematodes. These patients were tested only once. In each of these groups of disease, normal as well as abnormal results of chemotaxis were found. In cases with normal chemotaxis, the proportion of nonsegmented PMN was found to be between 4 and 17%. In cases with abnormal chemotaxis, this proportion was found between 21 and 98%.

Discussion

One of the main drawbacks of the different variants of the filter technique lies in their lack of reproducibility. We think that this is partly due to the methods used for counting the migrating cells. Like KELLER et al. [11], we also observed that most of these cells detach from the filter. The ratio of detached cells is not same in different experiments and varies between broad limits. For this reason, we studied a new way of reading the migration. This enabled us to establish an acceptable lowest limit for normal migration in humans.

Our finding of a constant and reversible decrease of the PMN response to chemotactic stimulation in infections raises the question as to whether some of the reported deficiencies might not have been a transitory consequence of the infections [4, 5, 8, 10, 16, 22]. Reversible deficiencies in infection were already mentioned by Mowat and Baum [17]. Clark and coworkers [4, 5, 25] suspected that impaired PMN migration could be secondary to infections; however, they did not retain this hypothesis, since they did not find such an impairment in several cases of infections, apart from their three cases of Chediak-Higashi syndrome [4] and one case of chronic infection with mucocutaneous candidiasis [5] in which the defect seems to be considered as primary. The differences between these results and ours could perhaps be explained by the fact that PMN were counted on the filter only.

The finding of a reversible intrinsic defect in PMN migration, along with an increase in nonsegmented cells, in infections (table II) as well as in other conditions (table III, IV), also raises the question as to whether younger cells are less capable of moving. They might be lacking in enzymes (myeloperoxydase?). Since their migration was also impaired when the plasma was not activated, it might rather be a problem of mobility than of 'understanding' of the chemotactic signal. Impaired random motility along with impaired chemotaxis has already been described [8, 16].

In testing 40 adults with recurrent infections, 16 with and 24 without specific immune deficiencies, we could not find any cellular impairment of chemotaxis, provided the patients were tested when completely free of infection [2, 3]. This, of course, does not rule out the possibility of finding a primary cellular deficiency of chemotaxis in adults, on studying a larger group of patients. Among these same 40 patients, a deficiency of chemotaxis due to plasma factors was found in 5 of them; when recontrolled later, these factors were found to be normal. This shows that one should also be cautious with the clinical interpretation of plasma deficiencies of chemotaxis some of which may be merely transitory.

Summary

The classical filter method for the assessment of chemotaxis was found to lack reproducibility in the case of human PMN and normal values of reference, which are necessary for any clinical interpretation, could not be established nor found in the literature. Therefore, a new modification of the filter procedure was investigated until reproducible results could be obtained.

The experimental pattern used allowed the separate assessment of (1) plasma factors of chemotaxis (cytotaxins), and (2) PMN ability to respond to these substances. Acceptable normal values of reference could be established both for unstimulated migration and for chemotaxis of human PMN.

It was shown that bacterial infection was associated with a clear-cut defect of the intrinsic PMN response without decrease in activity of plasma factors. The values promptly returned to normal when the infection was eliminated. A correlation was found between this defect and the increase in less mature nonsegmented PMN. When the proportion of these cells was increased for another reason, such as an acute hemorrhage or leukemia, deficiency of PMN migration was also observed. It might, thus, be a general phenomenon that only segmented PMN could respond to chemotactic stimulation. In case of infection or inflammation somewhere in the organism, most of the mature PMN are probably attracted to that site and so mainly younger cells, unable to respond, are collected from the vein. Therefore, meaningful investigations of suspected primary deficiency of chemotaxis can probably only be performed in the absence of any infection or inflammation at the time the blood is taken.

This paper strongly suggests that chemotaxis data in disease must be interpreted with caution. Some of the published cases of chemotaxis deficiency might not have been the cause of infections but, perhaps, a transitory consequence of them, since the patients might not have been completely free from infection at the moment of the investigation.

References

1   ALPER, C. A.; ABRAMSON, N.; JOHNSTON, R. B.; JANDL, J. H., and ROSEN, F. S.: Increased susceptibility to infections associated with abnormalities of complement-mediated functions and of the third component of complement. New Engl. J. Med. *282:* 249 (1970).

2   BAISERO, M. H.: Chimiotactisme des polynucléaires humains *in vitro*. III. Etude dans l'infection aiguë et chronique de l'adulte. Schweiz. med. Wschr. *103:* 1599 (1973).

3   BAISERO, M. H.; REGAMEY, A. et FREI, P. C.: Chimiotactisme des polynucléaires humains *in vitro*. Etude dans les infections bactériennes de l'adulte. Ann. Inst. Pasteur *123:* 120 (1972).

4   CLARK, R. A. and KIMBALL, A. R.: Defective granulocyte chemotaxis in the Chediak-Higashi syndrome. J. clin. Invest. *50:* 2645 (1971).

5   CLARK, R. A.; ROOT, R. K.; KIMBALL, H. R., and KIRKPATRICK, C. H.: Defective neutrophil chemotaxis and cellular immunity in a child with recurrent infections. Ann. intern. Med. *78:* 515 (1973).

6   CORNELY, H. P.: Reversal of chemotaxis *in vitro* and chemotactic activity of leucocytes fractions. Proc. Soc. exp. Biol. Med. *122:* 831 (1966).

7   DEMEO, A. M. and ANDERSEN, B. R.: Defective chemotaxis associated with a serum inhibitor in cirrhotic patients. New Engl. J. Med. *286:* 736 (1972).

8 EDELSON, P. J.; STITES, D. P.; GOLD, S., and FUDENBERG, H. H.: Disorders of neutrophil function. Defects in an early stage of the phagocytic process. Clin. exp. Immunol. *13:* 21 (1973).

9 FREI, P. C.; BAISERO, M. H., and OCHSNER, M.: Chemotaxis of human polymorphonuclears *in vitro*. II. Technical study (in preparation).

10 HIGGINS, G. R.; SWANSON, V., and YAMAGAKI, J.: Granulocytasthenia. A unique leukocyte dysfunction associated with a decrease resistance to infection. Clin. Res. *18:* 209 (1970).

11 KELLER, H. U.; BOREL, J. F.; WILKINSON, P. C.; HESS, M. W., and COTTIER, H.: Reassessment of Boyden's technique for measuring chemotaxis. J. immunol. Meth. *1:* 165 (1972).

12 KELLER, H. U. and SORKIN, E.: Studies on chemotaxis. I. On the chemotactic and complement-fixing activity of γ-globulins. Immunology, Lond. *9:* 441 (1965).

13 MILLER, M. E.: Deficiency of chemotactic function in a human neonate. A previously unrecognized defect of the inflammatory response (abstract). Pediat. Res. *3:* 497 (1969).

14 MILLER, M. E.: Cell movement and host defenses. Ann. intern. Med. *78:* 601 (1973).

15 MILLER, M. E. and NILSSON, U. R.: A familial deficiency of the phagocytosis-enhancing activity of serum related dysfunction of the fifth component of complement. New Engl. J. Med. *282:* 354 (1970).

16 MILLER, M. E.; OSKI, F. A., and HARRIS, M. B.: Lazy leukocyte syndrome. A new disorder of neutrophil function. Lancet *i:* 665 (1971).

17 MOWAT, A. G. and BAUM, J.: Polymorphonuclear leucocytes chemotaxis in patients with bacterial infections. Brit. med. J. *iii:* 617 (1971).

18 MOWAT, A. G. and BAUM, J.: Chemotaxis of polymorphonuclears from patients with diabetes mellitus. New Engl. J. Med. *284:* 621 (1971).

19 MOWAT, A. G. and BAUM, J.: Chemotaxis of polymorphonuclear leukocytes from patients with rheumatoid arthritis. J. clin. Invest. *50:* 2541 (1971).

20 REGAMEY, A.: Chimiotactisme des polynucléaires humains *in vitro*. I. Etude de la méthode chez l'adulte normal. Ann. Immunol. (Inst. Pasteur) *124 C:* 493 (1973).

21 SMITH, C. W.; HOLLERS, J. C.; DUPREE, E.; GOLDMAN, A. S., and LORD, R. A.: A serum inhibitor of leukotaxis in a child with recurrent infections. J. Lab. clin. Med. *79:* 878 (1972).

22 STEERMAN, R. L.; SNYDERMAN, R.; LEIKIN, S. L., and COLTEN, H. R.: Intrinsic defect of the polymorphonuclear leukocytes resulting in impaired chemotaxis and phagocytosis. Clin. exp. Immunol. *9:* 939 (1971).

23 WARD, P. A.: Insubstantial leucotaxis. J. Lab. clin. Med. *79:* 873 (1972).

24 WARD, P. A. and SCHLEGEL, J. R.: Impaired leukotactic responsiveness in a child with recurrent infections. Lancet *ii:* 3446 (1969).

25 WOLFF, S. M.; DALE, D. C.; CLARK, R. A.; ROOT, R. K., and KIMBALL, H. R.: The Chediak-Higashi syndrome. Studies of host defenses. Ann. intern. Med. *76:* 293 (1972).

Authors' address: P. C. FREI, M. H. BAISERO and M. OCHSNER, Département d'Immunologie et d'Allergie, Service de Médecine, Hôpital Cantonal Universitaire, *CH-1011 Lausanne* (Switzerland)

Antibiotics and Chemotherapy, vol. 19, pp. 362–368
(Karger, Basel 1974)

The Chemotactic Activity of Leucocytes Related to Blood Coagulation and Fibrinolysis

V. J. Stecher and E. Sorkin

Research Division, Ciba-Geigy Corporation, Ardsley, N.Y. and Department of Medicine, Swiss Research Institute, Davos Platz

A multiplicity of vascular and tissue injuries results in the deposition of fibrin. The subsequent history of such fibrin depends upon the balance achieved between the rates of new fibrin formation and fibrin resolution. Fibrinolysis represents a major repair mechanism at the site of fibrin deposition. It is well established that fibrin and its split products (FDP) are present during inflammatory processes of both immunologic and non-immunologic origin.

During our work on the chemotaxis of leucocytes we consistently noted that coagulation of normal blood from various species leads to the appearance of leucotactic factors. The fact that sera prepared from cell-free plasma lacked chemotactic activity for neutrophils provided a clue as to their evolution in serum. When rabbit blood was prevented from clotting by adding one part sodium oxalate to 9 parts blood and the cells were centrifuged off, the resulting plasma when clotted by recalcification or thrombin (4 NIH U/ml) gave rise to a serum containing practically no intrinsic chemotactic activity. This series of experiments provided not only a useful method for obtaining low background serum, but clearly implicated the involvement of blood cells in the generation of chemotactic factors [1].

Table I demonstrates that the chemotactic activity of normal rabbit serum is determined by the presence of buffy coat cells. In this series of experiments blood was taken from each of 3 rabbits in the presence or absence of citrate. In the case of the plasma samples, the cells were separated out by centrifugation, and then the peripheral white blood cells were added back to one sample and not to another, prior to the addition of thrombin.

The chemotactic activity of serum from a normal clot was equivalent to that found in serum obtained from plasma clotted in the presence of peri-

Table I. Chemotactic activity (neutrophils/field) of normal rabbit serum prepared by different methods

Method	Rabbit 1	Rabbit 2	Rabbit 3
Serum from normal blood clot	64	73	33
Serum from cell-free plasma clot	11	19	5
Citrated plasma	3	3	3
Serum from plasma clot with peripheral white blood cells	61	56	46

Chemotaxis studies were performed using the chambers described by BOYDEN [17]. The concentration of polymorphonuclear leucocytes in the chamber was 2×10^6/ml, and a 3-μm pore size 'Millipore' filter membrane was used. Normal and oxalated blood were taken sequentially from the same rabbit by means of heart puncture. Sterile conditions were maintained throughout. Serum and plasma were inactivated by incubation for 30 min at 56°C and used at a final concentration of 10% in Gey's balanced salt solution. Neutrophils were obtained from the peritoneal cavities of rabbits 18 h after injection of 50 ml of a 7-percent solution of sodium caseinate, and incubated in the Boyden chambers for 3 h at 37°C. The values represent an average of 4 microscopic fields of the lowermost surface of each of three individual filters. By permission of STECHER *et al.* [1].

pheral white blood cells. Serum from cell-free plasma clotted with thrombin possesses very low chemotactic attraction for neutrophils, but it was nevertheless consistently higher than that found for plasma not treated with thrombin. The appearance of this minimal chemotactic activity in plasma could be accounted for by the recent report by KAY *et al.* [2] demonstrating that human leukocytes are attracted *in vitro* toward a clot supernatant prepared from thrombin and fibrinogen. Plasmin cleaves not only fibrin (fibrinolysis), but also has the ability to cleave fibrinogen (fibrinogenolysis).

Fibrinogenolysis and fibrinolysis are not mutually exclusive processes, and both events might occur simultaneously *in vivo*. However, PLOW and EDGINGTON [3] recently published convincing experimental evidence that fibrin derivatives containing plasma reflect fibrinolysis after *in vivo* coagulation, whereas the fibrinogen derivatives containing plasma reflect primary fibrinogenolysis. The 'terminal' digestion products of both fibrin and fibrinogen are fragment D and fragment E [4]. Following electrophoresis, fragment D is near the cathode and fragment E near the anode. The molecular weights are approximately 83,000 and 50,000 respectively [5]. In a recent report on the effect of plasmin on the subunit structure of human fibrin, PIZZO et al. [6] found that while fragment E from fibrinogen and fragment E from fibrin were structurally very similar, fragment D from fibrinogen and fibrin differed in structure.

WEKSLER and COUPAL [7] recently confirmed that blood centrifuged to eliminate all cellular elements prior to allowing clot formation was free of chemotactic activity. Upon incubation of platelet protein extracts with fresh serum, chemotactic activity was evolved. However, they reported that human plasma anticoagulated with heparin or with sodium citrate did not generate chemotactic activity by platelet protein extracts for human leucocytes. Unfortunately, the concentration and source of heparin or sodium citrate used in the preparation of plasma was not given in their publication. We have found, for example, that commercial heparin containing a preservative and used at a concentration of only 5 U/ml cell suspension, prevented the cells from migrating. Cell viability tests indicated that no cell death had occurred. On the other hand, heparin without preservative could be used at a concentration of 20 U/ml without inhibiting leucotaxis.

It would be interesting to determine if the addition of human peripheral white blood cells to human plasma prior to clotting can equally result in formation of chemotactic factors. Alternatively, the rabbit system may be exhibiting a species difference in allowing chemotactic factors to form consequent to the clotting of citrated plasma containing buffy coat cells.

Once fibrin has been deposited and neutrophils begin to arrive at the inflammatory site, there is abundant experimental evidence that the reaction could be of a self-perpetuating nature. Studies by SABA et al. [8] demonstrated that intact granulocytes, but not lymphocytes, from rabbit and man have potent clot-promoting activity. Furthermore, the granulocytes not only accelerated the clotting of normal human plasma, but also compensated for the prolonged clotting time of human plasma with certain clotting factor deficiencies. The procoagulant effect in vitro occurred using cell concentrations found normally in vivo. These same authors have shown that cationic proteins isolated from rabbit granulocyte lysosomes possess potent anticoagulant activity, indicating that the same cell may perform two opposing functions, depending upon its status in situ. It should be noted that the mechanical forces arising during blood coagulation are quite considerable [9] and could lead to cellular breakdown.

Fibrinolysis may be considered as a physiological repair mechanism which leads to the elimination of fibrin which was deposited in response to cellular injury. FDP occur with varying frequency in clinical conditions, depending upon the severity of the reaction to the causative agent. These products are known to appear very dramatically in the case of acute thromboembolism as well as in the case of transplantation rejection.

NILSSON [10] reported an increase in serum FDP in patients who were undergoing rejection after kidney transplantation, as well as in several disease states, including cancer. BENNETT et al. [11] also supported the concept that elevation of serum FDP in association with kidney graft rejection indicates that fibrin deposition is involved in the pathogenesis of transplantation rejection. Following injury, fibrin may be deposited in inflamed synovial joints, but it is normally rapidly lysed by plasmin [12]. In order to have fibrinolytic activity, it is necessary to have local plasminogen activator present. BACH ANDERSEN and GORMSEN [13] have demonstrated that fibrinolytic activity is mainly localized around the venules and the capillaries of the stromal layer of synovial membranes. In rheumatoid arthritis, fibrinogen entering the joint is precipitated as fibrin and is often not removed, but instead forms fibrin bodies or is incorporated in the inflammatory pannus. This increased deposition of fibrin in joints and on synovial membranes has led to the hypothesis that fibrin has an important position in perpetuating the chronic inflammation of rheumatoid arthritis.

Recently fluorescent labeling methods were used to provide direct evidence that the deposition of fibrin is a consistent characteristic of both allergic contact dermatitis and classic delayed hypersensitivity skin reactions in man. Immunoglobulin and C3 deposits were not found in these delayed reactions [14].

It has now been established that enzymatic splitting of fibrin [15] as well as fibrinogen [2] can lead to the generation of chemotactic factors in vitro. There is in vivo evidence as well for a link between fibrinolysis, chemotaxis, and accumulation of cells in inflammatory lesions [16]. While crucial to an understanding of the inflammatory response, one must excercise caution in the interpretation of in vivo experiments. Numerous agents that appear to be chemotactic themselves may in fact be inducing the formation of chemotactic factors from the ubiquitous plasma. For this reason the Boyden chamber method [17] is being used with increasing frequency to study systems which may influence cell migration.

Figure 1 illustrates the chemotactic activity of human FDP for neutrophils plotted as a function of concentration. Although human fibrinogen was used to obtain this particular activity curve, preparations of bovine, horse and sheep FDP gave essentially the same results. Such a bell-shaped plot of

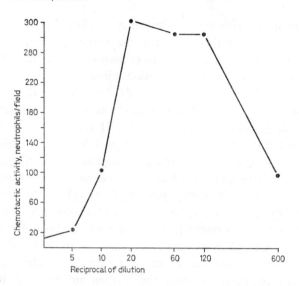

Fig. 1. Chemotactic activity of Human FDP as a function of concentration.

concentration versus cellular response is typical for chemotaxis reactions [18] and illustrates the importance of testing a dilution series in order to achieve maximal chemotactic activity.

It is of utmost importance that one work with pure preparations. WISSLER *et al.* [19] have shown, for example, that in the case of the anaphylatoxin-related leucotactic peptide system one can detect chemotactic activity for neutrophils at a concentration of 10^{-8} M. Indeed, the use of commercial preparations of 5 mg/ml fibrinogen yielded FDP possessing varying amounts of chemotactic activity. The purity of these fibrinogen preparations is normally quoted as containing 90 or 95% of clottable protein.

Fibrinogen from 4 species used in a recent study in our laboratory [15] was purified from whole plasma according to the method of POLSON [20]. The protein concentration of each sample was determined using a spectrophotometer, and prior to testing the final concentration was adjusted to 5 mg protein per ml. This figure was chosen because it is within the normal human plasma concentration range for fibrinogen.

Three different methods were used to ascertain the purity of our fibrinogen preparations. Sedimentation diagrams obtained by analytical ultracentrifugation yielded only one symmetrical peak. Likewise, analysis of the fibrinogen using cellulose acetate membranes and a Beckman microzone electrophoresis system showed one uniform peak. Finally, one precipitate line was obtained when human fibrinogen was subjected to immuno-electrophoresis and developed with a rabbit antiserum to human plasma proteins and rabbit anti-human fibrinogen antiserum.

The fibrinogen of four different species, when incubated alone for 24 h at 37°C, exhibited no chemotactic activity. Likewise, thrombin and plasmin, either alone or in combination, were not chemotactically active. However, when the fibrinogen was first converted to fibrin by the addition of thrombin and then incubated with plasmin for 24 h at 37°C, the lysis products formed were highly chemotactic, as illustrated in figure 1. Immunoelectrophoresis demonstrated the presence of the classic 'terminal' fibrin lysis products, fragments D and E. These results provide experimental evidence for a direct correlation between the process of fibrinolysis and chemotactic attraction of leucocytes.

Although the inflammatory response occurs in many different types of tissue, the reaction has certain general characteristics. One such aspect is the deposition of fibrin. Using a skin window chamber technique with dogs, BARNHART et al. [16] found that FDP-D resulting from proteolysis by plasmin attracted granulocytes. Furthermore, it was suggested that the delayed vascular permeability phase in acute inflammation may be mediated by FDP-E produced by fibrinolysis at the site of injury.

Summary

There is presently abundant experimental evidence to illustrate that the role of fibrinogen and its derivatives is not limited to hemostasis but quite likely also determines the direction of migratory cells *in vivo*. Blood coagulation and subsequent fibrinolytic activity clearly influence the course of the inflammatory response by the evolution of biologically active endogenous end products.

References

1   STECHER, V. J.; SORKIN, E., and RYAN, G. B.: Relation between blood coagulation and chemotaxis of leucocytes. Nature new Biol. *233:* 95 (1971).
2   KAY, A. B.; PEPPER, D. S., and EWART, M. R.: Generation of chemotactic activity for leukocytes by the action of thrombin on human fibrinogen. Nature new Biol. *243:* 56 (1973).
3   PLOW, E. F. and EDGINGTON, T. S.: Discriminating neo-antigenic differences between fibrinogen and fibrin derivatives. Proc. nat. Acad. Sci., Wash. *70:* 1169 (1973).
4   NUSSENZWEIG, V.; SELIGMANN, M.; PELMONT, J. et GRABER, P.: Les produits de

dégradation du fibrinogène humain par la plasmine. I. Séparation et propriétés physico-chimiques. Ann. Inst. Pasteur *100:* 377 (1961).

5 MARDER, V. J.: Identification and purification of fibrinogen degradation products produced by plasmin: considerations on the structure of fibrinogen. Scand. J. Haemat., suppl. 13, p. 21 (1971).

6 PIZZO, S. V.; SCHWARTZ, M. L.; HILL, R. C., and McKEE, P. A.: The effect of plasmin on the subunit structure of human fibrin. J. biol. Chem. *248:* 4574 (1973).

7 WEKSLER, B. B. and COUPAL, C. E.: Platelet-dependent generation of chemotactic activity in serum. J. exp. Med. *137:* 1419 (1973).

8 SABA, H. I.; HERION, J. C.; WALKER, R. I., and ROBERTS, H. R.: The procoagulant activity of granulocytes. Proc. Soc. exp. Biol. Med. *142:* 614 (1973).

9 RUBENSTEIN, E.: Blood clotting: the force of retraction. Science *138:* 1343 (1962).

10 NILSSON, I. M.: Determination of fibrinogen degradation products in serum and urine as a new diagnostic tool. Scand. J. Haemat., Suppl. 13, p. 317 (1971).

11 BENNETT, N. M.; BENNETT, D.; HOLLAND, N. H., and LUKE, R. G.: Serum fibrin degradation products in the diagnosis of transplantation rejection. Transplantation *14:* 311 (1972).

12 KWANN, H. C. and ASTRUP, T.: Fibrinolytic activity of reparative connective tissue. J. Path. Bact. *87:* 409 (1964).

13 BACH ANDERSEN, R. and GORMSEN, J.: Fibrinolytic and fibrin stabilizing activity of synovial membranes. Ann. rheum. Dis. *29:* 287 (1970).

14 COLVIN, R. B.; JOHNSON, R. A.; MIHN, M. C., jr., and DVORAK, H. F.: Role of the clotting system in cell-mediated hypersensitivity. J. exp. Med. *138:* 686 (1973).

15 STECHER, V. J. and SORKIN, E.: The chemotactic activity of fibrin lysis products. Int. Arch. Allergy *43:* 879 (1972).

16 BARNHART, M. I.; SULISZ, L., and BLUHM, G. B.: Role for fibrinogen and its derivatives in acute inflammation; in FORSCHER and HOUCK Immunopathology of inflammation, pp. 59–65 (Excerpta Medica, Amsterdam 1971).

17 BOYDEN, S. V.: The chemotactic effect of mixtures of antibody and antigen on polymorphonuclear leucocytes. J. exp. Med. *115:* 453 (1962).

18 WISSLER, J. H.; STECHER, V. J., and SORKIN, E.: Biochemistry and biology of a leucotactic binary serum peptide system related to anaphylatoxin. Int. Arch. Allergy *42:* 722 (1972).

19 WISSLER, J. H.; STECHER, V. J., and SORKIN, E.: Chemistry and biology of the anaphylatoxin related serum peptide system. III. Evaluation of leucotactic activity as a property of a new peptide system with classical anaphylatoxin and cocytotoxin as components. Europ. J. Immunol. *2:* 90 (1972).

20 POLSON, A.: Mechanism of cryoprecipitation. Prep. Biochem. *2:* 53 (1972).

Authors' addresses: V. J. STECHER, Research Division, Ciba-Geigy Corporation, *Ardsley, NY 10502* (USA); E. SORKIN, Department of Medicine, Swiss Research Institute, *CH-7270 Davos Platz* (Switzerland)

Antibiotics and Chemotherapy, vol. 19, pp. 369–381
(Karger, Basel 1974)

Necrotaxis

Chemotaxis towards an Injured Cell[1]

M. BESSIS

Institut de Pathologie Cellulaire (INSERM, U. 48), Hôpital de Bicêtre,
Le Kremlin-Bicêtre

I. Introduction

It was during experiments on micro-irradiation that I first noticed that dying cells attracted phagocytes. In 1960, when using an ultraviolet (UV) microbeam to damage certain cell organelles containing nucleic acids, I happened one day to irradiate one white cell too strongly. The cell became swollen and after a few minutes burst. Immediately leukocytes appeared from all directions, surrounded the dying cell and phagocytized it. At the time I did not pay much attention to this phenomenon because it seemed to me natural that cell corpses be phagocytized by macrophages. I thought that this was common knowledge: millions of cell corpses formed every minute in the organism must be cleared by phagocytes. Then, a few years later, we began working on laser microbeams. This time the aim was to destroy certain organelles which had been vitally stained (such as mitochondria), but to keep the cell alive. Also in this case, when a cell was killed by an overdose of radiation, the macrophages, even those at a distance of 500 microns, were attracted by the cell corpse and came to phagocytize it. While I was compiling the bibliography for a note on this subject, I found out that far from being a known fact, there were differences of opinion between authors as to whether

1   This work has been partly supported by the contract DAJA 37-70-C-0678, European Research Office, US Army.

or not a dead tissue can attract leukocytes. According to HARRIS [1953a, b] dead granulocytes and autolysed tissues, or tissues damaged by various enzymes could not induce chemotaxis. In observing that the injured tissues were infiltrated by leukocytes after a certain period of time, he remarked that this could be caused by locally developed bacteria. This remark could also be made about certain experiments with tissue fragments introduced into living bodies. However, HURLEY [1963], RYAN and HURLEY [1966] using the same methods that HARRIS had previously used, demonstrated that tissue damaged under sterile conditions and maintained sterile, became chemotactic after incubation in fresh plasma (this was not so after incubation in serum or in a saline solution). These experiments were made with either Harris' photographing tracing technique, or Boyden's millipore filter technique. BUCKLEY [1963] studied leukocyte chemotaxis following focal aseptic heat injury *in vivo*, in the 'ear chamber' of the rabbit. He avoided the secondary effect due to bacterial infection by making a lesion with a radio-frequency generator. Between 4 and 24 h after the lesion, leukocytes appeared in the area which had been injured by heat and moved around at random in that zone. FLOREY [1964], also having observed movement at random of leukocytes, came to the conclusion that the action of possible substances on the migration of leukocytes in inflammation was purely speculative. But BUCKLEY remarked that before moving at random in the area of the lesion, leukocytes moved in a definite direction and he recorded the phenomenon by time-lapse photography. The following criticism can be made of this observation: it is obvious that if one finds an accumulation of leukocytes at the injured spot, they had to travel there. There are at least two means by which they may have come: they could have been either attracted or trapped. In both cases you can see the photographic tracks of their passage. The technique of micro-irradiation gives an answer to the question in showing clearly that the leukocytes are *attracted* by the dying cell.

I wrote two short notes with BURTÉ [BESSIS and BURTÉ, 1964, 1965] on chemotaxis observed after the destruction of a cell by UV and laser microbeams. But it is only recently that we have come back to these studies in my laboratory. Preliminary results will be presented and the main questions arising from this work will be discussed.

One last remark: it is questionable as to whether or not a new term was necessary to mark this particular species of chemotaxis. A new term is necessary when it can avoid resorting to a long phrase. Necrotaxis is the 'phenomenon which determines the migration of cells towards cells (or molecules) altered in a way which would seem irreversible'. The term 'necrotaxis' summarizes what we want to say in one word.

II. Material and Techniques

Human blood or that of experimental animals (rats and rabbits) is drawn from a vein with an anticoagulant (sodium citrate or heparin). The blood is allowed to sediment by gravity for approximately 1 h. Then the buffy coat (containing leukocytes and a few red blood cells) is removed and a drop deposited between slide and coverslip. The preparation is then sealed with paraffin and examined with the phase-contrast microbeam device.

For the evaluation of necrotaxis, a field including some leukocytes and a few red blood cells is selected. One of the cells located in the middle of the field is destroyed by means of the microbeam and the movement of the surrounding leukocytes is observed and recorded by time-lapse cinematography (approximately one picture every 4 sec). The distance over which the necrotactic effect is sensed is measured and the various cell types found within the field and their velocity are noted. Phagocytosis, which eventually follows necrotaxis, is also recorded.

Initial observations are made at low magnification (100×) so that the distances of necrotactic action can be measured. Thereafter, a higher magnification (1,000×) is used to allow identification of the various cell types and the study of internal cell modifications.

To destroy the cells by UV we used a very short wave length of 3,700 Å (cadmium spark). To destroy the cells by laser we used a ruby laser (wavelength 6,943 Å, duration of flash 500 μsec. To destroy the white cells with the ruby laser it is necessary to use vital staining. This was done with Janus green (final concentration 1/5,000) or methylene blue (final concentration 1/2,000). Under these conditions the dye cannot be seen with the naked eye, but is sufficient for the laser effect [Infrastaining, BESSIS and STORB, 1965].

The same infrastaining was used to obtain a small coagulum by burning pure albumin solutions. Pure hemoglobin solutions are burned by the ruby laser without the need of additional stain.

The devices for micro-irradiation by UV and laser beams have been described already [BESSIS and NOMARSKI, 1960; BESSIS and TER-POGOSSIAN, 1965].

III. Analysis of the Stages of Necrotaxis

Nothing can replace the direct vision of a phenomenon, especially when it is seen by time-lapse motion picture[2].

In this section we are going to describe the phenomenon as seen with the usual technique: later we can see the variations that can be made. The results observed were the same whether the destroyed cell was a polymorph, a lymphocyte, a red blood cell or a tissue culture cell. Approximately 1 sec after the destruction of one of these cells it was observed that the leukocytes directed themselves towards the damaged cell. Polymorphonuclear neutrophils were briskly attracted, eosinophils to a lesser degree, monocytes even less, whilst lymphocytes were not attracted at all. (This is not surprising since these cells do not manifest classical chemotaxis.)

It should be mentioned that COMANDON [1917] observed that a parasitized cell produced a chemotactic phenomenon. (The author said that the leukocyte is directed towards the parasitized cell in a straight line in spite of contrary currents which may exist in the ambient medium.) COMANDON attributed this chemotaxis to substances which the parasite releases, but perhaps it is also pertinent to assign a role to necrotactic substances leaking out of the damaged red cell.

The leukocytes approach their target in such a way that can be anything between a completely straight line and a tortuous path. When there are a lot of white corpuscles surrounding the damaged cell, and what is known in pathology as a 'rosette' appears. The formation of 'rosettes' can be seen clearly through time-lapse film observations. After the irradiated cell has shown signs of agony in its death [the period of irreversible cell damage *preceding* death; BESSIS, 1964, see below], neighboring healthy cells begin to travel towards the corpse from all directions. In a few minutes a 'foreign body rosette' of classical pathology is formed. These cells literally tear the corpse apart. Phagocytosis usually takes place morsel by morsel, each phagocyte carrying off part of the prey [POLICARD and BESSIS, 1953]. The newly arrived polymorphs can be seen to push the phagocytes aside vigorously so as not to miss their portion. At a cellular level, the phenomenon evokes a violent orgy of feeding sharks in which blood escaping from a wound causes one to attack another of its kind. After a few moments, no traces of the dead cell remain.

---

2   A film on this subject is available from Sandoz, Motion Picture Dept., CH-4000 Basel, Switzerland.

The fact that leukocytes are not attracted by cells which have been dead for some time is an important observation. POLICARD and COLLET [1953] have already indicated the fact that macrophages do not bother about dead cells. It would seem that it is only within a few seconds after a cell has been subjected to cytolysis that surrounding cells are directed towards it. Sometimes polymorphs can be seen to abandon a morsel of the corpse having phagocytosized part of the dead cell. This rejected portion will not attract another cell, even a close passer-by. Such observations would suggest the existence of a gradient in the environment. Leukocytes sensitive to this gradient react, directing themselves actively towards the highest concentration of this hypothetical necrotactic substance. Dead cells probably do not attract leukocytes because their necrotactic substances have already been exuded.

In order to follow the experiments it is important to bear in mind one particular point concerning the shape of the cell during its locomotion. All cells assume a spherical form in free suspension in a liquid. Progression implies a point of attachment. *In vitro* attachment takes place on the supporting surface of the preparation; *in vivo* it takes place on fibrils, sheets of connective tissue or on fixed neighboring cells. The movement of a cell may take place at random in an apparently disorganized manner, or alternatively it may be oriented in a precise direction. In the latter case leukocytes crawl along in a peculiar way: the front part develops a cytoplasmic veil in the form of a fringe, the *metapode* (from a Greek word meaning front) whilst the back part tapers out and gels, the *uropode* (from a Greek word meaning tail). The metapode therefore indicates the direction in which the cell has oriented itself.

In a drop of blood, white cells move about at random and nothing particularly attracts them towards the red cells. After probing red cells a bit they turn around and move off in another direction. But as soon as a red cell is hit by the microbeam, the white cell develops a large metapode and changes its direction heading towards the target. The first cells to arrive on the spot encircle the prey and phagocytize it. However, it should be noted that this attraction does not necessarily always end in phagocytosis; in some cases the cell moves off somewhere else after a little while.

It has already been remarked that most of the questions relating to necrotaxis are common to all chemotaxis. For instance: how do the molecules which leave the destroyed red cell diffuse and reach the leukocyte? What receptors allow the leukocyte to recognize these messenger molecules?

What is the mechanism by which the leukocyte follows the concentration gradients to its target? Once these messages arrive at the cell surface, how do they subsequently act on the contractile system to make the cell move in the right direction? The microbeam technique is easy to operate and permits direct microscopic observation of this phenomenon which can be of considerable importance to the investigator.

IV. Changes in Experimental Procedures

A. Change in the Target

1. *Red cells* taken from the same human subject as the phagocytes, red cells from other human subjects, from mammals and amphibians, were used. The necrotaxis was always of the same magnitude. However, there was some variation according to the size of the lesion made on the target and, therefore, the size of the red cell. Red cells taken from frogs gave the most dramatic results.

When incompatible red cells were used it was necessary to carefully wash the white and red cells in heated plasma so as to inactivate the complement system. A small number of red cells was used to avoid agglutination.

2. *White cells and cultured cells.* If leukocytes or cells taken from tissue culture are used as targets they must be artificially colored (as indicated in 'Material and Techniques') unless they have natural absorption in the ruby laser wave length. The time-lapse of appearance of the veils and the average speed are about the same as for red cells. Human cells and those from the different species of animals used gave comparable results.

3. *Hemoglobin in solution.* We chose a concentration of hemoglobin which, after irradiation, gave a coagulum of approximately equal volume to that of a coagulated red cell. The hemoglobin concentration should be about 10% for this purpose. We used crystallized commercial hemoglobin from horses and nonpurified human hemoglobin. Its presence does not interfere with the survival of the leukocytes, at least for the time of the experiment (15–30 min). After coagulation of a volume of 5–10 μm in diameter, necrotaxis can be seen, similar to when erythrocytes are destroyed [BARAT and BOISFLEURY, 1970].

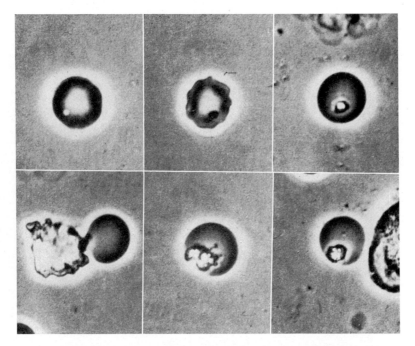

Fig. 1. Hemolysis of erythrocytes produced by a laser beam. Progressive intensity determines larger and larger lesions. Necrotaxis is maximal with more extensive coagulation.

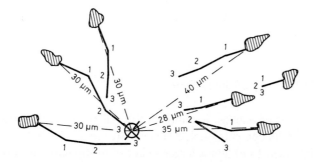

Fig. 2. Necrotaxis. A diagram of leukocyte movements after destruction of an erythrocyte by the laser. The hatched areas indicate the position of the leukocytes at the outset; the cross shows the site at which the red cell was destroyed, and each number designates the minute-by-minute location of the leukocytes.

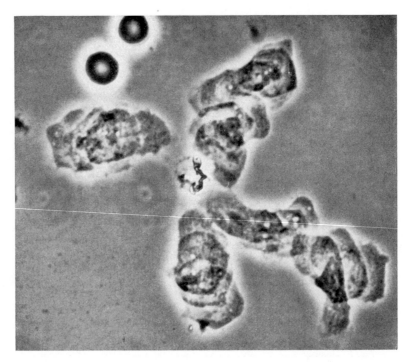

Fig. 3. Necrotaxis. Superimposed sequential photographs illustrating the course of leukocytes attracted by a red cell that has been destroyed with a laser microbeam.

4. *Albumin in solution.* Similar results are observed with albumin. To achieve a coagulum, albumin must be stained. We used a solution of Janus green (0.1/1,000) and the results were identical with those with the hemoglobin solutions.

B. Change in Intensity of the Lesion

When the amount of damage is minimal, the cell shows only a small coagulation spot and its contents discharge slowly (in about 30 sec). This experiment was intended to obtain a larger quantity of discharge of intracellular content by piercing a small hole in the cell membrane. But the results were just the opposite of what was expected. Necrotaxis was not as distinct as when the target cell was more completely coagulated. The fastest necrotaxis is obtained by completely coagulating the entire cell (e.g., with a red cell from a frog which is fairly voluminous, the phagocytes come from a greater distance and in larger numbers). It is not rare to see 100 cells surrounding this coagulum within 15 min. This phenomenon is not easy to explain. Perhaps heat denaturation of proteins has an important part to play in it. We shall come back to this question in the discussion.

C. Change in the Environment

1. Effects of a Nonplasmatic Environment
To find out if necrotaxis was influenced by a plasma factor as suggested by RYAN and HURLEY [1966], we substituted the plasma with synthetic media. Different media used in tissue culture were tried and the one that preserved leukocytes the most effectively seemed to be [BARAT and DE BOIS-FLEURY, 1970] the mixture of equal quantities of 199 [MORGAN et al., 1950] and NCT 109. The red and white cells were carefully washed three times in this solution. They were not in a very healthy state but remained alive long enough to clearly show necrotaxis towards the irradiated cell. The presence of plasma does not seem to be essential.

2. Effect of a Hyperviscous Environment
This subject will be brought up in the discussion.

V. Discussion

I do not intend to go into the characteristics common to necrotaxis and chemotaxis, but to bring forward briefly three specific points about necrotaxis: (1) What is the nature of necrotactic substances? (2) What is the mechanism of recognition of damaged cells and molecules? and (3) Does necrotaxis have a role *in vivo*?

A. Nature of Necrotactic Substances

First of all, the question arises as to whether necrotaxis is the result of a diffused substance. Other mechanisms could well be imagined: such as an electric field, or radiation gradient. The heat given off by the impact certainly does not play any role. For a local thermic elevation of 60–80 °C (necessary to coagulate proteins) and for a flash duration of 500 μsec, the increase in temperature at 100 μm from the target is only about 0.01 °C and for less than 1 sec [SALET et al., 1970]. Laser irradiation of a carbon particle (in a solution containing no protein) does not produce necrotaxis.

We have also seen that necrotaxis does not depend on the nature of the target as long as it is a protein (we have not yet tried targets of a non-protein nature). It would not seem to be the cellular contents alone because cell 'ghosts' which have lost practically all their contents become necrotactic when hit by the microbeam. Thus, it appears that proteins denatured by heat and/or the liberation of enzymes due to the death of the cell induce necrotaxis after microbeam treatment.

To obtain indications as to the molecular weight of this hypothetical necrotactic substance, on the suggestions of BARNES [1969] the viscosity of the solution was changed. The viscosity of the saline solution is about 0.90 centistock and that of normal plasma about 1.43 centistock. The viscosity was increased by adding polyvinyl-pyrrolidone or methocel (Dow Chemical Co., Midland, Mich.) until concentrations of 1.92 centistock were reached [BARAT and DE BOISFLEURY, 1970]. In high concentrations leukocytes can no longer move about, but one can be sure that they perceive the death of the cell after micro-irradiation because they develop metapodes (the characteristic cytoplasmic veils) in the direction of the irradiated cell. Moreover, the time-lapse of appearance of these fringes is about the same as in plasma or saline solution. We can conclude accordingly that the molecular weight of such necrotactic substance is very low, or at any rate much lower than that of a protein such as hemoglobin, for example. Polypeptides seem to have a suitable molecular weight. Perhaps they are related to those recently exposed by SORKIN's team [SORKIN et al., 1970; WISSLER et al., 1972]. Work along these lines would seem most interesting and promising. I would like to mention at this point that WISSLER et al. isolated a serum-derived leuco-tactic peptide system composed of two small molecules, one of which is anaphylotoxin. This substance, amongst other properties, induces muscular fibrils to contract. The idea occurred to me to observe the effect of the destruction of a red cell on neighboring cells of the heart muscle in tissue culture and beating ciliate cells. The results were stunning. The contractions accelerated each time and their amplitude increased. Cells which appeared to be dying, and had stopped moving, came back to life within a few seconds. It is debatable as to whether this same substance (1) attracts leukocytes, and (2) activates the movement of leukocytes in the same way as it activates contractions, thus contributing to the efficiency of necrotaxis.

B. Mechanism of Recognition of Dead and Dying Cells

First of all, it is essential to separate the 'recognition' of cells after contact from the taxis which attracts phagocytes. For instance, certain cells adhere to substances like glass, quartz and silicone. They also adhere to other cells, dead or altered by heat, glutaraldehyde, osmic acid, etc. [RA-BINOVITCH, 1967a, b], to altered cells in general and not to normal cells. This phenomenon is perhaps not related to necrotaxis which is chemotaxis or, in other words, locomotion towards a target. There are, however, relationships between chemotaxis, adhesiveness and phagocytosis. For example, lymphocytes have never been seen to phagocytize and are probably not capable of chemotaxis. It is true, however, that they accumulate around some specific tissue lesions, and so the question arises as to how they came to be there: perhaps at random, and then they were caught in a trap. In the same way, it would seem that a leukocyte is not oriented towards a foreign red cell, but if it is touched by one, it phagocytizes it. The point is the attraction towards a dying cell. It is difficult to define the moment of death of a cell. The death agony (prior to death) has to be distinguished from the cell necrosis following death [BESSIS, 1964]. Yet, we have already seen that it is not death itself that is recognized because cells that have been dead for some time do not attract phagocytes.

C. The Role of Necrotaxis *in vivo*

A layman would not be astonished by anything in necrotaxis. It all happens as in nature: a creature is struck, it dies and is eaten. A teleologist who knows that thousands of cells die every day and *must* be eliminated[3], would think it quite logical (or teleological) that such a mechanism exists favoring the hunt for corpses. He would even find a cascade of finalities in this. The phagocyte, seeking and trapping its 'food' serves the community of cells which makes up the organism.

So, as chemotaxis has a function in the organism in the fight against infection or in reproduction (chemotaxis of spermatozoids), necrotaxis attracts cells to the particular spot where they have a part to play.

3   E.g., in one day there are 200 billion red cells and 15 billion white cells to be destroyed. Even if a phagocyte can consume these dead globules in 4 min, therefore admitting that it can consume 300 in 24 h, a thousand million phagocytes for the red cells alone would be necessary.

Reactions of living cells in the presence of dying or dead cells vary considerably according to the tissue, the number of cells rapidly or slowly transforming into corpses, and the cause of death. It is quite probable that the mechanism of necrotaxis is not the same as that of 'immunological recognition' of self and not-self, and that of 'immuno-surveillance' [Burnet, 1970]. It does not need the presence of plasma molecules, and lymphocytes are not concerned. It would appear to be more analogous to the mechanism of metamorphosis of amphibians [Boyden, 1963].

Last, but not least, necrotaxis may have an important role in pathology: I will recall only two of the examples: (1) inflammatory reactions of various types that can be seen after pathologic destruction of a great many cells [Majno, 1964], and (2) accumulation of leukocytes in certain types of polyarthritis, which are perhaps kept going by a vicious circle due to the death of local leukocytes [Sorkin et al., 1970]. In this case suppression of necrotaxis could be beneficial to the patient.

Not enough attention has been paid to the phenomenon of the elimination of cells and cell debris after death by phagocytosis. The experimental circumstances in which a single cell can be damaged in the midst of others is particularly valuable for these studies.

Summary

The name 'necrotaxis' has been given to a special type of chemotaxis in which granulocytes and monocytes are attracted to cells in the process of dying. Micro-irradiation devices (classic UV and laser) have been used to destroy a target cell. One second after the destruction of this cell, it can be seen that leukocytes direct themselves towards the damaged cell. Hemoglobin and albumin coagulated by the microbeam are equally necrotactic. The presence of plasma does not seem necessary, as the phenomenon occurs also in a saline environment. The following specific points have been briefly discussed: the nature of necrotactic substances, the recognition mechanism of damaged cells and molecules and the eventual role of necrotaxis *in vivo*.

References

Barat, N. et Boisfleury, A. de: Etude de différents paramètres intervenant dans le nécrotactisme des granulocytes après destruction d'une cellule par faisceau Laser. Nouv. Rev. franç. Hémat. *10:* 739 (1970).

Barnes, F. S.: Personal commun. (1969).

Bessis, M.: Studies on cell agony and death. An attempt at classification; in De Reuck and Knight Cellular injury. CIBA Foundation Symposium, p. 287 (Churchill, London 1964).

BESSIS, M. et BURTÉ, B.: Chimiotactisme après destruction d'une cellule par microfaisceaux Laser. C.R. Soc. Biol. *158:* 1995 (1964).

BESSIS, M. and BURTÉ, B.: Positive and negative chemotaxis as observed after destruction of a cell by ultra-violet and Laser microbeams. Tex.Rep.Biol.Med.*23:* suppl.1, p. 204 (1965).

BESSIS, M. et NOMARSKI, G.: Irradiation UV des organites cellulaires avec observation continue en contraste de phase. J. biophys. biochem. Cytol. *8:* 777 (1960).

BESSIS, M. et STORB, R.: Sensibilité des leucocytes au Laser à rubis après différentes colorations vitales. Nouv. Rev. franç. Hémat. *5:* 459 (1965).

BESSIS, M. and TER-POGOSSIAN, M.: Micropuncture of cells by means of a Laser beam. Ann. N.Y. Acad. Sci. *122:* 689 (1965).

BOYDEN, S.: Cellular recognition of foreign matter. Int. Rev. exp. Pathol. *2:* 311 (1963).

BUCKLEY, I. K.: Delayed secondary damage and leucocyte chemotaxis following aseptic heat injury *in vivo*. Exp. molec. Path. *2:* 402 (1963).

BURNET, SIR M.: Cellular immunology, 2nd ed., p. 628 (University of Cambridge Press, London 1970).

COMANDON, J.: Phagocytose *in vitro* des hématozoaires du calfat. C.R. Soc. Biol. *80:* 314 (1917).

FLOREY, H. W.: Chemotaxis, phagocytosis and the formation of abcesses; in FLOREY General pathology, 3rd ed., p. 98 (Lloyd-Luke, London 1964).

HARRIS, H.: Chemotaxis of granulocytes. J. Path. Bact. *66:* 135 (1953a).

HARRIS, H.: Chemotaxis of monocytes. Brit. J. exp. Path. *34:* 276 (1953b).

HURLEY, J. V.: Incubation of serum with tissue extracts as a cause of chemotaxis of granulocytes. Nature, Lond. *198:* 1212 (1963).

MAJNO, G.: Interactions between dead cells and living tissue; in DE REUCK and KNIGHT Cellular injury, p. 87, CIBA Foundation Symposium (Churchill, London 1964).

MORGAN, J. F.; MORTON, H. J., and PARKER, R. C.: Nutrition of animal cells in tissue culture. I. Initial studies on a synthetic medium. Proc. Soc. exp. Biol. Med. *73:* 1 (1950).

POLICARD, A. et BESSIS, M.: Fractionnement d'hématies par les leucocytes au cours de la phagocytose. C.R. Soc. Biol. *147:* 982 (1953).

POLICARD, A. et COLLET, A.: Recherches par microcinématographie en contraste de phase sur la phagocytose des particules minérales. Rev. Hémat. *8:* 132 (1953).

RABINOVITCH, M.: The dissociation of the attachment and ingestion phases of phago-cytosis by macrophages. Exp. Cell Res. *46:* 19 (1967a).

RABINOVITCH, M.: Attachment of modified erythrocytes to phagocytic cells in absence of serum. Proc. Soc. exp. Biol. Med. *124:* 396 (1967b).

RYAN, G. B. and HURLEY, J. V.: The chemotaxis of polymorphonuclear leucocytes to-wards damaged tissue. Brit. J. exp. Path. *47:* 530 (1966).

SALET, C.; LUTZ, M. et BARNES, F. S.: Paramètres physiques caractérisant le dommage thermique sélectif de mitochondries en micro-irradiation par Laser. Photochem. Photobiol. *11:* 193 (1970).

SORKIN, E.; STECHER, V. J., and BOREL, J. F.: Chemotaxis of leukocytes and inflammation. Ser. Haemat. *3:* 111 (1970).

WISSLER, J. H.; STECHER, V. J., and SORKIN, E.: Biochemistry and biology of a leucotactic binary serum peptide system related to anaphylatoxin. Int. Arch. Allergy *42:* 722 (1972).

Author's address: Dr. MARCEL BESSIS, Institut de Pathologie Cellulaire (INSERM, U. 48), Hôpital de Bicêtre, *F-94270 Le Kremlin-Bicêtre* (France)

Antibiotics and Chemotherapy, vol. 19, pp. 382–408
(Karger, Basel 1974)

Granuloma Formation and its Relation to Chemotaxis

A Discussion of the Role of Two Groups of Micro-Organisms: The Mycobacteria and the Anaerobic Coryneforms[1]

R. G. WHITE

University Department of Bacteriology and Immunology, Western Infirmary, Glasgow

The migration of phagocytic cells into sites of infection forms an essential feature of the tissue response in bacterial disease. Despite the early recognition by LEBER [1891] and METCHNIKOFF [1893] of bacterial chemotactic forces, knowledge of the manner of their working *in vivo* has remained fragmentary and incomplete.

Some micro-organisms are termed *pyogenic;* they lead to pus formation when they infect animal tissues. In many species the usual pyogenic organisms are streptococci and staphylococci; in ruminants and pigs, so-called *Corynebacterium pyogenes* is the common pyogenic organism, whereas in horses the usual pyogenic organism is a group *C. streptococcus.* We would certainly expect that such organisms would provide some form of direct or indirect chemotactic influence on granulocytes.

The organisms which are central to this article produce an entirely different type of cellular response. One, *Mycobacterium tuberculosis*, is classically associated with the chronic granulomatous type of cell reaction. The term 'granuloma' covers a solid mass of accumulated cells which typically includes macrophages but also other cell types and a varying component of necrotic tissue. Granulomas are 'chronic' or persistent, and the relevant question therefore arises 'How is this collection of cells maintained?'

Granulomas are built up by emigration of circulating cells from the blood. These appear to be largely restricted in most early granulomata to monocytes which are of bone marrow origin [EBERT and FLOREY, 1939;

---

1   Departmental Publication No. 3722.

VOLKMAN and GOWANS, 1965a, b; SPECTOR and LYKKE, 1966]. However, although its participation is possibly minor in early lesions, there is compelling evidence that as the granuloma matures it becomes increasingly colonised by lymphocytes migrating from the bloodstream. These form typical lymphoid nodules in the later stages of a granuloma formed by injecting antigen in water-in-oil emulsion with added *M. tuberculosis* (complete Freund-type adjuvant) [WHITE *et al.*, 1955; SUTER and WHITE, 1954; SPECTOR and LYKKE, 1966]. Histological examination of such late granuloma gives the impression of a tissue organisation in which there is a through-put of cells from a modified vasculature and the development of functional compartments. Thus, the later stages of a granuloma may achieve an organisation with some features of a lymph-node or spleen with specific areas of differentiation of plasma cells, lymphoid nodules and in some instances even the development of germinal centres.

It is clear, therefore, that the continuing existence of a granuloma is likely to represent the complex response to a wide variety of concurrent stimuli. In the early stages, it is likely that macrophage chemotactic factors play an important part in recruiting cells from the circulation. Once the granuloma is established, the rate of entry of cells can be measured by transfusing monocytes labelled *in vivo* with tritiated thymidine to animals bearing granulomas of varying age. With this approach, a clear distinction becomes obvious between the different granulomata provoked by carrageenan and those provoked by *Bordetella pertussis* vaccine or complete Freund adjuvant (CFA) containing mycobacteria. With a carrageenan granuloma the daily entry falls, after the first week or so, to a low level of around 10,000–20,000 cells. This is a 'low turnover lesion' [SPECTOR and RYAN, 1970]. In high-turnover lesions, such as that provoked by local injection of *B. pertussis* vaccine or *M. tuberculosis* in mineral oil, the daily entry of monocytes remains high, i.e. around 250,000 cells, for months [RYAN and SPECTOR, 1969]. The high-turnover granuloma due to CFA is associated also with a significant entry of polymorphonuclear leucocytes.

Cell division in the granuloma can be measured by giving a single injection of tritiated thymidine 30 min before death of the animal and, in autoradiographs of histological sections, counting the radioactive label resulting from premitotic DNA synthesis. In high-turnover lesions the value remains high, i.e. 4.6%, whereas in the low-turnover lesion it starts high and soon falls to about 0.5–2.0%. The mycobacterial granuloma is, therefore, maintained by both immigration of cells into it from the blood and by cell-division.

Over a range of different types of granulomata the rates of sustained immigration and of macrophage proliferation were found to vary in parallel. Further work by RYAN and SPECTOR [1970] has clarified this relationship further. Using established high-turnover granulomata produced from inocula of *B. pertussis*, destruction of the bone marrow with shielding of the reaction site is followed within 24 h by virtual cessation of DNA synthesis in the macrophages of the reaction. Thus, even at 6 weeks, most of the macrophages of the granulomata were very recent arrivals from the bone marrow and their rate of proliferation was high. This presumably accounts for the parallelism which is constantly observed between extent of macrophage immigration and proliferation. Presumably, as suggested by SPECTOR and RYAN [1970], the stimulus for both immigration and for replication of macrophages is the systemic release of a chemotactic substance. Presumably this is, or is accompanied by, a cytotoxic substance which damages the macrophages of the granuloma after injection. By keeping up the numbers of locally available macrophages, the systemic release of the cytotoxin is prevented. The system therefore reaches homeostasis, increased damage to macrophages causes chemotaxis of further macrophages and mitotic proliferation of the cells after emigration.

The high-turnover granuloma therefore appears to depend for its continuing existence mainly on recruitment and local mitotic proliferation. Low-turnover granulomata depend more on the existence of long-lived macrophages. Thus, in the case of carrageenan granulomata, destruction of the bone marrow by X-rays causes little change in the already low DNA synthesis of macrophages in lesions of over 1 week old, and no fall at all in 6-week lesions [RYAN and SPECTOR, 1969]. In such a lesion the main basis for its persistence appears to depend upon the intrinsic longevity of its macrophages.

Evidence for Macrophage-Chemotactic Factors in Microbially Determined Granulomata

Two groups of micro-organisms will form the basis of the following discussion. Both (mycobacteria and the anaerobic coryneforms) have attracted attention as immunological adjuvants. The former are used universally as part of CFA and may be either virulent or attenuated strains of *M. tuberculosis*, of human, bovine, avian or atypical strains such as *M. kansaii*, saprophytic mycobacteria (*M. phlei, M. butyricum* or *M. smegmatis*), or members of the closely related *Nocardia* genus, such as *N. asteroides* or *N. rhodochrous (C. rubrum)*. The anaerobic corynebacteria, of which *C. parvum* is the best-known example, are a group of micro-organisms which induce macrophage granuloma formation which is largely restricted to the site, if this is subcutaneous or intramuscular, at which they are injected [PINCKARD *et al.*, 1968]. Mycobacteria, if injected in oil, can cause a systematised proliferation of macrophages in lymph-nodes throughout the body, spleen and lungs [WHITE *et al.*, 1955; SUTER and WHITE, 1954], an effect which can be reproduced by the chloroform-soluble wax D fraction or peptidoglycolipid extracted from the bacilli [WHITE, 1965].

It is necessary to consider what is meant by the term 'anaerobic coryneforms' as used here. An anaerobic coryneform organism referred to as *C. parvum* 936B was used originally by HALPERN *et al.* [1964] in order to increase the clearance of carbon from the blood of mice. In subsequent papers the same and other anaerobic coryneform bacteria have been used to inhibit tumour growth [HALPERN *et al.*, 1966; SMITH and WOODRUFF, 1968; FISHER *et al.*, 1970] without evidence for antigenic or taxonomic relationship between the organisms used. It seems clear from the work of JOHNSON and CUMMINS [1972] that the term *C. parvum* has very little taxonomic meaning, since bacteria bearing this name appear in each of several serologically and biochemically defined broad categories. My colleagues and I have, therefore, used the term 'anaerobic coryneforms' to include these diverse bacteria with macrophage-stimulating properties, following the taxonomic groups of JOHNSON and CUMMINS [1972] who postulate 4 groups on the basis of serology, cell wall analysis for sugars, amino sugars and amino acids and estimates of DNA homology. The type strains accepted for each of these groups are *C. acnes* 0208 (serological group I), *C. acnes* 0162 (serological group II), *C. granulosum* 0507 (serological group III) and *Propionibacterium avidum* 0575 (serological group IV) [JOHNSON and CUMMINS, 1972] (table I). Various organisms designated as *C. parvum* have proved to belong to each of the three serological groups I, II and III.

Of 21 strains of anaerobic coryneforms tested, 15 were able to increase substantially (by at least 50%) the phagocytic index of mice (table I); the bacteria effective in this way were derived from each of the 4 serological groups [O'NEILL *et al.*, 1973]. Nevertheless, 3 of 5 strains with the highest activity belonged to serological group I.

Attempts have been made by several authorities to include the anaerobic coryneform organisms in the genus *Propionibacterium* [DOUGLAS and GUNTER, 1946]. The type species on which this genus are based are organisms

Table I. Phagocytic indices and percentage increases in spleen weight produced by 4 serological groups of anaerobic coryneforms and by a group of 'classical' propionibacteria

Micro-organism	Phagocytic index K		Ratio of test K to control K	Spleen weights ratio test to control
	average of 4–5 mice	controls, average of 4–5 mice		
Anaerobic corynebacteria				
Serological group I				
C. parvum 0208	0.057	0.014	4.2	1.64
C. parvum 1383	0.061	0.015	4.1	1.21
C. liquefaciens 814	0.044	0.015	2.9	1.32
C. acnes 737	0.017	0.014	1.21	no increase
C. parvum 3085	0.041	0.015	2.7	1.9
C. parvum 6134	0.036	0.014	2.6	2.6
C. lymphophilum 6294	0.015	0.015	1.0	1.2
Serological group II				
C. parvum 10390	0.025	0.012	2.1	2.1
C. diphtheroides 2764	0.025	0.013	1.9	–
C. anaerobium 578	0.052	0.024	2.2	6.9
C. acnes 6280	0.020	0.013	1.5	1.1
C. granulosum 6292	0.015	0.017	0.9	1.4
C. liquefaciens 6290	0.017	0.017	1.0	1.7
Serological group III				
P. granulosum 0507	0.044	0.014	3.1	1.1
C. parvum 10387	0.021	0.014	1.5	1.2
C. parvum A	0.021	0.014	1.5	1.1
C. parvum B	0.019	0.015	1.3	1.5
C. parvum C	0.014	0.015	0.9	1.0
Serological group IV				
P. avidum 0575	0.044	0.024	1.8	1.4
P. avidum 4982	0.055	0.011	5.0	3.1
P. avidum 0589	0.035	0.015	2.3	1.7
'Classical' propionibacteria				
P. freudenreichii 10470	0.029	0.024	1.2	–
P. arabinosum 5958	0.022	0.024	0.9	1.0
P. jensenii 5960	0.015	0.014	1.1	1.0
P. rubrum 8901	0.014	0.014	1.0	1.0

K = phagocytic index.

(the classical propionibacteria) which are isolated from cheese, butter and other milk products [VAN NIEL, 1928]. We have included several examples of classical bacteria in our tests. In phagocytic clearance tests, these propionibacteria are unable to increase the phagocytic index of mice (table I). In other biological tests they have proved uniformly inactive (v.i.); these facts have, therefore, led us to prefer the designation anaerobic coryneforms to propionibacteria.

The chemotactic activity of a bacterial cell could, in general, become manifest in three ways; firstly, by a direct effect demonstrable *in vitro* and *in vivo* causing directional migration towards the bacteria; secondly, by an indirect effect also demonstrable *in vitro* whereby the bacteria activated chemotactic components of serum or plasma added to the system (cytotaxigen) and, thirdly, by indirect effects demonstrable additionally or only *in vitro*. There has been a number of attempts to prove the existence and workings of neutrophil chemotactic factors *in vivo* [BUCKLEY, 1963; DE SHAZO and COCHRANE, 1971] but, until recently, no reports of the production of macrophage chemotactic factors *in vivo*. SNYDERMAN *et al.* [1971a] showed that within 30 min of injection of bacterial endotoxin a neutrophil chemotactic factor, identifiable as the peptide C5a, appeared in the peritoneal exudate. This activity reached a maximum between 1 and 3 h after injection, thereafter declining. Recently my colleagues in this laboratory [WILKINSON *et al.*, 1973a] have shown that intraperitoneal injection of glycogen causes a rise in chemotactic activity for both neutrophils and macrophages. The early rise in neutrophils is attributable to C5a, which also probably accounts for the increase in macrophage-chemotactic factors since SNYDERMAN and his colleagues [SNYDERMAN *et al.* 1971b, 1972] have shown that C5a has the ability to attract macrophages as well as neutrophils. However, while the activity for neutrophils is transient, the activity for macrophages persists for the next 4 days. This persisting stimulus for macrophages is presumably a factor additional to C5a, but has not been identified further. Further experiments have shown that an intraperitoneal injection of heat-killed *M. tuberculosis* (human strain $H_{37}R_v$) reproduces an identical pattern of exudation of phagocytic cells as seen with glycogen. Presumably, therefore, the differences between mycobacteria and glycogen as granuloma-inducing agents must be due to mycobacterial properties which persist in exerting their effect on macrophage reproduction and recruitment from the circulation long after this initial chemotactic response.

When we look at the ability of mycobacteria and the anaerobic coryneforms to exert chemotaxis, we find interesting differences. Until recently, mycobacteria failed to provide evidence for a specific chemotactic action on macrophages. HARRIS [1953] reported that macrophages, in general, responded to the same bacterial stimuli as granulocytes and that *M. tuberculosis* strain $H_{37}R_a$ attracted both types of cells equally. Contrariwise, SCHÄR [1955] found that *M. tuberculosis* provided no direct chemotactic attraction for *M. tuberculosis*. She studied several constituents including phosphatides, polysaccharides and chloroform-soluble waxes. Phosphatide and wax fractions from a number of strains of *M. tuberculosis* showed weak leucotactic activity at high concentrations only. The polysaccharide fractions were inactive. In a later paper, MEIER and SCHÄR [1957] confirmed that polysaccharides from the bovine strain of *M. tuberculosis* and from *M. leprae* were not chemotactic. More recently, SYMON *et al.* [1972] found that *M. tuberculosis* had a strong chemotactic effect when tested *in vitro* by a modified Boyden chamber method in the presence of plasma, attracting both neutrophil leucocytes and peritoneal macrophages. Mycobacteria were found to attract macrophages more strongly than neutrophils. Parallel tests with bacterial endotoxin showed that it attracted neutrophils far more strongly than macrophages. When tested in the absence of plasma, *M. tuberculosis* had virtually no ability to attract macrophages. Presumably complement components are necessary for the chemotactic activity of *M. tuberculosis* on macrophages. Tests to determine which fractions of *M. tuberculosis* were cytotoxigenic are shown in table II. These findings indicate that the cytoplasmic fraction of *M. tuberculosis* was richer in activity than the cell wall; also that removal of the lipid from the bacterial cell did not remove the chemotactic activity. Several protein fractions which were extracted from *M. tuberculosis* were active, but polysaccharide fractions had a much weaker effect and the chloroform-soluble material, wax D, was inactive. *M. phlei* also contained an extractable cytotaxigen for macrophages in the presence of plasma, but its activity was weaker than that from *M. tuberculosis*.

It must be admitted that at this stage of our information it is difficult to see a clear relationship between the chemotactic factors described by SYMON *et al.* [1972] for *M. tuberculosis* and granuloma formation. Thus, chloroform-soluble wax D fractions of human strains of mycobacteria or the purified peptidoglycolipids derived therefrom are certainly able to induce a granuloma both at the site of their injection (which is usually in a vehicle of mineral oil) and in a wide field of lymph-nodes. This granuloma, macroscopically and microscopically, closely resembles that produced by the same weight (40 or 200 μg) of whole mycobacteria [WHITE *et al.*, 1955; WHITE,

Table II. Chemotactic activity of mycobacteria and mycobacterial fractions for guinea-pig peritoneal macrophages in the presence of plasma [data of SYMON *et al.*, 1972]

Test substances, 10 μg/ml	Mean chemotactic indices for all experiments in the presence of		
	no plasma	pre-heated plasma, 56°C 30 min (10%)	fresh plasma, 10%
M. tuberculosis strain C			
Whole cells	0	0	317
Cytoplasm	0	0	324
Cell walls	0	14	62
Wax D	0	7	10
M. tuberculosis strain $H_{37}R_a$			
Totally delipidated cells	0	11	357
Wax D	0	0	7
Protein A	0	0	206
Protein B	0	0	145
Protein C	0	6	232
Protein T	0	2	75
Glycopeptide	0	0	57
Polysaccharide II	0	0	118
M. phlei			
Totally delipidated cells	0	0	27
Cell walls	0	0	95
Glycopeptide	0	0	38
M. avium			
Wax D	0	0	6

Note that the chemotactic indices in this table are derived from the counts of macrophages reaching the lower surface of an 8-μm filter after 5 h at 37°C. The chemotactic index was derived by comparison with appropriate positive and negative controls by the following formula:

$$\text{chemotactic index} = \frac{\text{test count-control count (Gey's solution or plasma)}}{\text{positive (casein) control count-negative (Gey's count)}} \times 100.$$

1965] in including epithelioid cells, giant cells, lymphoid nodules and areas of necrosis in its make-up. Such a granuloma requires 10–14 days to achieve a typical appearance and include impressive areas of packed epithelioid cells. Possibly the cytoplasmic polysaccharide or protein factors of SYMON *et al.*

Table III. Chemotactic activity of anaerobic corynebacteria and propionibacteria for guinea-pig peritoneal macrophages

Strain	Chemotactic migration of macrophages[1]; microns migrated through 8-μm filter in 130 min; towards bacterial suspension at (mean migration to the nearest 10 μm)	
	$1,500 \times 10^6$ organisms/ml	$1,500 \times 10^4$ organisms/ml
Group I (*P. acnes* group)		
C. parvum 0208	20	40
C. parvum 3085	30	90
C. parvum 1383	20	120
C. liquefaciens 814	120	30
P. acnes 737 (VPI. No. 0389)	60	50
Group II		
C. diphtheroides 2764	100	10
C. anaerobium 578	90	40
C. parvum 10390[2]	30	30
Group III (*P. granulosum* group)		
C. parvum 10387	60	40
C. parvum A	30	50
C. parvum B	40	20
C. parvum C	30	60
P. granulosum 0507	40	40
Group IV (*P. avidum* group)		
P. avidum 0575	70	50
P. avidum 4982	40	120
P. avidum 0589	30	90
Classical propionibacteria		
P. freudenreichii	30	30
P. jensenii	20	not treated
P. rubrum	20	10
P. arabinosum	30	20

Negative control Gey's solution, 20. Positive control casein 5 mg/ml, 140.

No plasma or serum were added to the tests shown in this table.

The strain numbers of the bacteria in this table are numbers given by the Virginia Polytechnic Institute, by the Institut Pasteur or by Burroughs Wellcome Co. The original sources can be learned from table I in JOHNSON and CUMMINS [1972] and table I in O'NEILL *et al.* [1973].

1 Tests giving 10–30 μm migration are considered negative, 40–50 μm weakly positive and 60 μm and above, unequivocally positive.

2 A batch of *C. parvum* 10390 originally tested was strongly positive but this result could not be repeated with subsequent batches.

[1972] are responsible for an earlier phase of macrophage exudation. It is probable that, besides the cytotaxigenic influence of the mycobacteria other factors, e.g. macrophage chemotaxic factors released from damaged cells [CLARK and CLARK, 1930; JACOBY, 1937; BOREL, 1970] also have a role in granuloma formation.

All groups of the anaerobic coryneforms include representatives with a direct and specific chemotactic effect for macrophages (table III) [WILKINSON et al., 1973]. Most members of the anaerobic coryneforms (serological groups 1–4) produce a chemotactic factor which attracts macrophages directly and without the necessity for addition of plasma. Preliminary experiments indicate that the macrophage chemotactic factor from C. anaerobium 578 is a non-dialysable macromolecule which occurs in culture filtrates of the organism at the time of logarithmic growth. It is a labile substance being destroyed by moist heat at 100 °C and does not survive preservation at –20 °C. It is significant that propionibacteria derived from cheese or milk products (classical propionibacteria) had no chemotactic activity for macrophages. The anaerobic coryneforms failed to exert chemotaxis on neutrophils. Thus, they become quite outstanding for their strong macrophage specific chemotactic action. When human serum is added to the test system, these organisms can react with a serum component to cause chemotaxis of neutrophils.

It is of some interest to compare the ability of these organisms to secure chemotaxis of macrophages and their other biological properties such as the stimulation of phagocytosis, as measured by enhanced clearance of gelatin-stabilised carbon from the mouse circulation [test for increased phagocyte index: HALPERN et al., 1964]. The relationship between carbon clearance and chemotactic activity [O'NEILL et al., 1973] is shown in figure 1. A significant correlation is shown between the two functions ($p = 0.005$; correlation coefficient $r = 0.66$). It is also of interest that the most inactive organisms were all derived from the 'classical propionibacteria'. Mycobacteria also act to increase the phagocytic index of mice. In this case it has been argued that the stimulation of macrophages is an indirect effect of the bacteria in stimulating the thymus-dependent mechanisms of cell-mediated immunity [WHITE, 1971a]. This will be discussed further below.

It is further of interest to consider possible relationships between factors which exert chemotaxis and which stimulate the metabolism of macrophages. The energy requirements by macrophages were defined by KARNOVSKY et al. [1970] in respect of phagocytosis. Casein-induced guinea-pig peritoneal macrophages were shown to have their power of phagocytosis blocked by

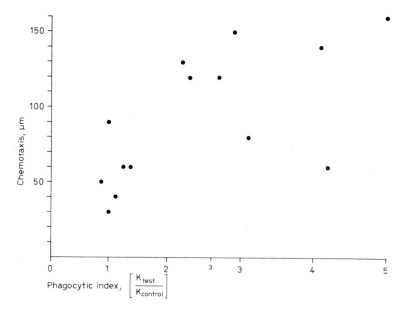

Fig. 1. Correlation of macrophage chemotaxis (ordinates in microns) with ratio of phagocytic indices (abscissae) for 19 different strains of anaerobic coryneforms and classical propionibacteria. The chemotactic activity taken is the sum of the activities at two different counts of bacteria 1.5×10^9 and 1.5×10^7 (see table III). The ratio of phagocytic indices is the ratio of mean value for K for a group of test mice to the mean of the group of controls. A good correlation between chemotaxis and K is shown for most results. Assuming the best straight line to be given by the equation $y = a + bx$, the plot of a against b yields a correlation coefficient (r) in this instance of 0.66. Tested statistically this result is found significant ($p = 0.005$).

inhibitors of anaerobic glycolysis but not by those for oxidative phosphorylation. Contrariwise, phagocytosis of alveolar macrophages (unstimulated) was blocked both by inhibitors of glycolysis and those of oxidative phosphorylation. The question of the energy requirement of stimulated *versus* unstimulated macrophages from various tissue sources was re-investigated by WILKINSON *et al.* [1973a], with the result that stimulated cells (whether peritoneal or alveolar) were found to depend on glycolysis for energy of chemotactic migration, whereas unstimulated cells used oxidative pathways in addition.

When a strong chemotactic agent, active for both neutrophils and macrophages [KELLER and SORKIN, 1967] is incubated for 3 h together with either of these cell types, derived from human blood or mouse or guinea-pig

Table IV. Production of lysosomal enzymes by human blood neutrophils and by guinea-pig peritoneal macrophages on incubation with chemotactic factors

Substance added to neutrophils, 2-hour incubation at 37°C (3×10^6 cells in 4 ml)	Increase in enzyme content[1]					
	human blood neutrophils			guinea-pig peritoneal macrophages		
	β-glucuronidase μg/ml	β-galactosidase μg/ml	acid phospha-tase, KAU/ml	β-glucuronidase μg/ml	β-galactosidase μg/ml	acid phosphatase KAU/ml
Casein 5 mg/ml	4.25	4.7	0.80	1.85	0	20.4
C. anaerobium 578 $1,500 \times 10^6$ organisms per ml	0.60	0.50	0	3.5	0	6.8

1 The figures represent enzyme content of cells incubated with the chemotactic substance minus enzyme content of cells incubated under the same conditions but without the chemotactic substance.
KAU = Kay-Armstrong units.

peritoneum, respectively, enzyme levels generally become raised [table IV: data of O'NEILL and colleagues, to be published]. In the experiment of table IV, the neutrophil levels for β-glucuronidase and β-galacturonidase were raised while the levels of acid phosphatase remained unchanged. This poor response in the case of acid phosphatase in neutrophils has been observed previously [BARKA and ANDERSON, 1962; BRAUNSTEINER and SCHMALZL, 1970]. When *C. anaerobium* 578 was used as the stimulus, none of the enzyme levels in neutrophils increased, but increases occurred in β-glucuronidase and acid phosphatase in peritoneal macrophages. These preliminary findings suggest that the anaerobic coryneforms may achieve their macrophage specificity by a selective stimulatory effect in certain enzyme systems, and that selective chemotaxis of different types of phagocyte may, therefore, depend on the activation of different metabolic pathways.

My colleagues and I have used a further method for detecting and measuring the enzyme activation of macrophages in granulomata formed within the lungs after intravenous injection of bacteria [CATER *et al.*, to be published]. The method is an adaptation from that of MYRVIK *et al.* [1961] for obtaining alveolar macrophages for *in vitro* study. In our method, 6 mg of heat-killed bacteria are injected intravenously into a 6- to 10-week-old chicken. This results in a rapid increase, reaching maximum at about 7 days, in the enzyme content of macrophages which can be washed out of the lung by tracheal perfusion *post mortem*. We have estimated acid phosphatase, β-D-glucuronidase, β-D-galactosidase and phospholipase A activity in the lysed macrophages. The increase in enzymes is partly accounted for by increase in numbers and partly by up to 2-fold increase in enzyme content per cell. In later experiments it was found simpler to homogenise the whole lung and estimate enzyme levels on an aliquot of this homogenate.

In such a test (table V) 6 mg of heat-killed *M. avium* will uniformly produce, after intravenous injection, about a 10-fold increase in acid phosphatase activity over that shown by saline-injected contemporaneous control birds. Other acid hydrolases, such as β-glucuronidase, β-galactosidase and phospholipase A have been shown to be increased in a similar ratio. The test provides a rapid method for surveying the ability of anaerobic coryneforms, mycobacteria and other bacteria to provoke the appearance of increased numbers of 'activated' macrophages. Table V also includes the results of injecting a suspension of carbon particles intravenously. Assuming that these are chemically inert and devoid of chemotactic activity it was assumed that they might provide an indication of the metabolic activation which is merely a consequence of phagocytosis. It can be seen that the responses to a variety

Table V. Pulmonary acid phosphatase levels in chickens 6 days after an intravenous injection of different heat-killed micro-organisms

Genus	Organism or particle	Serological group	Acid phosphatase KAU/ml	Mean
Mycobacterium	*M. avium*		220 235 260	238
Anaerobic coryneform	*P. avidum* 4982	4	186 195 209	196
	C. parvum 3085	1	160 165 174	166
	P. avidum 0589	4	156 161 170	162
	C. anaerobium 578	2	130 150 170	150
	C. parvum 0208	1	110 130 135	125
	C. liquefaciens 814	1	85 105 165	118
	C. parvum 1383	1	100 109 120	109
	C. parvum B	3	85 100 105	96
	C. parvum 6294	1	65 65 150	83
	P. granulosum 0507	3	68.7 74.3 65.3	69
	C. parvum 6292	2	52 59 72	61
	C. parvum C	3	20.2 33.4 36.1	30

Table V (continued)

Genus	Organism or particle	Serological group	Acid phosphatase KAU/ml	Mean
	Carbon suspension		58.93	
			54.51	57.9
			60.37	
Classical pro-pionibacteria	*P. freudenreichii* 10470		35	
			45	41.6
			45	
	P. jensenii 5960		30	
			24	28.6
			32	
	P. rubrum 8901		27	
			20	25.6
			30	
	P. arabinosum 5958		15	
			19	19.6
			25	
	Saline		20	
			16	19.6
			22	

of anaerobic coryneform organisms are scattered over a wide range, some such as *P. avidum* 4982 providing responses which approach that of *M. avium* and some, such as *C. parvum* C and *C. granulosum* 6292 which are only just above the 'carbon threshold' level. All of the classical propionibacteria yielded results at or below the 'carbon threshold' levels.

The degree of correlation which was found between the results of estimating enzyme increases (acid phosphatase) in the resultant pulmonary macrophages and phagocytic indices (ratio of test and control) is shown in figure 2 (solid dots). The degree of correlation which was found between the results of estimating chemotactic activity and enzyme increases (and phosphatase) in pulmonary macrophages is shown also in figure 2 (open squares).

Since the effect of the anaerobic coryneforms on chemotaxis of macrophages is specific and occurs without the activity of serum factors, it is tempting to regard these organisms as direct activators of enzymes such as acid phosphatase. As evidenced below, the anaerobic coryneforms do not stimulate cell-mediated immunity and indirect affects *via* thymus-dependent

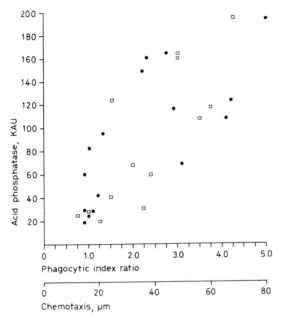

Fig. 2. (1) Correlation of macrophage chemotaxis (abscissae in microns) with levels of acid phosphatase (□) in alveolar macrophages (KAU = Kay-Armstrong units) after stimulation with 14 different strains of anaerobic coryneforms and classical propionibacteria. Correlation coefficient (r) = 0.79. When tested statistically this result was found highly significant (p = 0.0004). (2) Correlation of ratios of phagocytic indices (ordinates in microns) with levels of acid phophatase (•) in alveolar macrophages (KAU) after stimulation with 16 different strains of anaerobic coryneforms and classical propionibacteria. Correlation coefficient (r) = 0.73. When tested statistically this result was found highly significant (p = 0.0006).

lymphocytes are unlikely. Contrariwise, it is probable that the striking activation of pulmonary macrophages by mycobacteria is partly an indirect effect of their ability to stimulate cell-mediated immunity – although this has not been demonstrated, as yet, by experiments involving neonatally thymectomised mice.

The Influence of Cell-Mediated Hypersensitivity on Granuloma Formation

In the granuloma which has existed for more than 3 days it is possible that an increasing influence in determining the cellular composition, cytological and histological appearances would derive from the development of cell-mediated hypersensitivity. The presence of antigen in the granuloma would be expected to cause macrophage exudation and immobilisation within the lesion. Presumably, also, cell-mediated immunity will act as an important determinant for necrosis.

In this connection, the use of anaerobic coryneforms provided a sharp contrast to mycobacteria. The latter are well-known for their striking ability to facilitate the induction of cell-mediated immunity when antigen is injected in water-in-oil emulsions with added mycobacteria. On the contrary, the anaerobic coryneforms have virtually no ability to facilitate the induction of cell-mediated immunity [O'NEILL et al., 1973]. O'NEILL and colleagues used a range of anaerobic coryneforms in water-in-oil emulsion with added ovalbumin. Skin and corneal tests failed to show any increase in delayed-type hypersensitivity by guinea pigs receiving varying doses of anaerobic coryneforms (200 μg to 2.5 ml) with ovalbumin in water-in-oil emulsion over animals receiving ovalbumin in simple water-in-oil emulsion. In addition, the characteristic effect of injecting ovalbumin in water-in-oil emulsion with mycobacteria – the development of a γ_2-precipitin arc of anti-ovalbumin on immunoelectrophoresis [WHITE et al., 1963] was not observed following the use of anaerobic coryneforms.

It seems clear that the lack of any stimulatory effect of anaerobic coryneforms for cell-mediated hypersensitivity should lead to histological differences between their granulomatous reaction and that which follows the injection of mycobacteria. As described by O'NEILL et al. [1973] the local site of injection in the footpad of guinea-pigs receiving various anaerobic coryneforms in oily suspension becomes occupied by a relatively simple proliferation of macrophages. Control animals receiving the oil vehicle showed accumulation of some foamy-type macrophages around oil vesicles. Animals receiving in addition anaerobic coryneforms had a larger, more compact granuloma consisting of dense sheets of macrophages surrounding oil vesicles. The macrophages, although densely packed in masses of cells with polygonal outline, did not show epithelioid morphology. Giant cells were not seen and necrosis was absent. Contrasting these findings with those which follow injection of comparable doses of mycobacteria in oil, it is tempting to conclude that the epithelioid cellular change, the development of giant cells and the necrosis are all attributable to the enhanced cell-mediated hypersensitivity.

Support for this is provided by experiments in the guinea-pig [WHITE, 1971a] in which complete Freund-type adjuvant (containing heat-killed

M. tuberculosis and haemocyanin from *Helix pomatia* as antigen) was used as a primary sensitising injection into one hind footpad. Three weeks later the other hind foot was injected with a small dose of antigen (100 μg haemocyanin) in water-oil emulsion. The sensitising injection results 3 weeks later in a foot which is greatly swollen, with a cyanotic scaly surface and some ulceration. This enlargement is due to a granuloma of epithelioid type macrophages surrounding spherical oil vesicles. This granuloma has a characteristic firm, grey, translucent cross-section. The microscope shows areas of necrosis, areas of lymphocytes in the form of nodules and giant cells. This histology of the granuloma is mainly determined by the mycobacterial component of the complete Freund-type adjuvant since haemocyanin in water-oil emulsion (without mycobacterial component) injected into a normal guinea-pig fails to cause any such naked eye changes in the injected foot at 2–3 weeks later. Microscopically this control lesion contains scattered foamy macrophages but is devoid of epithelioid cells, giant cells or necrosis.

But with the same injection made into animals already sensitised to HCN, i.e. an injection into the opposite foot of a guinea-pig injected 3 weeks previously with haemocyanin in water-oil emulsion plus mycobacteria, a larger swelling results which comes to resemble in a remarkable way the opposite foot. Thus it has, at 2–3 weeks later, a scaly, blue surface, with ulcerative necrosis and a firm, translucent, grey cut surface on transection. Microscopically it is an epithelioid cell granuloma. These findings infer that the stimulus for formation of an epithelioid granuloma is an immunological one, which can develop without injection of tubercle bacilli directly into the area. The slow release of a protein antigen (haemocyanin) into the tissues of an animal with established cell-mediated hypersensitivity (and humoral immunity) to the same antigen is productive of such granuloma formation. It should be added that the resulting lesion is not an exact counterpart of the original granuloma. Giant cells are noticeably scarce within it. Thus, there is room for certain direct effects of mycobacteria.

The presumption is, therefore, that antigen within the tissue of animals with a high level of cell-mediated hypersensitivity release lymphokines which can transform the exudative lesion into a typical tuberculous epithelioid cell granuloma. Other components of adjuvant mixtures which are not generally regarded as immunological antigens can also produce analogous 'granulomatous hypersensitivity' [EPSTEIN, 1967], e.g. the mineral oil and the emulsifying agents. In experiments in which mycobacteria in oil were injected into the breast muscle of chickens, WHITE and HERBERT [unpublished information] found that a typical epithelioid and giant cell granuloma results 14–21 days later. Injection of either the mineral oil (Bayol or Drakeol) or the emulsifying agent (Arlacel A) at this time into a wattle resulted within 7–10 days in the development of an epithelioid granuloma with a remarkable resemblance to the original breast granuloma. Injection of either mineral oil or Arlacel into normal, unsensitised birds resulted in a minimal exudation of granulocytes and macrophages which differ qualitatively from the lesion in the sensitised birds. It is probable that the generation of such granulomatous hypersensitivity requires the presence of a persistent or indigestible antigen, and that the mineral oil and Arlacel possess the necessary antigenic determinants and the ability to persist *in vivo*. That the acyl side-chain of a catechol can act as a determinant for delayed contact sensitivity is shown by the work of BAER *et al.* [1967].

Further evidence that the epithelioid granuloma following injection of CFA is a consequence of the activation of a population of thymus-derived lymphocytes and the development of delayed-type hypersensitivity is obtained by deviation of this response by prior injection of protein antigen [BOYDEN, 1957], which ablates the increase of cell-mediated hypersensitivity and at the same time decreases focally the size and epithelioid cell component of the local granuloma [WILKINSON and WHITE, 1966]. In the chicken, the development of an epithelioid granuloma is not influenced by neonatal bursectomy and irradiation, but is greatly impaired by neonatal thymectomy and whole-body X-irradiation before injection of CFA mixture [AIYEDUN *et al.*, quoted by WHITE, 1971b].

Interaction of antigen with guinea-pig sensitised lymphocytes has been shown to release factors which are chemotactic for macrophages [WARD *et al.*, 1969, 1970]. This chemotactic factor, which may be of importance in the supply of cells to the chronic granuloma resulting from injection of CFA, can be separated from macrophage inhibition factor (MIF). It might be expected that the generation of this factor should depend upon antigen interaction with a thymus-dependent population of lymphocytes. Evidence for this was provided by ALTMAN and KIRCHNER [1972] by their demonstration that neonatally bursectomised and irradiated chickens could still generate this macrophage chemotactic factor. WILKINSON and WOLSTENCROFT [personal commun.] failed to confirm that the generation of such a factor in guinea-pigs sensitised to bovine γ-globulin was dependent upon antigen. However, primed lymphocytes were also shown by these investigators to release by contact with antigen a factor which inhibited the chemotaxis of macrophages.

Late Organisation of Cells in Granulomata

It has been emphasised above that in chronic inflammatory reactions caused by, for example, an injection of complete or incomplete Freund adjuvant, the entry of cells in the first 3 or 7 days is limited almost entirely to granulocytes and monocytes derived from the bone marrow. It was a remarkable result of the experiments of SPECTOR and WILLOUGHBY [1968] that injection of CFA or incomplete Freund adjuvant after whole-body irradiation of rats (600–750 rats) was followed by total suppression of the lesion seen at 3 and 7 days, and that whereas the granuloma could be re-established by intravenous injections of bone marrow cells, no reconstitution took place with lymph-node cells or thymic cells similarly injected by the intravenous route. In such irradiation chimeras, the architecture of the chronic inflammatory response to mineral oil was largely replaced within 3 days of its injection.

The next question to arise is: will lymphocytes migrate into an early granuloma if they are 'committed', i.e. sensitised against antigen present within the granuloma? Many workers, e.g. TURK [1967] have repeatedly failed to show any specific attraction of sensitised lymphocytes to tuberculin reactions. SPECTOR and WILLOUGHBY [1968] similarly were unable to restore the cellular accumulation of a tuberculin response in an irradiated rat injected intravenously with lymph-node cells or thymus cells from sensitised histocompatible donor rats. However, irradiated rats given lymph-node cells from a tuberculin-sensitive donor and bone marrow cells from a non-sensitised donor produced convincing tuberculin reactions at 48 h after challenge. These experiments show that lymph-node cells (or marrow cells) could transfer the cell-mediated hypersensitivity, but only marrow cells were capable of migrating into the reaction site to produce the typical cellular accumulation of a tuberculin reaction. By electron microscopy criteria, lymphocytic exudation in the tuberculin response in guinea-pigs is confined to the small veins in the sub-dermis [WIENER et al., 1967]. On the basis of these results it seems that cells of the lymphocyte series and even cells from animals sensitised to an antigen within the tissue site do not normally become recruited to an early granuloma in large numbers. However, in the late granulomatous reaction which follows the use of CFA, it is clear from the histology that infiltration by large numbers of lymphocytes has occurred, and that the granuloma is productive of substantial amounts of antibody. SMITH et al. [1970] studied the migration of lymphocytes in the granuloma which developed from CFA injected into the hind limbs of sheep. Large numbers of polymorphonuclear neutrophils appeared in the draining lymph in the first 24–48 h. Subsequently the numbers of lymphocytes in the lymph rose and the lymphocyte output increased until, 2–4 weeks after injection, the number of lymphocytes coming from the granuloma was about the same as from a normal lymph-node. Lymphocytes were identified within and migrating through the walls of blood capillaries in the granuloma. It was thought that these cells were passing from the blood stream into the lymph.

SPECTOR and LYKKE [1966] had previously pointed to the development of lymphocytes within a rabbit granulomatous response to CFA and claimed that progeny of dividing histiocytes and macrophages 'came to resemble lymphocytes'. However, later SPECTOR and WILLOUGHBY [1968] concluded that migrating lymphocytes did enter established granulomata in the rat (see below).

Local antibody production in the granuloma formed after injection of CFA has been demonstrated in the horse [FREUND, 1953], the rabbit [ASKONAS and HUMPHREY, 1958] and chicken [FRENCH et al., 1970] but hardly occurs in the guinea-pig [ASKONAS and WHITE, 1956]. The findings in the chicken are particularly interesting since an early anti-HSA response at 7–9 days appears to be the result of biosynthesis in the spleen, and the later prolonged and increased antibody response is due mainly to production in the local granuloma. Preliminary results with ^{3}H-uridine labelled autologous bursa and thymus cells have shown that both these cell types rapidly gain access to the granuloma tissue at 5 weeks [SMITHYMAN and WHITE, unpublished observations].

As indicated above, the granuloma induced by water-oil emulsion plus M. tuberculosis differs from that due to incomplete Freund adjuvant in the presence of large collections of cells which resemble small and medium lymphocytes. The lymphocytes in such lymphoid foci can become highly organised and in the granulomata resulting 1 month after an injection of a suspension of killed M. avium in mineral oil into the breast muscle of the chicken, well structured germinal centres are present (fig. 3) which resemble the germinal centres of chicken spleen in the distribution of B-lymphocytes and dendritic cells bearing complexes of antigen and antibody.

SPECTOR and WILLOUGHBY [1968] injected tritium-labelled bone marrow and lymph-node cells from both sensitive and normal rat donors, into recipient rats of the same inbred strain, into whose hind paws CFA had been injected 5 weeks previously. Autoradiographs of the granuloma prepared 24 h later showed the presence of labelled lymphocytes from sensitised, but

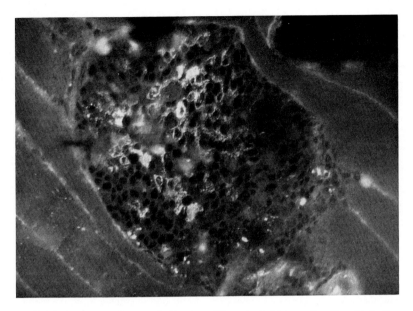

Fig. 3. Fluorescence micrograph. Section of breast of chicken: the site of injection 45 days after an injection of Freund's complete adjuvant (CFA) containing human serum albumin (HSA) as added antigen. Section stained by sandwich fluorescent antibody method for demonstration of HSA and anti-HSA. The section demonstrates complexes of HSA:anti-HSA in dendritic cells within a germinal centre in voluntary muscle at edge of granuloma (preparation of D. C. HENDERSON). × 600.

not from unsensitised donors. It is not clear what the migrant cells from bone marrow were, but they may have been lymphocytes in view of the fact that rat marrow contains up to 20% of these cells [RAMSELL and YOFFEY, 1961]. These experiments, therefore, indicate a supply of lymphocytes into late granulomas which histologically show the presence of lymphoid foci, and which are associated with the generation of delayed hypersensitivity. Presumably, here we have the explanation for the difference between the late granulomata induced by mycobacteria and anaerobic coryneforms, respectivelyly. Thus, the latter organisms fail to induce increased levels of cell-mediated hypersensitivity and have granulomas which lack the presence of lymphoid foci. Other observations such as those of SMITH *et al.* [1970] suggest a vigorous traffic of small and large lymphocytes through the established granuloma due to CFA injection in the sheep. It is suggested that the entry of lymphocytes into a granuloma in substantial numbers may depend upon the organisation of the blood vasculature with the provision of vessels functionally equivalent to the post-capillary vessels of the para-cortex of lymph-nodes. It has been suggested that lymphocytes migrate through the specialised high endothelium of these vessels [MARCHESI and GOWANS, 1964]. However, the increased traffic of lymphocytes within the lymph leaving a hydronephrotic sheep kidney developed without evidence of any specialised post-capillary venules in the kidney [SMITH et al., 1970].

At the moment of writing, the literature appears singularly bereft of findings supporting the role of chemotactic factors for lymphocytes with potential importance for recruiting these cells to chronic granulomata.

Summary

1. The granuloma is a tissue population of cells in dynamic equilibrium. The examples involving mycobacteria or pertussis organisms are high-turnover structures which maintain themselves by constant immigration of cells into themselves from the blood and by replication of macrophages. In this, they contrast with the low-turnover structure due to carrageenan, which depends for its continuing existence more on long-lived macrophages. Apparently, there is little or no evidence which allows an estimate of the manner and extent to which similar long-lived macrophages may contribute to the high-turnover structure also.

2. The *in vitro* chemotactic properties of mycobacteria differ sharply from those of the anaerobic coryneforms. The latter are a wide-flung group of bacteria which include 4 distinct groups on the basis of serology, cell-wall analyses for sugars, amino sugars and amino acids and estimates of DNA homology. The type strains for each are *Corynebacterium acnes* 0208 (group I), *C. acnes* 0162 (group II), *C. granulosum* 0507 (group III) and *Propionibacterium avidum* 0575 (group IV). Different organisms designated *Corynebacterium parvum* belonged to each of the groups I, II and III. Throughout the anaerobic coryneforms one finds, in varying degrees, the ability to activate macrophages directly and specifically in chemotaxis. Good correlations exist between ability to activate chemotaxis *in vitro*, ability to stimulate an increase of phagocytic index *in vivo*, and ability to increase the macrophage content of acid hydrolases *in vivo*. Contrariwise, mycobacteria fail to activate macrophages for chemotaxis directly *in vitro* but require addition of components of serum.

3. The granulomatous response to the anaerobic coryneforms is relatively simple, being mainly a population of normal macrophages (histiocytes). The mycobacterial granuloma is complex including necrosis, organized masses of epithelioid cells, giant cells and eventually an organisation which includes an intense through-put of lymphocytes and the presence of lymphoid nodules with germinal centres and antibody-synthesizing plasma cells. The dependence of these additional changes on the stimulation of thymus dependent lymphocytes (T cells) and cell-mediated hypersensitivity, processes which are associated with the use of mycobacteria, is discussed.

Acknowledgements

In this discussion I have made extensive use of the unpublished results of several members of my department, P. C. WILKINSON, G. O'NEILL, J. C. CATER, R. McINROY and D. N. K. SYMON. Their help and advice is gratefully acknowledged. I am also greatly indebted to Prof. CECIL CUMMINS of Blacksburg, Virginia, USA for help in understanding the difficulties which are involved in classifying the anaerobic coryneform bacteria, for the supply of many strains of these organisms and for serologically typing others.

This work was supported by grant No. 972/521/B from the Medical Research Council.

References

ALTMAN, L. C. and KIRCHNER, H.: The production of a monocytic chemotactic factor by agammaglobulinemic chicken spleen cells. J. Immunol. *109:* 1149 (1972).

ASKONAS, A. B. and HUMPHREY, J. H.: Formation of specific antibodies and γ-globulin *in vitro*. A study of the synthetic ability of various tissues from rabbits immunized by different methods. Biochem. J. *68:* 252 (1958).

ASKONAS, A. B. and WHITE, R. G.: Sites of antibody production in the guinea pig. The relation between *in vitro* synthesis of anti-ovalbumin and γ-globulin and distribution of antibody-containing plasma cells. Brit. J. exp. Path. *37:* 61 (1956).

BAER, H.; WATKINS, R. C.; KURTZ, A. P.; BYCK, J. S., and DAWSON, C. R.: Delayed contact sensitivity to catechols. III. Relationship of side-chain length to sensitizing potency. J. Immunol. *99:* 370 (1967).

BARKA, T. and ANDERSON, P. J.: Histochemical methods for acid phosphatase using hexazonium pararosanilin as coupler. J. Histochem. Cytochem. *10:* 741 (1962).

BOREL, J. F.: Studies on chemotaxis. Effect of subcellular leukocyte fractions on neutrophils and macrophages. Int. arch. Allergy *39:* 247 (1970).

BOREL, J. F. and SORKIN, E.: Differences between plasma and serum mediated chemotaxis of leukocytes. Experientia *25:* 1333 (1969).

BOYDEN, S. V.: The effect of a previous injection of tuberculoprotein on the development of tuberculin hypersensitivity following BCG vaccination in guinea-pigs. Brit. J. exp. Path. *38:* 611 (1957).

BRAUNSTEINER, H. and SCHMALZL, F.: Cytochemistry of monocytes and macrophages; in VAN FURTH Mononuclear phagocytes, p. 62 (Blackwell, Oxford 1970).

BUCKLEY, I.: Delayed secondary damage and leucocyte chemotaxis following focal aseptic heat injury *in vivo*. Exp. molec. Path. *2:* 402 (1963).

CLARK, E. R. and CLARK, E. L.: Observations on the macrophages of living amphibian larvae. Amer. J. Anat. *46:* 91 (1930).

DOUGLAS, H. C. and GUNTER, S. E.: Taxonomic position of *Corynebacterium acnes*. J. Bact. *52:* 15 (1946).

EBERT, R. H. and FLOREY, H. W.: The extravascular development of the monocyte observed *in vivo*. Brit. J. exp. Path. *20:* 342 (1939).

EPSTEIN, W. L.: Granulomatous hypersensitivity; in KALLÓS and WAKSMAN Progress in allergy, vol. 11, p. 36 (Karger, Basel 1967).

FISHER, J. C.; GRACE, W. R., and MANNICK, J. A.: Effect of non-specific immune stimulation with *Corynebacterium parvum* on patterns of tumour growth. Cancer, Philad. *26:* 1379 (1970).

FRENCH, V. I.; STARK, J. M., and WHITE, R. G.: The influence of adjuvants on the immunological response of the chicken. II. Effects of Freund's complete adjuvant in later antigen production after a single injection of immunogen. Immunology, Lond. *18:* 645 (1970).

FREUND, J.: Personal communication (1953).

HALPERN, B. N.; BIOZZI, G.; STIFFEL, C., and MOUTON, D.: Inhibition of tumour growth by administration of killed *Corynebacterium parvum.* Nature, Lond. *212:* 853 (1966).

HALPERN, B. N.; PRÉVOT, A. R.; BIOZZI, G.; STIFFEL, C.; MOUTON, D.; MORARD, J. C.; BOUTHILLIER, Y. et DECREUSEFOND, C.: Stimulation de l'activité phagocytaire du système reticulo-endothelial provoquée par *Corynebacterium parvum.* J. reticuloendothel. Soc. *1:* 77 (1964).

HARRIS, H.: Chemotaxis of monocytes. Brit. J. exp. Path. *34:* 276 (1953).

KARNOVSKY, M. L.; SIMMONS, S.; GLASS, A. E.; SCHAFER, A. W., and D'ARCY HART, P.: Metabolism of macrophages; in VAN FURTH Mononuclear phagocytes, p. 103 (Blackwell, Oxford 1970).

JACOBY, F.: Cannibalism and chemotaxis in the hen 'monocytes' *in vitro.* J. Physiol., Lond. *91:* 22 (1937).

JOHNSON, J. L. and CUMMINS, C. S.: Cell wall composition and desoxyribonucleic acid similarities among the anaerobic coryneforms, classical proprionibacteria and strains of *Arachnia proprionica.* J. Bact. *109:* 1047 (1972).

KELLER, H. U. and SORKIN, E.: Studies on chemotaxis. VI. Specific chemotaxis in rabbit polymorphonuclear leucocytes and mononuclear cells. Int. Arch. Allergy *31:* 575 (1967).

LEBER, T.: Über die Entstehung der Entzündung und die Wirkung der entzündungerregenden Schädlichkeiten (Engelmann, Leipzig 1891).

MARCHESI, V. T. and GOWANS, J. L.: The migration of lymphocytes through the endothelium of venules in lymph nodes. An electron microscope study. Proc. roy. Soc. B *159:* 283 (1964).

MEIER, R. und SCHÄR, B.: Vorkommen leukocytotaktischer Polysaccharide in bakteriellem, pflanzlichem und tierischem Ausgangsmaterial. Z. physiol. Chem. *307:* 103 (1957).

METCHNIKOFF, E.: Lectures on the comparative pathology of inflammation. Translated by F. A. STARLING and E. H. STARLING (Kegan-Paul, London 1893).

MYRVIK, Q. N.; LEAKE, E. S., and FARISS, B.: Studies on pulmonary alveolar macrophages from the normal rabbit. A technique to procure them in a high state of purity. J. Immunol. *86:* 128 (1961).

NIEL, C. B. VAN: The propionic acid bacteria (Boissevain, Haarlem 1928).

O'NEILL, G. J.; HENDERSON, D. C., and WHITE, R. G.: The role of anaerobic coryneforms on specific and non-specific immunological responses. I. Effect on particle clearance and humoral and cell-mediated immunological responses. Immunology, Lond. *24:* 977 (1973).

PINCKARD, R. N.; WEIR, D. M., and MCBRIDE, W. H.: Effects of *Corynebacterium parvum* in immunological unresponsiveness to bovine serum albumin in the rabbit. Nature, Lond. *215:* 870 (1967).

PINCKARD, R. N.; WEIR, D. M., and McBRIDE, W. H.: Factors influencing the immune response. III. The blocking effect of *Corynebacterium parvum* upon the induction of acquired immunological unresponsiveness to bovine serum albumin in the adult rabbit. Clin. exp. Immunol. *3:* 413 (1968).

RAMSELL, T. G. and YOFFEY, J. M.: The bone marrow of the adult male rat. Acta anat. *47:* 55 (1961).

RYAN, G. B. and SPECTOR, W. G.: Natural selection of long lived macrophages in experimental granulomata. J. Path. Bact. *99:* 139 (1969).

RYAN, G. B. and SPECTOR, W. G.: Macrophage turnover in inflamed connective tissue. Proc. roy. Soc. B *175:* 269 (1970).

SCHÄR, B.: Effects of constituents of tubercle bacilli on leucocyte migration.
cit. KRADOLFER, F.: Ciba Foundation Symposium on Experimental Tuberculosis, p. 65 (Churchill, London 1955).

SHAZO, C. V. DE and COCHRANE, C. G.: Discussion of paper by P. A. WARD, pp. 239–241; in AUSTEN and BECKER Biochemistry of the acute allergic reactions (Blackwell, Oxford 1971).

SMITH, J. B.; McINTOSH, G. H., and MORRIS, B.: The migration of cells through chronically inflamed tissues. J. Path. Bact. *100:* 21 (1970).

SMITH, L. H. and WOODRUFF, M. F. A.: Comparative effect of two strains of *Corynebacterium parvum* on phagocytic activity and tumour growth. Nature, Lond. *219:* 197 (1968).

SNYDERMAN, R.; ALTMAN, L. C.; HAUSMAN, M. S., and MERGENHAGEN, S. E.: Human mononuclear leukocyte chemotaxis. A quantitative assay for humoral and cellular chemotactic factors. J. Immunol. *108:* 857 (1972).

SNYDERMAN, R.; PHILLIPS, J. K., and MERGENHAGEN, S. E.: Biological activity of complement *in vivo*. Role of C5 in the accumulation of polymorphonuclear leukocytes in inflammation. J. exp. Med. *134:* 1131 (1971a).

SNYDERMAN, R.; SHIN, H. S., and HAUSMAN, M. S.: A chemotactic factor for mononuclear leukocytes. Proc. Soc. exp. Biol. Med. *138:* 387 (1971b).

SPECTOR, W. G. and LYKKE, A. W. U.: The cellular evolution of inflammatory granulomata. J. Path. Bact. *92:* 163 (1966).

SPECTOR, W. G. and RYAN, G. B.: The mononuclear phagocyte; in VAN FURTH Inflammation in mononuclear phagocytes, p. 219 (Blackwell, Oxford 1970).

SPECTOR, W. G. and WILLOUGHBY, D. A.: The origin of mononuclear cells in chronic inflammation and tuberculin reactions in the rat. J. Path. Bact. *96:* 389 (1968).

SUTER, E. and WHITE, R. G.: The response of the reticulo-endothelial system to the injection of the 'purified wax' and the lipopolysaccharide of tubercle bacilli. A histologic and an immunologic study. Amer. Rev. Tuberc. *70:* 793 (1954).

SYMON, D. N. K.; McKAY, I. C., and WILKINSON, P. C.: Plasma-dependent chemotaxis of macrophages towards *Mycobacterium tuberculosis* and other organisms. Immunology, Lond. *22:* 267 (1972).

TURK, J. L.: Delayed hypersensitivity, p. 71 (North-Holland, Amsterdam 1967).

VOLKMAN, A. and GOWANS, J. L.: The production of macrophages in the rat. Brit. J. exp. Path. *46:* 50 (1965a).

VOLKMAN, A. and GOWANS, J. L.: The origin of macrophages from the bone marrow in the rat. Brit. J. exp. Path. *46:* 62 (1965b).

WARD, P. A.: Chemotaxis of mononuclear cells. J. exp. Med. *128:* 1201 (1968).

WARD, P. A.; REMOLD, H. G., and DAVID, J. R.: Leukotactic factor produced by sensitized lymphocytes. Science *163:* 1079 (1969).

WARD, P. A.; REMOLD, H. G., and DAVID, J. R.: The production from antigen-stimulated lymphocytes of a leukotactic factor distinct from migration inhibition factor. Cell. Immunol. *1:* 162 (1970).

WHITE, R. G.: The role of peptido-glycolipids of *M. tuberculosis* and related organisms in immunological adjuvance; in STERZL Molecular and cellular basis of antibody formation, p. 71 (Czechoslovak Academy of Sciences, Prague 1965).

WHITE, R. G.: Factors controlling the adjuvant or immunosuppressive action of antigens in complete Freund-type mixtures. Ann. Sclavo *13:* 821 (1971a).

WHITE, R. G.: Adjuvant stimulation of antibody synthesis. In: 6th Int. Symp. Immunopathology, Grindelwald, p. 91 (Schwabe, Basel 1971b).

WHITE, R. G.; COONS, A. H., and CONNOLLY, J. M.: The role of a wax fraction of *M. tuberculosis* in adjuvant emulsions on the production of antibody to egg albumin. J. exp. Med. *102:* 83 (1955).

WHITE, R. G.; JENKIN, G. C., and WILKINSON, P. C.: The production of skin-sensitizing antibody in the guinea-pig. Int. Arch. Allergy *22:* 156 (1963).

WIENER, J.; LATTES, R. G., and SPIRO, D.: Electron microscopic study of leukocyte emigration and vascular permeability in tuberculin sensitivity. Amer. J. Path. *50:* 485 (1967).

WILKINSON, P. C.; O'NEILL, G. J.; MCINROY, R. J.; CATER, J. C., and ROBERTS, J. A.: Immunopotentiation by anaerobic coryneform bacilli; in WOLSTENHOLME and KNIGHT Ciba Symposium on Immunopotentiation, p. 121 (North-Holland, Amsterdam 1973a).

WILKINSON, P. C.; O'NEILL, G. J., and WAPSHAW, K. G.: Rôle of anaerobic coryneforms in specific and non-specific immunological reactions. II. Production of a chemotactic factor specific for macrophages. Immunology, Lond. *24:* 997 (1973b).

WILKINSON, P. C.; O'NEILL, G. J.; WAPSHAW, K. G., and SYMON, D. N. K.: Enhancement of macrophage chemotaxis by adjuvant-active bacteria. Immunology, Lond. (in press).

WILKINSON, P. C. and WHITE, R. G.: The rôle of mycobacteria and silica in the immunological response of the guinea pig. Immunology, Lond. *11:* 229 (1966).

Author's address: Dr. R. G. WHITE, University Department of Bacteriology and Immunology, Western Infirmary, *Glasgow G11 6NT* (Scotland)

Antibiotics and Chemotherapy, vol. 19, pp. 409–420
(Karger, Basel 1974)

Enzyme Activation and the Mechanism of Neutrophil Chemotaxis[1,2]

E. L. BECKER

Department of Pathology, University of Connecticut Health Center, Farmington, Conn.

The nature of the various factors chemotactic for polymorphonuclear leukocytes has been discussed in other chapters. Here, I wish to review our present knowledge of the biochemical mechanisms of the chemotactic response of the polymorphonuclear leukocyte paying particular attention to the nature and role of esterase activation in this response. It must be confessed at the outset of this endeavor that although it is evident that the chemotactic response is the result of a multistep, probably multibranched, sequence of biochemical reactions it is still impossible to give any coherent description of the sequence(s) which is (are) firmly based on well established, unequivocal experimental evidence.

The activation of a serine esterase (an esterase with serine in its active site) has been shown to be required in the chemotactic response of rabbit polymorphonuclear leukocytes to C5a, C3a, $C\overline{567}$ and a small molecular weight partially purified bacterial chemotactic factor found in *Escherichia coli* culture filtrates (called 'bacterial factor' from here on). The evidence for this conclusion is the following.

1. Various series of *p*-nitrophenyl ethyl phosphonates inhibit the chemotactic response of rabbit polymorphonuclear leukocytes when the chemotactic factor acts on the cell but not when they act separately on either the cell or the chemotactic factor [36]. The hypothesis offered in explanation of these findings is that the chemotactic agent activates either directly or indirectly a serine esterase, a so-called 'activatable esterase', existing in or

1 This is communication No. 47 from the Department of Pathology, University of Connecticut Health Center.
2 This work was partially supported by PHS grant 09648.

on the cell in an enzymatically inert form. This 'activatable esterase' is not susceptible to inhibition by phosphonate esters unless and until it is activated. When activated it is capable of enzymatically initiating the next (unknown) step in the sequence.

2. The relation between the structure of the various inhibitory phosphonates and their ability to inhibit chemotaxis, that is their 'inhibition profiles', shows a characteristic and specific form [36].

3. Leukocytes incubated with one or another of the complement-derived chemotactic factors, C3a, C5a or $C\overline{567}$ progressively lose their ability to give a chemotactic response; that is, they are 'deactivated' or 'desensitized' [33, 35]. The complement-derived chemotactic factors deactivate to themselves as well as to each other and the bacterial factor [33]. However, the bacterial factor does not deactivate to itself or to any of the complement-derived factors [23, 32, 33].

4. The deactivation is prevented by incubating the cells and the chemotactic factor with the phosphonate esters. The phosphonate inhibition profiles of the inhibition of deactivation are the same as the inhibition profiles of the inhibition of chemotaxis. This suggested that activation of the putative 'activatable esterase' is involved in deactivation as well as in chemotaxis [32].

5. Similarly, aromatic amino acid esters prevent deactivation and inhibit chemotaxis. This suggested the possibility that the 'activatable esterase' is the precursor of an esterase capable of hydrolyzing amino acid esters [35].

6. The rabbit polymorphonuclear leukocyte contains an enzyme, esterase 1, capable of hydrolyzing the aromatic amino acid ester, acetyl-DLl-phenylalanine-7-naphthyl ester [7].

7. Esterase 1 is inhibited by the p-nitrophenyl ethyl phosphonates in a manner characteristic of the inhibition of serine esterases [7].

8. The inhibition profiles obtained when the several series of p-nitrophenyl ethyl phosphonates inactivate esterase 1 are the same as those found for the 'activatable esterase' in both inhibition of chemotaxis and the prevention of the deactivation of chemotaxis [7]. This indicates that esterase 1 is the activated form of the 'activatable esterase'.

9. Esterase 1 is also found in a precursor form, proesterase 1. Proesterase 1 exists in or on the polymorphonuclear leukocyte in an enzymatically inert, phosphonate insusceptible form [34].

10. Proesterase 1 is transformed into esterase 1 capable of hydrolyzing acetyl-DL-phenylalanine-7-naphthyl esterase when the chemotactic factor interacts with the polymorphonuclear leukocyte [34].

11. Proesterase 1 is the 'activatable esterase' of chemotaxis.

Evidence from phosphonate inhibition profiles indicates that C3a, C5a, C$\overline{567}$ and the bacterial factor induce chemotaxis by activating proesterase 1 to esterase 1 [2]. However, only C3a, C5a, and C$\overline{567}$ activate the proesterase of rabbit peripheral blood neutrophils as measured by the increase of acetyl-DL-phenylalanine-ß-naphthyl esterase activity of the cell [2]. Attempts to detect, in a like manner, the activation of proesterase 1 by treating the same leukocytes with bacterial factor have consistently failed [2]. It is concluded that due to unknown reasons the bacterial factor is unable to activate proesterase 1 to the same extent as the complement-derived chemotactic factors.

The hypothesis that bacterial factor is less able to activate proesterase 1 than the complement-derived chemotactic factors can explain the observations alluded to previously, that the bacterial factor is not able to deactivate to itself or to the complement-derived factors, even though these latter factors can deactivate to themselves, to each other and to the bacterial factor. Since, by hypothesis, bacterial factor activates only a small proportion of the leukocyte proesterase 1, it always leaves sufficient proesterase 1 so that enough is available for a chemotactic response to either fresh bacterial factor, or to one or another of the complement-derived factors. Thus, it neither deactivates to itself nor to the other factors. If, on the other hand, sufficient proesterase 1 is used up through the interaction of the neutrophil with one or another of the complement-derived factors insufficient proesterase 1 will be left on the cell to interact with any of the chemotactic agents to give a chemotactic response. Thus, the cell deactivated to one of the complement-derived factors will also be deactivated to the others and to the bacterial factor.

The maximal chemotactic activity of cells stimulated by increasing concentrations of the bacterial factor is less than that of cells reacting to increasing concentrations of either C3a, C5a or C$\overline{567}$ [2]. This is also explicable on the hypothesis that the bacterial factor is able to activate less proesterase 1 than the complement-derived factors.

Direct biochemical demonstration of the activation of proesterase 1 by the complement-derived chemotactic factor is demonstrable only with peripheral blood neutrophils not with peritoneal polymorphonuclear leukocytes, indicating a lesser amount of proesterase 1 on the peritoneal cell compared to the blood leukocytes. The reason suggested for this difference is that the proesterase 1 of the blood leukocytes has been utilized in responding to the *in vivo* chemotactic stimulus which drew them into the peritoneal cavity [34]. The presence of the larger amount of proesterase 1 on the blood polymorphonuclear leukocyte than on the peritoneal leukocyte is associated with a demonstrably greater chemotactic responsiveness of the blood cell [34].

These results indicate that the activation of proesterase 1 is a general requirement for the chemotactic activity of rabbit polymorphonuclear leuko cytes with at least the bacterial and the complement-derived chemotactic factors and strongly suggest that under a number of circumstances the degree of such activation controls the level of chemotactic activity.

If this is so, and obviously more work directed to this point is required, some of the questions which arise are: How is proesterase 1 activated by the chemotactic factor? What is the nature of esterase 1, and its mode and place of action in the biochemical sequence(s) finally leading to the chemotactic response? Unfortunately, the answer to all of these questions at this time is that we not only do not know, but we have disappointingly little positive, firmly established evidence upon which to base any cogent speculations.

There is no information concerning the manner in which the several chemotactic factors activate proesterase 1. It is unknown whether this activation is a direct result of the interaction of the chemotactic factor with the cell or only occurs after a prior reaction or a sequence of reactions subsequent to this primary interaction. There is evidence from other systems where proesterase activation occurs, that it is apparently an early, if not the earliest, step in the given sequence [reviewed in ref. No. 5]. The only evidence that might be considered possibly bearing on the place which proesterase 1 occupies in the sequence leading to the chemotactic response is the heretofore unpublished findings shown in table I. The results demonstrate that the activation of proesterase 1 takes place just as well in the presence of EDTA as in its absence; that is, the activation of proesterase 1 does not require divalent cations. This suggests, although it does not prove, that the activation step occurs before the Ca^{2+} requiring step. In addition, there is even less direct evidence that proesterase 1 acts before the so-called 'activated esterase' [8]. (The latter is an enzyme different from esterase 1, so far identified only by its ability to be inhibited by phosphonate esters and DFP, and for which, as yet, no substrate has been found, see below.)

We also have no evidence in regard to the nature of the proesterase 1 and its site on the cell. One expects proesterase 1 to be on the cell membrane but this remains to be proved. The very low level of activity of proesterase 1 prevents approach to these problems using conventional techniques. Even in the blood neutrophil, 95% or more of the total esterase 1 activity is in the activated form, esterase 1. The data of table I showing the optical density

Table I. Activation of proesterase 1 in presence of EDTA

Experiment No.	Chelating agent	Treatment	Acetyl-DL-phenyl-alanine-β-naphthyl esterase activity, OD 485 mm	Number of replicate determinations	Δ OD attributable to proesterase 1 activation
42873	1 mM EDTA	control	0.101 ± 0.002	5	
		50 μg C5a	0.118 ± 0.002	2	0.017
		100 μg C5a	0.129 ± 0.010	2	0.024[1]
		control	0.095 ± 0.003	3	
		50 μg C5a	0.109 ± 0.007	2	0.017
		100 μg C5a	0.118 ± 0.003	2	0.0028861
5173	6.7 mM EDTA	control	0.130 ± 0.01	6	
		50 μg C5a	0.190 ± 0.015	2	0.060[1]
		control	0.148 ± 0.005	6	
		50 μg C5a	0.182 ± 0.029	2	0.034[1]
5572	5mM EDTA	control	0.360 ± 0.009	4	
		100 μg C5a	0.380 ± 0.004	2	0.020
		control	0.376 ± 0.004	4	
		100 μg C5a	0.416 ± 0.015	2	0.040[1]

1 Statistically significant difference from control, $p < 0.05$. Polymorphonuclear leukocytes were isolated from rabbit blood by the gelatin procedure [2]. The activation of proesterase 1 and the assay of acetyl-DL-phenylalanine-β-naphthyl esterase activity were performed as previously described on 6×10^6 neutrophils in 0.2 ml final volume OD = optical density.

differences attributable to the activation of proesterase 1 (last column, table I) are evidence for this low degree of activity of proesterase 1.

The activated form of proesterase 1, esterase 1, is found largely if not wholly in the microsomal fraction [13; BECKER and KEGELES, unpublished observations] and a part of the activity is associated with the membrane portion of this fraction [BECKER and KEGELES, unpublished observations]. On standing overnight in the cold in the presence of 1% Triton X-100, the molecular weight of esterase 1 drops to 40,000–80,000 [BECKER and KEGELES, unpublished observations] as ascertained by sucrose density gradient ultracentrifugation and Sephadex chromatography. Further purification has been hampered by the instability and tendency of the enzyme to aggregate.

Because esterase 1 is capable of hydrolyzing acetyl-DL-phenylalanine-ß-napthyl esterase [7], and its activity is inhibited by various aromatic amino acid derivatives, the present working hypothesis is that it is a protease. Even more indirect evidence for this latter hypothesis is the finding that known proteolytic enzymes such as kallikrein [22] and plasminogen activator are chemotactic for polymorphonuclear leukocytes. These substances are active through their enzymatic site, but whether they act in the chemotactic sequence as substitutes for proesterase 1, or to activate proesterase 1 or in some other manner, is unknown. Obviously, knowledge as to the nature of esterase 1 and its role in chemotaxis awaits further purification of the enzyme.

The fragmentary nature of our knowledge of the site and mode of activation of proesterase 1 is matched by the bits and pieces of which our knowledge consists in regard to the other events in chemotaxis as demonstrated in the brief review which follows.

As already mentioned, in addition to proesterase 1 and the activated form, esterase 1, there exists another serine esterase the so-called 'activated esterase of chemotaxis'. Unlike proesterase 1, this enzyme functions in chemotaxis in an already activated form. Unlike esterase 1, no substrate for it, either natural or synthetic, has been found, it being identified only by its ability to be inhibited by phosphonate esters and DFP [8]. The 'activated esterase' is inhibited by contact of the cell with phosphonate in the absence of the chemotactic factor, differing in this respect from proesterase 1. In addition, the profile of inhibition of the activated esterase is different from proesterase 1 [8]. Moreover, the inhibition of the activated esterase is prevented by neutral and acidic esters of acetic acid [8]. The only other information we have concerning the 'activated esterase' is essentially negative. It is not one of the recognized carboxylic acid esterases of the leukocyte; it apparently is not concerned with acid production by the cell [8], nor in the phosphonate inhibition of K^+ retention [3].

One can obtain inhibition of the activated esterase under conditions where there is no interference with the viability of the cell as measured by uptake of trypan blue [25, 36] or inhibition of phagocytosis. What the place is of the activated esterase in the sequence is not known but, as mentioned, there is evidence that the activated esterase acts after proesterase 1 [8].

Both Ca^{2+} and Mg^{2+} are required in the external medium for maximal chemotactic response of rabbit [6] and human neutrophils. In sufficient concentration, Mg^{2+} alone gives maximal spontaneous motility, although Ca^{2+} reduced the concentration of Mg^{2+} so required [6]. C5a enhances the influx of $^{45}Ca^{2+}$ into the cell and its efflux even more.

$\overline{C567}$ induces an increase in O_2 consumption of the polymorphonuclear leukocyte [WARD, information unpublished], as does C5a and the chemotactic enzyme, kallikrein [18]. In the latter instances, the increased O_2 consumption was shown to be due to an increase in the activity of the hexose monophosphate shunt. Oxidation by the shunt mechanism apparently does not provide the source of energy for chemotaxis [18]. Inhibitors of oxidative phosphorylation such as 2,4-dinitrophenol give only slight inhibition of chemotaxis at 1–10 mM [11, 31] and actually enhance the chemotactic response and cell motility at lower concentrations. These findings, together with the demonstration of the inhibitory activity of iodoacetate [10], suggest that anaerobic glycolysis rather than oxidation is the major if not sole source of energy involved in cell motility and chemotaxis of neutrophils.

As described in more detail below, I have previously postulated chemotaxis might involve the contractile machinery of the cell. The inhibition of chemotaxis by the microtubular disaggregating agents colchicine [3, 9] and vincristine and vinblastine [31] might be considered support for this suggestion. In addition the mold metabolite, cytochalasin B, which is reported to depress processes involving microfilaments [reviewed in ref. No. 38] reversibly inhibits chemotaxis and cell motility at concentrations of 0.5 μg/ml or higher [6]. Below this concentration, it actually enhances the chemotactic response [6]. Unfortunately, these agents apparently have more than one mode of action [5, 12, 14, 19, 30, 39] making the interpretation of their action in chemotaxis and cell motility difficult.

SENDA et al. have shown that an actomyosin-like protein can be isolated from horse neutrophils [28]. Szent-Györgyi, in 1949, demonstrated that muscle fibers extracted with glycerol contract on the addition of ATP and this finding formed the basis for subsequent work on the relation of actomyosin interaction to muscle contraction. ATP-induced contraction of a number of isolated glycerinated cells has been used as evidence for a contractile mechanism similar to that of muscle in the cells. Recently, we have shown that rabbit neutrophils after extraction with glycerol are capable of being contracted by ATP, and this microscopically observed contraction is associated with a decrease in cell volume [20]. The characteristics of the volume decrease are similar to those found in the ATP-induced contraction of glycerinated muscle fibers strengthening the suggestion that there exists in the neutrophil, as in other cells, a contractile system similar to actomyosin. Obviously, the relationship, if any, of this system to chemotaxis or other cell functions of the neutrophil has still to be determined.

Cyclic AMP has been reported to modify the chemotactic activity of

classical anaphylatoxin [38]. In addition, following the work of LEAHY et al.
[24] a number of investigators [1, p. 252, 17, 26, 27] have reported that
cyclic AMP is weakly chemotactic for neutrophils. Dibutyryl cyclic AMP
is also chemotactic but ATP, ADP and AMP are not [27]. PGE1 is also
slightly chemotactic under the same circumstances [21; RIVKIN and BECKER,
unpublished information]. Dibutyryl cyclic AMP, ATP and ADP, but not
AMP, inhibit chemotaxis when placed in the upper compartment of a
modified Boyden chamber; the inhibition by ATP appears to take place by
a different mechanism than that of the inhibition by ADP. Epinephrine and
isoproternol, agents which increase intracellular cyclic AMP (see below),
inhibit chemotaxis and spontaneous motility but norepinephrine does
not [26]. Theophylline also inhibits chemotaxis and exerts a synergistic effect
on the inhibition of chemotaxis by epinephrine. The β-adrenergic blocker,
propranolol, prevents the action of epinephrine [26]. However, the ability
of epinephrine and isoproterenol to inhibit the chemotactic response of
the neutrophils of different rabbits varies greatly, and there in no parallelism
between the ability of epinephrine to inhibit the chemotactic response of
neutrophils from different rabbits and the ability of this drug to raise the
cyclic AMP levels of the same cells [RIVKIN and BECKER, unpublished in-
formation]. Prostaglandin PGE, but not PGF_2a inhibit chemotaxis and
spontaneous motility [26]; the first substance increased intracellular cyclic
AMP levels but the last does not, or does so very much more weakly.
Cholera enterotoxin at concentrations of 50–1,250 ng/ml after a characteristic lag in time inhibits the chemotactic response up to a maximum of
about 60–70%; after the same time lag there is a rise in intracellular cyclic
AMP [RIVKIN and BECKER, unpublished results]. Thus, there is a general
qualitative parallelism between the rise in intracellular cyclic AMP and
inhibition of chemotaxis. However, the parallelism is not quantitatively
exact and inhibition of chemotaxis by any of the agents raising cyclic AMP
is not complete. These discrepancies suggest the presence of different intracellular compartments for the cyclic AMP with only the cyclic AMP in
one or some of these compartments being capable of affecting chemotaxis.
Moreover, there is a suggestion that only part of the mechanism of
chemotactic response may be affected by increases in intracellular cyclic
AMP. Preliminary evidence indicates that cells treated with widely varying
concentrations of bacterial chemotactic factor show no change in their
cyclic AMP levels. Confirmation of these very preliminary results would
suggest that changes in adenyl cyclase activity may modulate the chemotactic
response, or at least a major portion of it, but that adenyl cyclase is not

itself part of the direct biochemical sequence in the chemotaxis of rabbit neutrophil.

A hypothesis [3–6] which attempts to explain some of the above results is that the chemotactic factor(s) induces cell movement by stimulating the contractile machinery of the cell, presumably both the microtubules and microfilaments. A further concept is that the contractile apparatus is similar in nature and is under similar control to the actomyosin of muscle, particularly smooth muscle actomyosin. In line with this suggestion, activation of proesterase 1 might induce membrane changes leading to the influx of extracellular Ca^{2+} and a translocation of intracellular Ca^{2+} from a bound to an unbound site and its subsequent diffusion from the cell. The Ca^{2+}, in this view provides the excitation-response coupling.

Any view of the mechanism of the chemotactic response must take into account the recent finding that the bacterial and complement derived chemotactic factors induce secretion of enzymes from the lysosomal granules [29]. From present work it appears that the two responses share portions of the same steps in the biochemical sequences underlying the two activities but at least one critical step in each process might be different or have a different importance. It is possible that interaction of the cell with either the same receptors or with two different kinds of receptors is responsible for the triggering of the two functions. There is evidence that an 'activatable esterase' might also be involved in the lysosomal enzyme secretion induced by the chemotactic factors. Study of whether this putative activatable esterase is identical with proesterase 1 should help differentiate between these two possibilities. Another possibility, of course, is that the chemotactic factors do not bind to defined receptors, in the pharmacologic sense, on the cell membrane, but merely perturb the membrane without irreversibly binding to it. This perturbation, depending on attendant circumstances could then result in either chemotaxis or lysosomal enzyme secretion or both. The availability of purified chemotactic factor should help decide among the various alternatives.

Summary

The complement-derived chemotactic factors C567, C3a and C5a and a low molecular weight chemotactic factor obtained from *E. coli* culture filtrates induce a chemotactic response in rabbit polymorphonuclear leukocytes by directly or indirectly activating a cell-bound serine esterase, esterase 1. Proesterase 1 exists in or on the cell in an enzymatically inert form incapable of being inhibited by *p*-nitrophenyl ethyl phosphonates but, when activated to esterase 1, hydrolyzes acetyl-DL-phenylalanine-β-naphthyl ester and is inhibited by 10^{-8} M to 2.5×10^{-9} M phosphonate esters. The activation of proesterase 1 by the chemotactic stimulus does not require Ca^{2+}. The nature of esterase 1 and its mode of action in chemotaxis is presently unknown.

The optimal chemotactic response requires Ca^{2+} and Mg^{2+} in the external medium, there is a requirement for energy probably met by glycolysis. There is very indirect evidence that the cell movement involved in the chemotactic response involves the contractile mechanism of the cell.

The effect on chemotaxis of using agents capable of raising intracellular cyclic AMP and the ability of chemotactic factors to change intracellular cyclic AMP levels were tested. The results indicate that activation of adenyl cyclase is not in the main biochemical sequence responsible for chemotaxis. However, if adenyl cyclase is activated, the consequent increase of intracellular cyclic AMP decreases the chemotactic response probably through decreasing the cell motility.

References

1   AUSTEN, K. F. and BECKER, E. L.: Biochemistry of the acute allergic reactions. 2nd Int. Symp. (Blackwell, Oxford 1971).

2   BECKER, E. L.: The relationship of the chemotactic behavior of the complement derived factors C3a, C5a and C$\overline{567}$, and a bacterial chemotactic factor to their ability to activate the proesterase 1 of rabbit polymorphonuclear leukocytes. J. exp. Med. *135:* 376–387 (1972).

3   BECKER, E. L.: Phosphonate inhibition of the accumulation and retention of K^+ by rabbit neutrophils in relation to chemotaxis. J. Immunol. *106:* 689–697 (1971).

4   BECKER, E. L.; DAVIS, A.T.; ESTENSON, R. D., and QUIE, P. G.: Cytochalasin B. Inhibition and stimulation of chemotaxis of rabbit and human polymorphonuclear leukocytes. J. Immunol. *108:* 396–402 (1972).

5   BECKER, E. L. and HENSON, P. M.: *In vitro* studies of immunologically induced secretion of mediators from cells and related phenomena. Adv. Immunol. (in press).

6   BECKER, E. L. and SHOWELL, H. J.: Effect of Ca^{2+} and Mg^2_+ on the chemotactic responsiveness of rabbit polymorphonuclear leukocytes. Z. ImmunForsch. *143:* 466–476 (1972).

7   BECKER, E. L. and WARD, P. A.: Esterases of the polymorphonuclear leukocyte capable of hydrolyzing acetyl DL phenylalanine β naphthyl ester. Relationship to the activatable esterase of chemotaxis. J. exp. Med. *129:* 569–589 (1967).

8   BECKER, E. L. and WARD, P. A.: Partial biochemical characterization of the activated esterase required in complement dependent chemotaxis of rabbit polymorphonuclear leukocytes. J. exp. Med. *125:* 1021–1038 (1967).

9   CANER, J. E. Z.: Colchicine inhibition of chemotactic migration of human polymorphonuclear leukocytes. Arth. Rheumat. *7:* 297–302 (1964).

10 CARRUTHERS, B. M.: Leukocyte motility. I. Method of study, normal variation, effect of physical alterations in environment and effect of iodoacetate. Canad. J. Physiol. Pharmacol. *44:* 475–485 (1966).

11 CARRUTHERS, B. M.: Leukocyte motility. II. Effect of absence of glucose in medium.

Effect of presence of deoxyglucose, dinitrophenyl puromycin, actomycin D and trypsin on the response to chemotactic substance. Effect of segregation of cells from chemotactic substance. Canad. J. Physiol. Pharmacol. *45:* 269–280 (1967).

12 COHN, R. D.; BANERGEE, S. D.; SHELTON, E. H., and BERNFIELD, M. R.: Cytochalasin B. Lack of effect on mucopolysaccharide synthesis and selective alteration in precursor uptake. Proc. nat. Acad. Sci., Wash. *69:* 2865–2869 (1972).

13 DAVIES, P.; KRAKAUER, K., and WEISSMANN, G.: Subcellular distribution of neutral proteases and peptidases in rabbit polymorphonuclear leukocytes. Nature, Lond. *228:* 761–762 (1970).

14 ESTENSON, R. D. and PLAGEMANN, P. G. W.: Inhibition of glucose and of glucose amino transport. Proc. nat. Acad. Sci., Wash. *69:* 1430–1434 (1972).

15 GABBAY, K. N. and TZE, W. J.: Inhibition of glucose-induced release of insulin by aldolase reductase inhibitors. Proc. nat. Acad. Sci., Wash. *69:* 1435–1439 (1972).

16 GALLIN, J. I. and ROSENTHAL, A. S.: Divalent cation requirements and calcium fluxes during human granulocyte chemotaxis (abstract). Fed. Proc. *32:* 819 (1973).

17 GARROW, R. I.; BÖTTGER, B., and BARNES, F. S.: Analysis of chemotaxis in white blood cells. Biophys. J. *11:* 860–867 (1971).

18 GOETZEL, E. J. and AUSTEN, K. F.: Stimulation of the human neutrophil hexose monophosphate shunt (HMPS) by purified chemotactic factor (abstract). Fed. Proc. *32:* 973 (1973).

19 HASLAM, R. J.: Inhibition of blood platelet function by cytochalasin. Effects on thrombosthenin and on glucose metabolism. Biochem. J. *127:* 34 (1972).

20 HSU, L. S.; BECKER, E. L., and WEISEL, A.: Contraction and lysosomal enzyme release induced by ATP in rabbit polymorphonuclear leukocytes (PMNs) treated with glycerol (abstract). Fed. Proc. *32:* 820 (1973).

21 KALEY, G. and WEINER, R.: Effect of prostaglandin E_1 on leukocyte migration. Nature New Biol. *234:* 114–115 (1971).

22 KAPLAN, A. P.; KAY, A. B., and AUSTEN, K. F.: A prealbumin activator of prekallekrein. III. Appearance of chemotactic activity for human neutrophils by the conversion of human prekallikrein to kallekrein. J. exp. Med. *135:* 81–97 (1972).

23 KELLER, H. U. and SORKIN, E.: Chemotaxis of leukocytes. Experientia *24:* 641–672 (1968).

24 LEAHY, D. R.; MCLEAN, E. J., jr., and BONNER, J. T.: Evidence for cyclic 3′, 5′ adenosine monophosphate as chemotactic agent for polymorphonuclear leukocytes. Blood *36:* 52–54 (1970).

25 PEARLMAN, D. C.; WARD, P. A., and BECKER, E. L.: The requirement of serine esterase function in complement dependent erythrophagocytosis. J. exp. Med. *130:* 745–764 (1969).

26 RAMSEY, W. S.: Analysis of individual leucocyte behavior during chemotaxis. Exp. Cell Res. *70:* 129–139 (1972).

27 RIVKIN, I. and BECKER, E. L.: Possible implication of the adenyl cyclase system in the chemotaxis of rabbit peritoneal polymorphonuclear leukocytes (PMNs) (abstract). Fed. Proc. *31:* 657 (1972).

28 SENDA, N.; SHIBATA, N.; TATSUMI, N.; KONDO, K., and HAMADA, K.: A contractile protein from leukocytes. Its extraction and some of its properties. Biochim. biophys. Acta *181:* 191–195 (1969).

29 Showell, H.; Hsu, L. S., and Becker, E. L.: Chemotaxis, lysosomal enzyme release, and volume expansion of rabbit polymorphonuclear leukocytes (PMN) induced by chemotactic factors (abstract). Fed. Proc. 32: 973 (1973).

30 Trefaro, J. M.; Collier, B.; Lastowecka, A., and Stern, D.: Inhibition by colchicine and vinblastine of acetyl choline-induced catecholamine release from the adrenal gland. An anticholenergic action not an effect on microtubules. Molec. Pharmacol. 8: 264–267 (1972).

31 Ward, P. A.: The chemosuppression of chemotaxis. J. exp. Med. 124: 209–226 (1966).

32 Ward, P. A.: Leukotactic factors in health and disease. Amer. J. Path. 64: 521–530 (1971).

33 Ward, P. A.: Complement-derived chemotactic factors and their interaction with neutrophilic granulocytes; in Ingrahm Proceedings of the international symposium on the biological activities of complement, pp. 108–116 (Karger, Basel 1971).

34 Ward, P. A. and Becker, E. L.: Biochemical demonstration of the activatable esterase of the rabbit neutrophil involved in the chemotactic response. J. Immunol. 105: 1057–1067 (1970).

35 Ward, P. A. and Becker, E. L.: The deactivation of rabbit neutrophils by chemotactic factor and the nature of the activatable esterase. J. exp. Med. 127: 693–710 (1968).

36 Ward, P. A. and Becker, E. L.: Mechanisms of the inhibition of chemotaxis by phosphonate esters. J. exp. Med. 125: 1001–1020 (1967).

37 Wessels, N. K.; Spooner, J. F.; Ash, J. F.; Bradley, M. O.; Ludena, M. A.; Wrenn, J. T., and Yamada, K. M.: Microfilaments in cellular and developmental processes. Science 171: 135–143 (1971).

38 Wissler, J. F.; Stecher, V. J., and Sorkin, E.: Regulation of chemotaxis of leukocytes by the anaphylatoxin-related peptide system; in Peeters Proc. 20th Coll. Protides of the Biological Fluids, pp. 411–416 (Pergamon Press, Oxford 1973).

39 Zigmond, S. H. and Hirsch, J.: Effects of cytochalasin B on polymorphonuclear leukocyte locomotion, phagocytosis and glycolysis. Exp. Cell Res. 73: 383–393 (1972).

Author's address: Dr. Elmer L. Becker, Department of Pathology, University of Connecticut Health Center, *Farmington, CT 06032* (USA)

Antibiotics and Chemotherapy, vol. 19, pp. 421–441
(Karger, Basel 1974)

Recognition in Leucocyte Chemotaxis

Studies with Structurally Modified Proteins[1, 2]

P. C. WILKINSON and I. C. McKAY

Department of Bacteriology and Immunology, University of Glasgow, Glasgow

The phagocytic cell system of the animal kingdom is undoubtedly a more primitive immunity mechanism in the phylogenetic sense than the specific immune system, as was first pointed out by METCHNIKOFF [1892], yet there is little information at the molecular level on how phagocytic cells recognise and remove foreign material or how they distinguish this material from normal constituents of the body. Slightly damaged red cells [JANDL and TOMLINSON, 1958; JACOB and JANDL, 1962; CROME and MOLLISON, 1964] or native proteins, which have been linked to small molecules such as *p*-amino-benzoic acid [DIXON *et al.*, 1951] or fluorescein isothiocyanate [NAIRN, 1969] under mild conditions, are removed from the circulation of non-immune hosts more rapidly than analogous but unmodified cells or protein molecules injected into control animals. This type of experiment suggests that phago-cytic cells may be able to recognise small abnormalities in the structure of the macromolecules of the body, but does not offer any explanation for this recognition in structural terms.

We have recently become interested in the molecular characteristics which allow certain substances to excite a locomotor response in leucocytes under circumstances in which other related substances evoke no response. The number of chemotactic substances documented in the literature is now considerable and is growing. If any generalisation may now be made about them, it may be stated that almost all of the chemotactic agents for leucocytes so far described have been proteins or polypeptides. No polysaccharide, lipid or nucleic acid has been shown to attract leucocytes directly. Small

1 Departmental publication No. 7307.
2 Part of this work was supported by a grant from the Medical Research Council.

molecules such as amino acids, fatty acids or monosaccharides are also without effect. Those reports which have suggested that small molecules such as cyclic AMP were cytotaxins generally have not stood the test of time. On looking through a list of proteins with chemotactic activity, it becomes obvious that many of them are not in their native conformation but have become modified by environmental influences. For instance, chemotactic activity is frequently generated from serum proteins by the action of proteolytic enzymes. Recently described examples include the complement peptides C3a and C5a [BOKISCH et al., 1969; SNYDERMAN et al., 1969], the conformationally altered IgG subtypes produced by the action of a tissue sulfhydril-group (SH) dependent protease or of papain [YOSHINAGA et al., 1970, 1972], peptides derived from collagenase-treated collagen [CHANG and HOUCK, 1970; HOUCK and CHANG, 1971], or kallikrein generated by the action of Hageman factor fragments on prekallikrein [KAPLAN et al., 1972]. These observations suggest that proteins which have been changed structurally by the action of enzymes or other factors are recognized as 'abnormal' by leucocytes. The work described here explores the molecular basis for this recognition. From the outset of this work, we realised that there could be two possible approaches to such a problem. One would be to examine the molecular characteristics of known chemotactic factors for common features which would provide a basis for recognition. However, this approach was not likely to be fruitful since too little is known about the structure of these factors for useful conclusions to be drawn. The second approach would be to attempt to confer chemotactic activity on non-chemotactic, native proteins by various structural manipulations, and thus to try to find a pattern of structural changes which could be related to chemotactic activity. The studies described here are based largely on this second approach.

Effects of Chemotactic Factors on Directional and on Non-Directional Migration of Leucocytes

Migration of leucocytes in Boyden chambers has generally been regarded as a directional response to a concentration gradient, and there is evidence that, in some situations, this is indeed so [KELLER and SORKIN, 1968]. However, cells may also accumulate in a locus by responding merely to the local concentration of the test substance, the substance thus influencing the rate of random migration. Sufficient attention has not been paid to

Table I. The migration of human neutrophils in varying absolute concentrations and in varying gradients of a chemotactic factor (casein)

Concentration of casein above filter (mg/ml) in cell compartment	Migration of neutrophils (μm) in 65 minutes (mean for 3 filters). Concentration of casein below filter, mg/ml				
	0	0.2	1.2	2.2	3.2
0	23				
0.2		46	82	76	89
1.2		43	72	87	94
2.2		42	61	84	100
3.2		40	49	56	64

this possibility in the measurement of chemotaxis of mammalian phagocytes. Since BOYDEN [1962] introduced the micropore filter method, it has been possible to measure leucocyte migration fairly accurately *in vitro*, but it must be emphasised that the migration of cells into filters in response to a gradient of any substance is not *per se* evidence of directional migration or chemotaxis, and the same can be said about the accumulation of cells in a locus *in vivo*.

This problem has been explored using the type of experiment shown in table I and first used by ZIGMOND and HIRSCH [1973]. In this type of experiment, it is necessary to use the 'leading-front' method of measuring cell migration which was described by WEKSLER and HILL [1969] and ZIGMOND and HIRSCH [1972, 1973] and which we use routinely in all experiments. In the experiment shown in table I, cells have been exposed to a stimulus (casein) under conditions (1) of positive gradient, where the cells migrate from a low concentration of casein towards a high concentration; (2) of negative gradient, where the cells migrate from a high concentration of casein towards a low concentration, and (3) of no gradient, where the concentration of casein is uniform in each chamber, but the absolute concentration varies from chamber to chamber. These latter conditions are shown along the diagonal from upper left to lower right in table I.

From this type of experiment, the following conclusions can be drawn. Firstly, the migratory response of leucocytes, i.e. neutrophils in table I, but also macrophages in other experiments, is sensitive to differences in the absolute concentration of factors such as casein, under conditions in which there is no gradient. As concentration is increased, the migration of leucocytes is enhanced, but above an optimal concentration, it decreases again. Secondly, under negative gradient conditions, leucocytes are 'trapped' at the upper surface of the filter and migrate poorly (figures below the diagonal in table I). Thirdly, under conditions of positive gradient, leucocyte migration is enhanced. This is best seen in table I by looking along the diagonals from lower left to upper right, along each of which the absolute concentration is uniform, but along which the gradient changes from negative at lower left to positive at upper right. Therefore, leucocytes respond, not only (1) directionally in a concentration gradient of the chemotactic factor, but also (2) non-directionally to absolute concentration.

The important conclusion from this type of experiment is that chemotaxis of leucocytes is not a simple directional response, as has often been assumed, but a mixed response in which the chemotactic factor simultaneously enhances non-directional locomotion, independent of the presence of a concentration gradient, and directional migration if there is a positive gradient. These two components of the response must both be taken into account in understanding how chemotactic factors are recognised by cells. We have studied several types of chemotactic factor using the system shown in table I, viz. casein, endotoxin-activated serum, bacterial chemotactic factors and denatured proteins, and the above conclusions can be shown to apply generally to the responses of leucocytes to proteins. These conclusions have been further strengthened by studying the final distribution of the cells in the filter, as suggested by ZIGMOND and HIRSCH [1973]. In the presence of a small or zero concentration gradient the cells attain a normal distribution, i.e. they appear to obey the equation for linear diffusion, but in the presence of a positive gradient the cell-distribution is distorted, indicating a directional component in the migration.

It may, indeed, be an advantage to an organism if its phagocytes respond both to the absolute concentration and to the concentration gradient of a cytotaxin, since the chemotactic or directional stimulus may act over a very limited range: at a distance from the source of cytotaxin the concentration gradient will be very weak and will be disturbed by the continual movement of cells and tissues. The cytotaxin, however, by stimulating the random migration of phagocytes, will still be able to increase the rate at which phagocytes arrive at the inflammatory locus. Thereafter the chemotactic or directional component of the response will be important, particularly since, once the cells arrive at the source of the cytotaxin, they will be trapped by the gradient and unable to move out again.

Sources of Leucocytes Used in Studies of Recognition

In the studies to be described, the human peripheral blood neutrophil leucocyte was the cell used in the great majority of experiments. Blood neutrophils behaved with great reproducibility in these experiments. We have also used guinea-pig oil- or glycogen-induced peritoneal neutrophils. The response of these cells is weaker than that of the blood neutrophil. Human peripheral blood monocytes migrated consistently towards the structurally altered proteins described below. Oil-induced guinea-pig macrophages also responded to them, but very unpredictably. The macrophages of certain guinea-pigs showed vigorous and clear-cut responses, whereas those of other animals treated in the same way failed to respond at all. We do not know the reason for this variability of response of induced exudate cells, but it may be related to the chemotactic 'deactivation' described by WARD and BECKER [1970] in similar cells in rabbits. Guinea-pig peritoneal eosinophils were found not to respond to the proteins discussed in this study (experiments of Mr. T. D. SCOTT).

Protein Conformation and Leucocyte Recognition

Three proteins have been studied in detail for the effect of conformational alterations on their chemotactic activity. Two of these, serum albumin and haemoglobin, are ubiquitous constituents of normal tissues and are not chemotactic in their native states. The third, casein, is the major milk protein which, in physiological salt solutions, is strongly chemotactic and which also has unusual structural features, discussed later, which may be relevant to its chemotactic activity.

Studies of Serum Albumin

In the initial studies, human serum albumin (HSA) was used as a model protein, since this protein is available commercially as a highly purified, lyophilised preparation without evidence of denaturation (Behringwerke, Marburg, Germany). Serum albumin consists of a single polypeptide chain, therefore denaturing agents cause alteration in three-dimensional structure of the monomeric units and secondary polymerisation, but not dissociation into subunits. A disadvantage of serum albumin is that its detailed structure is not well understood. Sequence studies and details of secondary and tertiary structure are not yet available.

Our first approach was to study the effects of denaturing agents on the chemotactic activity of HSA for human neutrophils. These studies have been reported fully [WILKINSON and MCKAY, 1971] and only a summary is given here. The use of strong denaturing agents such as guanidine HCl gave little useful information since, on the return of HSA treated with such agents to physiological conditions, the protein aggregated strongly and was no longer diffusible. More information was obtained from the use of milder procedures such as alkali-denaturation (pH 12), acid denaturation (pH 2), reduction-alkylation or mild heat-denaturation. These procedures varied in effectiveness in the order, alkali-denaturation>reduction-alkylation>acid-denaturation>heat-denaturation, as judged by the chemotactic activity of the denatured HSA. The change in chemotactic activity was reproducible but the products were never such potent cytotaxins, as for instance, endo-toxin-treated serum, or casein.

It was necessary to establish whether the essential requirement for recognition in chemotaxis was a change in the three-dimensional structure of the HSA monomer, or whether the secondary changes of polymerisation were also required. This was investigated by studies of polymerised and monomeric fractions of HSA separated from one another by permeation chromatography on Sephadex G-100 or G-200 (Pharmacia, Uppsala, Sweden). These studies showed that the monomer fractions had equal or greater chemotactic activity than the polymer fractions. Polymerisation was, therefore, not essential for recognition since neutrophils were capable of recognising changes in the structure of the HSA monomer.

For chemotactic studies of denatured proteins, the denaturing agent must be removed and the protein returned to physiological conditions. The unpredictable degree of refolding which occurs during this procedure makes it difficult to produce protein samples in which a reproducible alteration of three-dimensional structure has been achieved. We therefore estimated the degree of denaturation achieved in any given sample by various measure-ments of the physical state of the protein after return to physiological conditions. The following properties were studied: (1) viscosity; (2) surface activity, and (3) the difference spectrum of the native and denatured protein in the ultra-violet (UV) region between 280 and 290 nm. The correlation between viscosity, surface activity and chemotactic activity of samples of denatured, monomeric HSA was reported by WILKINSON and MCKAY [1971] and it was shown in that study that these properties increased in parallel. Table II shows the relationship of chemotactic activity of various samples of HSA to their difference spectrum at 287 nm. Again the relationship between the two functions was quite good.

The parallel between increase in viscosity and increase in chemotactic

Table II. Chemotactic activity of different samples of denatured human serum albumin (HSA) and the change in their absorbance at 287 nm as measured by difference spectroscopy

Preparation at 2 mg/ml and pH 7.2 in Gey's solution	Chemotactic migration of neutrophils (μm in 75 min)	Difference in absorbance between sample HSA and native HSA at 287 nm
HSA (native)	34	nil
Acid-denatured HSA (sample I)	44	nil
Acid-denatured HSA (sample II)	48	nil
Urea-denatured HSA (6 M urea)	52	0.010
Reduced-alkylated and acid-denatured HSA	68	0.008
Alkali-denatured HSA (sample I)	70	0.026
Reduced-alkylated and alkali-denatured HSA	76	0.042
Alkali-denatured HSA (sample II)	79	0.040

activity of denatured HSA suggested that chemotactic activity might be associated with unfolding of the protein and with a change in its shape from a globular to a more linear form. The parallel between increase in surface activity and increase in chemotactic activity suggested that the nature of the change which occurred during unfolding and which leucocytes detected might be the exposure of non-polar sites normally concealed in the interior of the molecule. Such exposure would cause the protein to show increased surface activity.

To explore these possibilities, we turned from the use of denaturing agents to the use of small molecules which could be linked under mild conditions to a protein molecule. By using a variety of such synthetic side-groups with selected properties, e.g. polar or non-polar character, negative or positive charge, etc., it was possible to explore in more detail the molecular basis of chemotactic recognition. These studies have been reported in full by WILKINSON and McKAY [1972]. Table III shows the chemotactic activity of some representative HSA conjugates from those reported in that paper.

Table III. Chemotactic activity for human neutrophils of some HSA preparations conjugated to different side-groups

Reagent conjugated to HSA	Conditions of reaction			Moles reagent per mole protein in reaction mixture	Chemotactic ratio[1]
	pH	temperature °C	time		
Butyric anhydride	9.0	20	30 min	111	2.81
Acetic anhydride	9.0	20	30 min	342	2.60
Propionic anhydride	9.0	20	30 min	134	2.49
Succinic anhydride	9.0	20	30 min	175	2.08
Tosyl chloride	9.0	20	30 min	92	2.29
Propylene oxide	5.3	20	3 days	175	1.31
Propylene oxide	7.2	20	3 days	175	1.70
Propylene oxide	9.0	20	3 days	175	2.21
Picryl chloride	9.0	20	30 min	56	1.66
Formaldehyde	7.2	20	3 days	257	1.00
Iodoacetamide	7.2	20	3 days	90	1.44
Iodine + KI	7.2	20	20 min	55	1.29
O-methyl isourea (guanidylation)	10.0	0–4	3 days	very high	1.18
Aniline[2]	9.0	0	60 min	30	2.10
Sulphanilamide[2]	9.0	0	60 min	2	1.95
Sulphanilic acid[2]	9.0	0	60 min	2	1.07

1 Chemotactic ratio = migration (μm) towards test protein/migration (μm) towards unconjugated HSA.
2 Conditions of reaction after diazotisation of the reagent.
3 A fuller version of this table is to be found in WILKINSON and McKAY [1972].

The most active side-groups were those which added non-polar groups to the protein, such as butyryl, propionyl or acetyl groups. The chemotactic activity of butyryl HSA is shown in figure 1. Aromatic groups of non-polar character, e.g. tosyl or benzylazo groups were also active, but aromatic rings bearing polar substituents, as in the picryl group, were less active. The addition of charged groups to proteins made little difference to their chemotactic activity. Guanidylation of HSA did not increase its chemotactic activity. Reagents such as sulphanilic acid and succinic anhydride, which introduce negatively charged groups, gave conjugates with rather less activity than their uncharged analogues prepared by treating HSA with

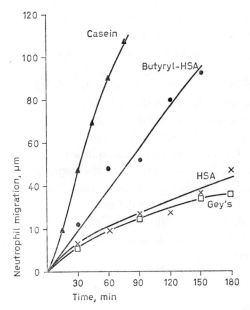

Fig. 1. Migration of human blood neutrophils towards casein (▲———▲), butyryl HSA (•———•), unconjugated HSA (x———x) and Gey's solution (☐———☐) after different times of incubation [from WILKINSON and McKAY, 1972, by courtesy of Verlag Chemie].

sulphanilamide and butyric or propionic anhydride. Neither poly-L-lysine nor poly-L-glutamic acid was chemotactic.

The degree of conjugation of many small molecules to proteins can be measured spectrophotometrically, e.g. azoconjugates in the UV and visible regions, or conjugates bearing alkyl groups in the infra-red region. It was thus possible to relate the degree of conjugation achieved to the acquisition of chemotactic activity and to show that these two parameters increased in parallel [WILKINSON and McKAY, 1972].

The Nature of the Leucocyte Receptor for Chemotaxis

Chemotactically active HSA conjugates were used to investigate the nature of the receptor for chemotactic factors on the neutrophil. If the combination of cytotaxin and cell receptor were stereospecific, in the same sense that the reaction of antigen with antibody is stereospecific, then it would be expected that the chemotactic activity of HSA conjugated with a small organic molecule would be easily inhibited by the presence of the same organic molecule in free solution, competing for binding sites, but not by slightly different molecules, since these could not compete for a truly stereo-specific receptor. This was investigated using tosyl-arginine methyl ester in free solution to study its influence on the chemotactic activity of tosyl HSA. Both protein and ester carry the tosyl group linked in the same way to a basic amino acid. In fact tosyl arginine methyl ester at a concentration between 10^{-4} and 10^{-3} M had no effect on the chemotactic activity of tosyl HSA at a concentration of about 10^{-5} M. From this, and experiments with other compounds, we concluded that there was not a stereospecific chemo-taxis receptor on the neutrophil and that it was more likely that the bond between cell and protein was of a more general type independent of the shape of the reactants, and that the strength of bonding was derived from reactions at multiple sites between the protein and the cell, rather than from a high affinity reaction at a single site.

Studies of Haemoglobin

Haemoglobin has certain advantages over serum albumin as a model protein for the study of the relationship between structural change and chemotactic recognition. Firstly, its structure is understood in detail. Secondly, it is a coloured protein with a complex spectrum in the visible region as well as in the UV. Many small changes in conformation can be monitored precisely by changes in the spectrum. Finally, haemoglobin, like serum albumin, is a protein present in large amounts in the body, and there-fore its study should be directly relevant to the activities of phagocytic cells in the body.

Certain physico-chemical properties of haemoglobin are relevant to the studies described below of its chemotactic activity. Haemoglobin undergoes reversible allosteric transitions on the binding of ligands such as oxygen or carbon monoxide. These transitions involve changes in the quaternary structure such that the 4 chains of the protein molecule realign themselves in relation to each other. This realignment involves a rotation of the chains through about 7–10° [MUIRHEAD et al., 1967]. There are also changes in tertiary structure on ligand binding, but these are very minor compared with those which follow exposure to a denaturing agent. The allosteric changes in haemoglobin are discussed in extenso by ANTONINI and BRUNORI [1971].

In addition, haemoglobin undergoes a major conformational change when the haem is liberated from globin. This may be achieved by the method of ROSSI-FANELLI *et al.* [1958] using acid acetone at –20 °C or with acidified ethyl methyl ketone as described by TEALE [1959]. After removal of haem, the tetrameric protein dissociates into αβ-chain dimers. The change in conformation on formation of globin (apohaemoglobin) from haemoglobin has been studied by measurements of circular dichroism [JAVAHERIAN and BEYCHOK, 1968]. The helical content of native ferrihaemoglobin decreases from 65±3% to 52±3% on removal of haem. It has been calculated that globin has 15–20 fewer residues in segments of right-handed α-helix than native ferrihaemoglobin [JAVAHERIAN and BEYCHOK, 1968]. Haem is embedded in a cleft, the 'hydrophobic pocket', in the haemoglobin molecule and is held in position by hydrophobic bonds between non-polar amino acids in the protein molecule and vinyl and methyl side-chains on the haem. Therefore, once haem is removed, these amino acids become freed and a major realignment of the globin molecule ensues. Haemoglobin has a high content of amino acids with non-polar side chains. Valine, alanine, leucine, phenylalanine, proline and tryptophan form 47% of the total amino acid content of human haemoglobin A [BRAUNITZER *et al.*, 1964]. It is unlikely that, following the loss of the prosthetic group, such a large number of non-polar side-chains can be accommodated in the interior of the molecule, so that the content of exposed non-polar groups must be higher in globin than in native haemoglobin. Globin is less soluble than haemoglobin in aqueous solutions, tends to aggregate at room temperature and is much more easily denatured or digested by proteolytic enzymes [ANTONINI and BRUNORI, 1971]. The structural changes in globin are fully reversible on readdition of haem or haemin, when the original conformation of haemoglobin is fully recovered. We have taken advantage of this reversible unfolding of globin in the studies of its chemotactic properties discussed below [WILKINSON, 1973].

Chemotactic studies of haemoglobin were carried out using human haemoglobin A prepared by the method of ROSSI-FANELLI *et al.* [1961]. The results are summarised in table IV. There was no difference in chemotactic activity between reduced haemoglobin (Fe^{II}), oxyhaemoglobin (Fe^{II}), carbon monoxide haemoglobin (Fe^{II}) and ferrihaemoglobin (Fe^{III}). None of these preparations had significant chemotactic activity. Therefore, the conformational changes associated with ligand binding are not sufficient or not of the right type to be recognised by leucocytes. On the other hand, the major conformational change which accompanies the transition, haemoglobin → globin, is accompanied by the acquisition of strong chemotactic

Table IV. Chemotactic migration of human blood leucocytes towards various preparations of human haemoglobin and globin (apohaemoglobin)

Sample at 1–2 mg/ml	Neutrophils[1], migration (μm) in 75 min in 3-μm filters	Monocytes[1], migration (μm) in 130 min in 8-μm filters
Oxyhaemoglobin	31	NT
CO haemoglobin	38	NT
Reduced haemoglobin ($Na_2S_2O_4$)	28	NT
Ferrihaemoglobin	37	33
Butyryl-ferrihaemoglobin	70	58
Globin (prepared by acid-acetone method)	93	70
Trypsin-treated globin (above preparation incubated overnight with trypsin)	44	NT
Controls		
Gey's solution	22	18
Casein, 5 mg/ml	105	110

1 Separated from dextran-sedimented plasma by centrifugation over Ficoll-Triosil.
NT = not tested.

activity. The chemotactic activity of globin can be abolished by renaturation. Haemin (B. D. H., Poole, England) prepared by the method of JAVAHERIAN and BEYCHOK [1968] was added in varying amounts to a solution of globin. The re-formation of haemoglobin from globin was monitored by measurement of the recovery of absorption intensity at 406 nm. Table V shows that, as globin reassociates with haemin to form haemoglobin, the chemotactic activity disappears. These experiments showed, therefore, that unfolding of the globin chain caused the protein to become chemotactic and that refolding of the chain was associated with loss of chemotactic activity. Note that, although each globin dimer has two binding sites for haemin, the molecule reassumes its native conformation upon binding of *one* haemin per dimer [JAVAHERIAN and BEYCHOK, 1968]. Likewise, chemotactic activity is abolished when one haemin molecule is bound per globin dimer (table V).

One reason for turning from HSA to haemoglobin in these studies was that none of the HSA preparations was more than moderately chemotactic.

Table V. Chemotactic activity of globin. The effect of adding haemin

Preparation (protein at 1.5 mg/ml)	Moles added haemin bound per mole globin dimer[1]	Neutrophil migration (micrometers in 75 min)
Globin	–	73
Globin + haemin	0.15	64
Globin + haemin	0.27	65
Globin + haemin	0.73	48
Globin + haemin	1.0	28
Ferrihaemoglobin	–	36

1 Calculated from the recovery of absorbance at 406 nm on addition of varying doses of haemin to a standard preparation of globin.

On the other hand, globin preparations were often strongly chemotactic. If non-polar residues are important for chemotactic recognition, this result would be expected, since haemoglobin contains 47% of such residues whereas serum albumin contains only 34% (including those listed above and isoleucine), therefore more such residues can be exposed on denaturation of haemoglobin than of serum albumin.

Studies of Casein

Casein is strongly chemotactic for neutrophils and macrophages [KELLER and SORKIN, 1967] but does not attract eosinophils. It is also frequently used *in vivo* as a stimulus to elicit the migration of neutrophils (during the first 24 h after injection) or of macrophages (3–4 days after injection) into the peritoneal cavity in various species of animal.

Casein, the acid-precipitable fraction of milk, is not a simple substance since it contains a number of structurally different proteins of which α_s-, β- and $\varkappa$-caseins are the major ones. In the presence of high concentrations of calcium ions, these casein proteins form micelles of high molecular weight [NOBLE and WAUGH, 1965]. Casein therefore exists in high molecular weight micellar form in milk where the Ca^{++} concentration is much higher than in other biological fluids.

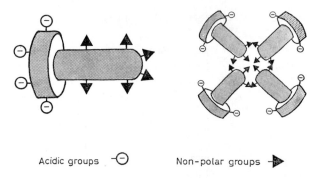

Acidic groups ⊖   Non-polar groups ➤

Fig. 2. Schematic model of the molecule of α_s- or β-casein showing the monomer on the left and its association into micelles, limited by the presence of the mutually repulsive acidic peptides, on the right [after the model of WAUGH *et al.*, 1970, and reproduced from 'Chemotaxis and inflammation', WILKINSON, 1974, by courtesy of Churchill-Livingstone].

Investigations in this laboratory [WILKINSON, 1972] were directed to the characterisation of the chemotactic activity of casein. From fractionation studies by calcium chloride precipitation and from Sephadex G-100 chromatography, it appeared that the major chemotactic acitivity was in the α_s- and β-caseins and that $\varkappa$-casein had little activity. This was of interest since α_s- and β-caseins resemble one another structurally but are quite different from $\varkappa$-casein. The chemotactic activity of casein was Ca^{++} ion sensitive. By Sephadex G-100 chromatography of whole casein in distilled water, it was shown that the most active fraction was a monomer fraction of about 20,000 molecular weight [WILKINSON, 1972]. As Ca^{++} ion concentration was increased, the chemotactic activity decreased. In milk, casein is not chemotactic. The decrease in chemotactic activity in high Ca^{++} concentrations, and its loss in whole milk, is presumably due to micelle formation, loss of diffusibility and perhaps masking of active sites.

WAUGH *et al.* [1970] have described unusual structural features of α_s- and β-casein which may give clues to their powerful chemotactic activity. However, it must be remembered that Hammarsten acid-precipitated casein must be dissolved at pH 12 in order to obtain a preparation which is soluble under physiological conditions, and that this step may affect the conformation of the protein. Although the intrinsic viscosities of α_s- and β-casein are both high, they are both globular proteins. In the model (fig. 2) proposed by WAUGH *et al.* [1970], both are formed from a major peptide with a high content of non-polar residues (40–50%) covalently linked to a smaller hydrophilic peptide which is situated at one end of the molecule and which is highly negatively charged since it contains a high proportion of serine phosphate monoesters and of free carboxyl groups. WAUGH *et al.* [1970] suggest that the non-polar side-chain content of the major peptide is too high for all these side-chains to be packed internally. Some are present at the surface of the molecule and can interact with similar groups on other casein molecules to form polymers. On the other hand, the acidic peptides on different molecules limit the degree of association by coulombic repulsion (fig. 2). Hence, binding of positive ions to these peptides reduces this repulsion and encourages intermolecular association.

The model of WAUGH *et al.* [1970] might explain some attributes of casein as a chemotactic protein. Thus, in low molecular weight form, the exposed non-polar groups are recognised by leucocytes. On polymer formation, these groups become buried, hence the protein loses chemotactic activity. Dephosphorylation of casein with acid or alkaline phosphatases also causes a reduction in its chemotactic activity, presumably due to polymerisation following the loss of negatively charged phosphate groups from the terminal peptide. Ca^{++} may also reduce the chemotactic activity of casein by its association with this acidic peptide.

The casein model indicates that a high content of exposed non-polar side-chains is probably insufficient on its own to make a protein chemotactic. Such a protein would polymerise rapidly, concealing the non-polar groups, unless some other groups were present, such as mutually repulsive acidic or basic groups, which would limit the degree of intermolecular association.

Chemotactic Factors as Activators of Lysosomal Hydrolases in Leucocytes

Is recognition in leucocytes simply a mechanism for activating directional locomotion, or do chemotactic factors have a more general action on leucocyte function? The fact that chemotactic factors simultaneously enhance non-directional as well as directional migration (table I) suggests that they may have an enhancing action on cell metabolism. Casein has, in fact, been shown to stimulate glycolysis, respiration and lysosomal enzyme activity in macrophages after injection *in vivo* [KARNOVSKY *et al.*, 1970].

We have made preliminary studies of the capacity of preparations of the structurally altered proteins described above to enhance lysosomal enzyme activity in leucocytes *in vitro* in comparison with native protein controls.

These experiments were carried out in collaboration with Mr. GEOFFREY O'NEILL and Mr. JEROME CATER. The enzymes measured were acid phosphatase, β-glucuronidase and β-galactosidase. They were measured in human peripheral blood granulocytes separated by centrifugation in 'Ficoll-Triosil', and in guinea-pig peritoneal macrophages. The results showed that denaturation or linkage of synthetic side-groups to proteins enhances their capacity to activate lysosomal enzymes in phagocytes, and therefore suggested that the same molecular recognition system described for chemotaxis may also serve in lysosomal activation. A further point of interest is that cell-specific cytotaxins only activate lysosomal enzyme activity in the cell-type for which they are specific in chemotaxis. This was studied using various strains of anaerobic coryneform bacteria, including *Corynebacterium parvum*, which produce a chemotactic factor specific for mononuclear phagocytes. These bacteria increased the lysosomal enzyme levels in macrophages only and not in neutrophils [WILKINSON *et al.*, 1973a]. Further evidence that the anaerobic coryneform bacteria are general stimulants of mononuclear phagocyte activity is provided by the close parallel between the ability of individual strains from this group to produce a macrophage chemotactic factor and their capacity to enhance carbon clearance in mice [O'NEILL *et al.*, 1973; WILKINSON *et al.*, 1973b], a function largely dependent on the activation of hepatic Kupffer cells. It is possible, therefore, that the recognition system described here may also serve to activate phagocytic cell functions other than chemotaxis.

General Observations and Conclusions

The experiments described above show clearly that conformational alterations in proteins are recognised by leucocytes and that this recognition activates a chemotactic response and possibly other cellular activities important for defence. We have suggested that non-polar groups play an important part in this recognition. These groups have a major role in determining the tertiary structure of globular proteins in aqueous solution by their mutual hydrophobic bonding in the interior of the protein, isolated from the solvent. On denaturation, these hydrophobic bonds are broken and the non-polar groups become exposed to solvent. The presence of non-polar groups in exposed sites on proteins may, therefore, provide a general marker for recognition by leucocytes by means of which the cell can monitor its environment for evidence of tissue damage without requiring receptors for stereospecific recognition. To draw an analogy, protein molecules might be likened to the coats worn by passers-by in a street. Each coat has an inner lining but this is not normally visible in a new and well-made coat. To recognise old or worn coats, it is not necessary to notice the pattern on the coat of every passer-by. All that is required is to pick out those coats on which the lining is showing through or hanging loose.

While the hypothesis presented here suggests that chemotactic recognition ensues from hydrophobic bonding between protein and cell, other features of protein structure which might be recognised by migrating leucocytes should also be considered. There is no evidence that this recognition is stereospecific. So many chemotactic factors have been described that it is highly unlikely that leucocytes could carry specific recognition sites for them all. Other more general alterations in proteins should therefore be considered, such as alterations in charge. Since all the charged groups are superficially placed in proteins such as haemoglobin, myoglobin or trypsin, denaturation cannot expose new charged groups as a signal for chemotactic recognition. Nor is net charge likely to be important, since the chemotactic factors already described fall into a wide range of net charge, and there are also many acidic, basic and uncharged proteins which are not chemotactic. Could leucocytes recognise random coil conformation in proteins? Can they recognise changes in protein shape? We were attracted by this idea initially, but have discarded it. Random coil proteins such as reduced-alkylated ribonuclease or polyglutamic acid or polylysine (at neutral pH) were found not to be chemotactic. A different possibility has been suggested by Wissler and Sorkin [1973], namely that leucocytes recognise the energy changes which occur in proteins on the transition from one conformation to another, transformations such as might occur in a protein in equilibrium between two or more conformational states. At present there is only circumstantial evidence for this interesting suggestion.

One important aspect of chemotactic recognition which the work discussed here does not touch upon is the cell-specificity of certain chemotactic factors. Wissler et al. [1972] have recently presented evidence for control of cell-specificity by multifactorial systems. In their experiments, two peptide factors were shown to act synergistically so that, depending on the absolute concentration and molar ratios of the two factors, either neutrophils or eosinophils were attracted. This work may provide an explanation for cell specificity which is compatible with recognition based on conformational changes in proteins or peptides.

The experimental pathologist, concerned with the response of tissues to injury, might validly ask what relevance the studies reported here have to the response of leucocytes in inflammation *in vivo*. The answer to this question is that the proteins used in these studies have been used as models to examine a recognition system which may be of general relevance to tissue injury. Denaturation of serum albumin or haemoglobin may or may not itself play a role in acute inflammation. It is certainly important for the clearance of these proteins from the tissues by the mononuclear phagocyte system. However, the point of interest here is not the activity of these proteins *in vivo* but their use as models of chemotactic factors. Many of the chemotactic factors which have been reported in inflammatory lesions are structurally altered forms of precursor proteins which are chemotactically inert in their native form but which acquire activity after enzymatic splitting or conformational changes. The present work offers a hypothesis on some of the physicochemical properties necessary for chemotactic activity of a polypeptide. It should be possible to test the hypothesis critically when the known chemotactic factors are more completely characterised so that it is possible to correlate their structure with their function.

Summary

The molecular basis for recognition of chemotactic factors by migrating leucocytes (neutrophils or monocytes) has been explored. Using model proteins such as serum albumin or haemoglobin, which were exposed to denaturing agents or conjugated to synthetic side groups, the following conclusions could be drawn. These proteins in their native state do not attract leucocytes. However, on denaturation, they may acquire chemotactic activity. Recognition of denatured proteins by leucocytes follows the change in the three-dimensional structure of the protein. Secondary changes such as polymerisation are not essential. Studies of proteins with conjugated side-groups indicated that the addition or exposure of non-polar groups was probably important in recognition. Added non-polar groups made the protein more chemotactic than added polar groups.

Since the chemotactic activity of conjugated proteins is not specifically inhibited by the presence of the conjugated substance in free solution, it is unlikely that recognition of such proteins depends on a close stereospecific bonding between a cell surface receptor and the side-group on the protein.

Substantial evidence that recognition in leucocyte chemotaxis follows unfolding of a protein chain is provided by studies of haemoglobin. When haemoglobin is converted to globin it unfolds and also becomes chemotactic. When haemin is added to globin and haemoglobin is re-formed, the chemotactic activity disappears.

The locomotor response of leucocytes to these substances is complex. The cells respond both by enhanced random movement to the absolute concentration of a chemotactic factor such as casein, and also by directional migration in a concentration gradient.

References

ANTONINI, E. and BRUNORI, M.: Hemoglobin and myoglobin in their reactions with ligands (North-Holland, Amsterdam 1971).

BOKISCH, V. A.; MÜLLER-EBERHARD, H. J., and COCHRANE, C. G.: Isolation of a fragment (C3a) of the third component of human complement containing anaphylatoxin and chemotactic activity and description of an anaphylatoxin inactivator of human serum. J. exp. Med. *129:* 1109–1130 (1969).

BOYDEN, S. V.: The chemotactic effect of mixtures of antibody and antigen on polymorphonuclear leukocytes. J. exp. Med. *115:* 453–466 (1962).

BRAUNITZER, G.; HILSE, K.; RUDLOFF, V., and HILSCHMANN, N.: The hemoglobins. Adv. Protein Chem. *19:* 1–71 (1964).

CHANG, C. and HOUCK, J. C.: Demonstration of the chemotactic properties of collagen. Proc. Soc. exp. Biol. Med. *134:* 22–26 (1970).

CROME, P. and MOLLISON, P. L.: Splenic destruction of Rh-sensitized and of heated red cells. Brit. J. Haemat. *10:* 137–153 (1964).

DIXON, F. J.; BUKANTZ, S. C., and DAMMIN, G. J.: The effect of sensitization and X-radiation on the metabolism of I[131] labeled proteins. Science *113:* 274–276 (1951).

HOUCK, J. and CHANG, C.: The chemotactic properties of the products of collagenolysis. Proc. Soc. exp. Biol. Med. *138:* 69–75 (1971).

JACOB, H. S. and JANDL, J. H.: Effects of sulfhydryl inhibition on red blood cells. II. Studies *in vivo*. J. clin. Invest. *41:* 1514–1523 (1962).

JANDL, J. H. and TOMLINSON, A. S.: The destruction of red cells by antibodies in man. II. Pyrogenic, leukocytic and dermal responses to immune hemolysis. J. clin. Invest. *37:* 1202–1228 (1958).

JAVAHERIAN, K. and BEYCHOK, S.: Subunit interactions in the conformational change of horse apohemoglobin on binding of hemin. J. molec. Biol. *37:* 1–11 (1968).

KAPLAN, A. P.; KAY, A. B., and AUSTEN, K. F.: A prealbumin activator of prekallikrein. III. Appearance of chemotactic activity for human neutrophils by the conversion of prekallikrein to kallikrein. J. exp. Med. *135:* 81–97 (1972).

KARNOVSKY, M. L.; SIMMONS, S.; GLASS, E. A.; SHAFER, A. W., and D'ARCY HART, P.: Metabolism of macrophages: in VAN FURTH Mononuclear phagocytes, pp. 103–120 (Blackwell, Oxford 1970).

KELLER, H. U. and SORKIN, E.: Studies on chemotaxis. VI. Specific chemotaxis in rabbit polymorphonuclear leucocytes and mononuclear cells. Int. Arch. Allergy *31:* 575–586 (1967).

KELLER, H. U. and SORKIN, E.: Chemotaxis of leucocytes. Experientia *24:* 641–652 (1968).

METCHNIKOFF, E.: La pathologie comparée de l'inflammation (Masson, Paris 1892); (English translation: Kegan Paul, London 1893).

MUIRHEAD, H.; COX, J. M.; MAZZARELLA, L., and PERUTZ, M. F.: Structure and function of haemoglobin. III. A three-dimensional Fourier synthesis of human deoxyhaemoglobin at 55 Å resolution. J. molec. Biol. *28:* 117–150 (1967).

NAIRN, R. C.: Fluorescent protein tracing, 3rd ed., pp. 97–99 (Livingstone, Edinburgh 1969).

NOBLE, R. W. and WAUGH, D. F.: Casein micelles. I. Formation and structure. J. amer. chem. Soc. *87:* 2236–2245 (1965).

O'NEILL, G. J.; HENDERSON, D. C., and WHITE, R. G.: Role of anaerobic coryneforms in specific and non-specific immunological reactions. I. Effect on particle clearance and humoral and cell-mediated responses. Immunology 24: 977–995 (1973).

ROSSI-FANELLI, A.; ANTONINI, E., and CAPUTO, A.: Studies on the structure of hemoglobin. I. Physicochemical properties of human globin. Biochim. biophys. Acta 30: 608–615 (1958).

ROSSI-FANELLI, A.; ANTONINI, E., and CAPUTO, A.: Studies on the relations between molecular and functional properties of hemoglobin. I. The effect of salts on the molecular weight of human hemoglobin. J. biol. Chem. 236: 391–396 (1961).

SNYDERMAN, R.; SHIN, H. S.; PHILLIPS, J. K.; GEWURZ, H., and MERGENHAGEN, S. E.: A neutrophil chemotactic factor derived from C'5 upon interaction of guinea pig serum with endotoxin. J. Immunol. 103: 413–422 (1969).

TEALE, F. W. J.: Cleavage of the haem-protein link by acid methylethylketone. Biochim. biophys. Acta 35: 543 (1959).

WARD, P. A. and BECKER, E. L.: Biochemical demonstration of the activatable esterase of the rabbit neutrophil involved in the chemotactic response. J. Immunol. 105: 1057–1067 (1970).

WAUGH, D. F.; CREAMER, L. K.; SLATTERY, C. W., and DRESDNER, G. W.: Core polymers of casein micelles. Biochemistry 9: 786–795 (1970).

WEKSLER, B. B. and HILL, M. J.: Inhibition of leukocyte migration by a staphylococcal factor. J. Bact. 98: 1030–1035 (1969).

WILKINSON, P. C.: Characterization of the chemotactic activity of casein for neutrophil leucocytes and macrophages. Experientia 28: 1051–1052 (1972).

WILKINSON, P. C.: Recognition of protein structure in leukocyte chemotaxis. Nature, Lond. 244: 512–513 (1973).

WILKINSON, P. C.: Chemotaxis and inflammation (Churchill-Livingstone, Edinburgh 1974).

WILKINSON, P. C. and McKAY, I. C.: The chemotactic activity of native and denatured serum albumin. Int. Arch. Allergy 41: 237–247 (1971).

WILKINSON, P. C. and McKAY, I. C.: The molecular requirements for chemotactic attraction of leucocytes by proteins. Studies of proteins with synthetic side groups. Europ. J. Immunol. 2: 570–577 (1972).

WILKINSON, P. C.; O'NEILL, G. J.; McINROY, R. J.; CATER, J. C., and ROBERTS, J. A.: Chemotaxis of macrophages. The role of a macrophage-specific cytotaxin from anaerobic corynebacteria and its relation to immunopotentiation in vivo; in WOLSTENHOLME and KNIGHT Ciba Foundation Symposium on Immunopotentiation (Associated Scientific Publishers, Amsterdam 1973a).

WILKINSON, P. C.; O'NEILL, G. J., and WAPSHAW, K. G.: Role of anaerobic coryneforms in specific and non-specific immunological reactions. II. Production of a chemotactic factor specific for macrophages. Immunology 24: 997–1006 (1973b).

WISSLER, J. H. and SORKIN, E.: Mechanism of cellular recognition and chemotactic activity for neutrophil leucocytes. Abstr. 9th Int. Congr. Biochem., Stockholm 1973, p. 318 (1973).

WISSLER, J. H.; STECHER, V. J., and SORKIN, E.: Biochemistry and biology of a leucotactic binary peptide system related to anaphylatoxin. Int. Arch. Allergy 42: 722–747 (1972).

Yoshinaga, M.; Mayumi, M.; Yamamoto, S., and Hayashi, H.: Immunoglobulin G as possible precursor of chemotactic factor. Nature, Lond. *225:* 1138–1139 (1970).

Yoshinaga, M.; Yamamoto, S.; Kayota, S., and Hayashi, H.: The natural mediator for PMN emigration in inflammation. IV. *In vitro* production of a chemotactic factor by papain from immunoglobulin G. Immunology, Lond. *22:* 393–399 (1972).

Zigmond, S. and Hirsch, J. G.: Effects of cytochalasin B on polymorphonuclear leucocyte locomotion, phagocytosis and glycolysis. Exp. Cell Res. *73:* 383–393 (1972).

Zigmond, S. H. and Hirsch, J. G.: Leukocyte locomotion and chemotaxis. New methods for evaluation, and demonstration of a cell-derived chemotactic factor. J. exp. Med. *137:* 387–410 (1973).

Authors' address: Dr. P. C. Wilkinson and Dr. I. C. McKay, Department of Bacteriology and Immunology, University of Glasgow, *Glasgow* (Scotland)

Antibiotics and Chemotherapy, vol. 19, pp. 442–463
(Karger, Basel 1974)

Regulation of Serum-Derived Chemotactic Activity by the Leucotactic Binary Peptide System

J. H. Wissler, E. Sorkin and Vera J. Stecher

Schweizerisches Forschungsinstitut, Medizinische Abteilung, Davos-Platz

Introduction

Accumulation of leucocytes is a basic feature of the process of inflammation [1–9, 25, 26]. In the various kinds of inflammatory processes different types of leucocytes are accumulating either alone or together at the reaction sites. This accumulation is time-dependent and proceeds most often in well-defined sequences [1–5, 7–9, 25 ,26]. Thus, cell-specific accumulation of different types of leucocytes or accumulation of a mixed cell population can occur *in vivo*. The task of attracted leucocytes is to take up dead tissue, to liberate other mediators for further activation of the defense system and to catalyze the healing process [1–6, 16, 18, 19, 21, 25, 26]. In addition, scavenging leucocytes (the phagocytes) have to remove by phagocytosis and pinocytosis foreign and host's own modified particulate and soluble material [22, 23], with resulting 'suicide' for most of them. On the other hand, failure of leucocyte accumulation and of their scavenging functions around tissue grafts and tumours, allows the latter to grow and to threaten the host [15]. There is some evidence *in vivo*, that tumour survival or its destruction is dependent on the ratio of the number of accumulated leucocytes to tumour size and tumour cell number present [10–12, 24]. In addition, malfunction of regulatory principles governing leucocyte accumulation and liberation of mediators often lead to excessive allergic reactions of the immediate or delayed type, to shock phenomena and other pathological tissue changes [6, 15, 25, 26] from which a considerable proportion of people are suffering [18].

We shall summarize in this paper the work concerning the molecular characteristics of a natural serum-derived chemotactic mediator system, the anaphylatoxin-related binary peptide leucotactic system with CAT[1] and CCT[1] as peptide components [30–40, 42, 45–47]. By activity mapping, regulatory principles governing selectivity and specificity of cell attraction are disclosed [35, 39].

Molecular Homogeneity and Biological Activities of CAT

Chemotactic Activity for Phagocytes and Spasmogenic Activity of CAT

Contact activation by preformed immune complexes, dextran or yeast of normal, fresh serum induces formation of spasmogenic and permeability-enhancing, histamine-liberating and shock-producing activity in the guinea pig and its organs, due to the presence of CAT [27, 30, 31, 36, 37], shown to be a split product of the complement component C5 [50]. These contact-activated sera may also display chemotactic activity for phagocytes, namely for neutrophils, eosinophils and macrophages the magnitude of which may be dependent on the type of contact agent, on the cell batch and on the species used, and, furthermore, on the concentration of activated serum used in the assay system [17, 38].

When CAT is purified [36–39] and the congruence of the distribution of spasmogenic activity and chemotactic activity for neutrophils, eosinophils and macrophages is compared in the sequence of six purification steps, including ion exchange, hydroxyapatite distribution and exchange chromatography and molecular sieve chromatographic steps, it was demonstrated [34, 38, 39] that there is a close relationship in the distributions of all of these activities (table I) [37–39]. This logically may lead to the assumption that all activities are intrinsic molecular properties of a single peptide which, by classical definition, is CAT [27, 35, 37], detectable through its spasmogenic activity. However, at this stage 1.6.0 (table I) of the purification operation, chemotactic activity for macrophages has been lost, although this activity followed the fractions with spasmogenic activity along the preceding five purification steps [34]. Electrophoretic analysis of spasmogenically active

1   Abbreviations: CAT=classical anaphylatoxin. CCT=cocytotaxin. ACL-system: anaphylatoxin-cocytotaxin binary leucotactic peptide system. Complement components and split products according to the recommendations of Bull. WHO 39: 935 (1968), also given in Immunochemistry 7: 137 (1970). Cytotaxigen: substance generating a cytotaxin in a distinct medium. Cytotaxin according to the proposal for a general nomenclature concerning agents eliciting tactic reactions of cells, as given in SORKIN et al. [41].

Table I. Schematic representation of the sequence and results of purification steps for CAT and CCT from rat serum, contact-activated with performed immune complexes, dextran, or yeast as cytotaxigens[1]

Purification step	Operation running combination 1.0.0 = anaphylatoxin running combination 2.0.0 = cocytotaxin	Activities[2]				Purification -fold	Yield, % CAT[2]
		CAT	CA-N	CA-M	CA-E		
1.0.0	contact reaction: serum + immune complex (or yeast, dextran)	+	+	+	+	1.0	100
1.1.0	batch-positive absorption on CM-Sephadex C-50	+	+	+	+	6.7	28
1.2.0	batch-negative absorption on calcium phosphate	+	+	+	+	12	24
1.3.0	chromatography on hydroxyapatite (distribution)	+	+	+	+	390	19
1.4.0	gel chromatography on Sephadex G-75	+	+	+	+	3,900	50[3]
1.5.0	analytical gel chromatography on Sephadex G-50	+	+	−	+	4,350	51
1.6.0	chromatography on hydroxyapatite (ion exchange)	+	+	−	+	4,350	33
1.7.0	differential gel chromatography (Sephadex G-75)	+	−	−	+	4,350	10
1.8.0	differential gel re-chromatography (Sephadex G–50)	+	−	−	+	4,900	10
1.9.0	crystallization of anaphylatoxin (isothermal precipitation with $[NH_4]_2SO_4$)	+	−	−	+[4]	4,900	10
2.8.0	differential gel re-chromatography (Sephadex G–50)	−	−	−	+[4]	1,900[5]	10
2.9.0	crystallization of cocytotaxin with $(NH_4)_2SO_4$ (precipitation in temperature gradient)	−	−	−	+[4]	1,900[5]	10
1.9.0 + 2.9.0	recombination of anaphylatoxin and cocytotaxin	+	+[4]	−	+[4]		

fractions at stage 1.6.0 (table I) reveals one major stained protein band with about three obvious minor contaminants which may be closely related to the major component [37–39]. Peak distribution analysis of chromatographic patterns, however, discloses oligodispersity of the preparation with about three major bands, beside minor contaminants [37–39]. Further extensive purification under refined conditions at physiological ionic strength and pH revealed that during these operations, chemotactic activity for neutrophils is also lost from the fractions comprising the smooth muscle-contracting and permeability-enhancing, as well as the shock-producing and histamine-liberating activity (table I) [34, 36, 38, 39]. Qualitatively, these latter activities remain unchanged as being intrinsic properties of the molecularly homo-geneous, crystallizable CAT[6] which, in this state, is devoid of significant chemotactic activity for neutrophils and macrophages (table I; fig. 1, 3, 5) [30, 31, 35–40, 45–47]. Hence, the classical activities of CAT, elaborated

---

Footnotes to table I

CAT and CCT represent the two independent peptide components of the leucotactic binary peptide system (ACL-system)[1]. The same scheme is applicable and has been used for purification of CAT and CCT from contact-activated pig and guinea pig serum. Note that chemotactic activity for neutrophils is lost during the purification within the last steps of the purification of CAT and CCT and is restored by recom-bination of these two peptides[6] (ACL-system). For display of this activity, no *stable* complex is formed by CAT and CCT, although interaction occurs. In contrast, chemotactic activity for macrophages is lost early in the purification operation prior to CAT-CCT separation and cannot be restored by the recombination of CAT and CCT. This purification behaviour already reflects a regulatory function of the ACL-system, namely cell-discriminatory chemotactic attraction of granulocytes (compare to table III and the text). CAT (spasmogenic) activity itself remains as an intrinsic molecular property of this one peptide of contact-activated serum[5], whereas chemo-tactic activity for eosinophils has been found to be displayed by the CAT and CCT preparations, although strongly differing in their concentration threshold (fig. 4, 6, 7). Methods and results on which these data are based, are given in WISSLER *et al.* [33–40, 45–47], STECHER *et al.* [28] and BERNAUER *et al.* [30, 31]. Table reprinted in comprehensively altered form from WISSLER *et al.* [39] with kind permission of the publisher (Karger, Basel).

1 Expression defined in 'Abbreviations'.
2 CA-N, CA-M, CA-E = chemotactic activity for neutrophils, macrophages, eosino-phils, respectively. CAT = anaphylatoxin activity.
3 Activation during purification step.
4 Activity strongly dependent on peptide concentration.
5 For calculation and definition see WISSLER [33].
6 Findings are subject to misquotation by LIEFLÄNDER *et al.* [51].

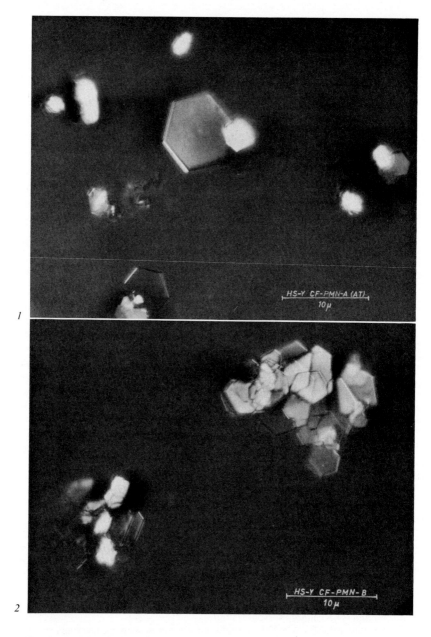

Fig. 1. Double-refractive crystals in polarized light of molecularly homogeneous CAT from yeast-treated pig serum. Similar crystalline preparations of CAT have been obtained from rat and guinea pig serum, activated with dextran [36, 37]. Methods for preparing the crystals are reported by WISSLER [37] and WISSLER *et al.* [46, 47]. Reprinted from WISSLER *et al.* [39] with kind permission of the publisher (Karger, Basel).

during the six decades by numerous investigators [for review, see 27] with crude CAT-containing sera, can now be attributed to the action of solutions of the crystallized CAT peptide. Furthermore, since these activities of solutions of the crystalline preparation have been qualitatively indistinguishable from the action of the crude and semi-purified CAT-containing serum [30, 31, 36, 37, 39], no artifacts have been induced in the CAT peptide in the process of its purification and crystallization. Nevertheless, chemotactic activity for eosinophils of the guinea pig remained unseparable from CAT, the mechanistic reason of which has to be elucidated [39, 40, 45]. Similar results have been recently obtained by PARISH [13, 29].

Nature of CCT, and its Role in Chemotactic Activity for
Phagocytic Leucocytes of Contact-Activated Serum or Plasma

To elucidate why chemotactic activity for neutrophils and macrophages is lost during the purification of CAT, we recombined the chemotactically *inactive* fractions of spasmogenically active CAT preparations with fractions of other chemotactically *and* spasmogenically inactive peptides, obtained within the sequence of purification operations (table I). Specifically, when a peptide, termed CCT [33, 36, 38, 39], was added to the CAT preparation, chemotactic activity for neutrophils[1], *but not for macrophages* is fully restored in comparison to the activity of the original peptide mixture prior to separation steps of CAT and CCT (stage 1.7.0 and 1.8.0 in table I and III). These mixtures display chemotactic activity for neutrophils, *but not for macrophages* [34, 36, 38, 39]. Reseparation of CCT from CAT again led to loss of chemotactic activity for neutrophils, demonstrating that the process of *acquisition* of chemotactic activity for neutrophils of the peptide mixture is fully reversible. Similar operations with fractions separated earlier from CAT and CCT in the sequence of purification steps (table I) have thus

---

1 Findings are subject to misquotation by LIEFLÄNDER *et al.* [51].

Fig. 2. Double-refractive crystals in polarized light of molecularly homogeneous CCT from yeast-treated pig serum. Similar crystalline preparations of CCT have been obtained from rat and guinea pig serum, activated with dextran [33, 36, 39]. Methods for preparing the crystals are reported by WISSLER [33]. Reprinted from WISSLER *et al.* [39] with kind permission of the publisher (Karger, Basel).

Table II. Comparison of some physicochemical and biological properties of CAT and CCT from pig, rat or guinea pig serum, which has been contact-activated either with preformed immune complexes, dextran or yeast as cytotaxigens

Physicochemical property, biological activity	Anaphylatoxin	Cocytotaxin
Molecular weight		
Peptides of hog and rat serum	9,500	8,500
Peptides of guinea pig serum	14,000	8,500
	−15,000	−12,000
Migration direction in electrophoresis (hog serum peptides)		
At pH 7.4 ⎫	cathode	cathode
At pH 9.0 ⎬ acrylamide gel	anode	cathode
At pH 11.0 ⎭	anode	anode
At pH 8.6 cellulose acetate	anode	cathode
At pH 8.6 agar.	anode	cathode
Extinction quotient at 280 and 260 nm	1.49	1.23
Carbohydrate (peptides of hog serum only)	negative	not done
Induction of lethal shock (guinea pig)	+	−[1]
Ileum contraction (guinea pig)	+	−
Coronary constriction (guinea pig)	+	−
Histamine liberation (guinea pig, blood level)	+	−
Chemotactic activity for neutrophils	−	−
Chemotactic activity for macrophages	−	−
Chemotactic activity for eosinophils	$+(2 \cdot 10^{-8} \text{M})$	$+(7 \cdot 10^{-7} \text{M})$

Note that CCT is a very strong basic peptide, whereas CAT is only very weakly basic, if at all. Due to endosmotic effects and dependent on its conformational state, the direction and the magnitude of electrophoretic migration of CAT at pH 7.4 from contact-activated rat and pig serum to either the cathode or the anode is influenced by the supporting matrix used. CAT from contact-activated guinea pig serum, irrespective of the matrix (polyacryl amide, cellulose acetate, agar, agarose), migrates only toward the anode at pH 7.4, but the magnitude of its electrophoretic migration is influenced by endosmosis and conformational alterations. Biological activities given in the table have been investigated over several orders of magnitude of the peptide concentrations (fig. 3–7). For eosinophils, the concentration thresholds of chemotactic activity is given in parentheses for both peptides (compare to figure 4, 6 and 7). Methods and results on which these data are based, are given in WISSLER et al. [33–40, 45–47], STECHER et al. [28] and BERNAUER et al. [30, 31]. Table reprinted in comprehensively altered form from WISSLER et al. [39] with kind permission of the publisher (Karger, Basel).

1 Up to 200–fold dosage of threshold of CAT.

far been unsuccessful to regain the lost chemotactic activity for macrophages.

Like CAT, CCT is no normal peptide constituent of plasma or serum of pig, guinea pig and rat, but is formed in serum or plasma by contact reactions with preformed immune complexes, dextran or yeast [35, 37]. One can assume that other natural or synthetic macromolecules may also be suitable for this purposes, as has been demonstrated for CAT formation [27, 35, 38]. The parent molecule of CCT is presently unknown. With yeast as contact agent, the amount of CCT to be formed under standard conditions (1 h incubation at 37 °C) [33, 37] in normal, fresh sera is about 3- to 6-fold the amount of generated CAT. It is mostly congruently purified to the spasmogenic activity of CAT (table I), and is separated from CAT in the last two purification steps which represent a sequence of analytical triple recycling gel permeation chromatography on Sephadexes of appropriate separation volume [33, 37–39]. The purification behaviour of CCT and CAT reflects their physicochemical similarity in some of their molecular parameters (table II), although some small differences exist in their molecular properties. They are detailed in table II.

In contrast to the physicochemical similarity, biologically, CCT differs strongly from CAT. Up to a 200-fold dosage of the minimum active dosis of CAT, CCT does not display spasmogenic activity on guinea pig organs, nor does it cause a fatal shock in the guinea pig under these conditions. The only biological activity, so far disclosed, is the strongly concentration-dependent chemotactic activity for eosinophils. In this property, it differs from CAT by its higher threshold, which is about 35-fold that of CAT (fig. 1, 4, 6; table II) [39, 40]. Since, by assaying spasmogenic activity, CAT contamination in CCT preparations can be evaluated to be less than 0.5% CAT, and since, physicochemically, no other contaminants could be detected so far, one may presently conclude that chemotactic activity for eosinophils is a molecular property of CCT, as it is of CAT [39, 40]. However, CCT does not display chemotactic activity for neutrophils and macrophages, nor does CAT. Hence, in contact-activated serum, there are present two agents chemotactic for eosinophils, one is an anaphylatoxin, the other one is not, neither of them attracts neutrophils significantly, as judged from the *in vitro* assay device [17, 36, 38, 39]. Table II compiles and compares the physicochemical and biological properties of CCT and CAT. Similar results in terms of number of agents chemotactically active for eosinophils in activated serum and the lack of anaphylatoxin to display chemotactic activity for neutrophils, have been recently reported by PARISH [13, 29].

Table III. Biological activities displayed by the ACL-system and by its two components, CAT and CCT, from different species and at different concentrations

Classical anaphylatoxin			Cocytotoxin			Chemotactic activity cell count/microscopic field			Anaphylatoxin activities	
serum source	mol. weight	molar conc.	serum source	mol. weight	molar conc.	CA-N[1]	CA-M[1]	CA-E[1]	lethal shock	ileum contraction
Hog	9,500	2.0×10^{-6}				0	0	52	+	+
Hog	9,500	1.1×10^{-7}				0	0	62	–	+
			hog	8,500	5.9×10^{-7}	0	0	0	–	–
			hog	8,500	5.9×10^{-8}	0	0	0	–	–
Hog	9,500	2.0×10^{-6} +	hog	8,500	5.9×10^{-7}	215	0	42	+	+
Hog	9,500	1.1×10^{-7} +	hog	8,500	5.9×10^{-8}	327	0	95	–	+
Rat	9,500	2.0×10^{-6}				0	0	nd	+	+
Rat	9,500	1.8×10^{-7}				0	0	nd	–	+
			rat	8,500	1.4×10^{-7}	0	0	nd	–	–
			rat	8,500	6.0×10^{-7}	0	0	nd	–	–
Rat	9,500	2.0×10^{-6} +	rat	8,500	1.4×10^{-7}	71	0	nd	+	+
Rat	9,500	1.8×10^{-7} +	rat	8,500	6.0×10^{-7}	107	0	nd	–	+
Guinea pig	14,000	1.8×10^{-6}				0	0	17	+	+
Guinea pig	14,000	7.0×10^{-8}				0	0	30	–	–
			guinea pig	8,500	1.0×10^{-6}	0	0	15	–	–
			guinea pig	8,500	1.0×10^{-7}	0	0	0	–	–
Guinea pig	14,000	1.8×10^{-6} +	guinea pig	8,500	1.0×10^{-6}	53	0		+	+
Guinea pig	14,000	7.0×10^{-8} +	guinea pig	8,500	1.0×10^{-7}	87	0	70	–	–

CAT (spasmogenic) activity is displayed exclusively by CAT and also in related oligodisperse systems through CAT. Chemotactic activity for neutrophils is displayed exclusively by the binary ACL-system, but not by its native components [34, 35, 43, 44, 48]. Chemotactic activity for eosinophils is displayed exclusively by the binary ACL-system and its two components. Chemotactic activity for macrophages is neither displayed by the binary ACL-system nor its two components. Note that the magnitude of all apparent activities is strongly dependent on the concentration of the peptides (compare to figures 3–7). Chemotactic activity for macrophages is, however, uniformly negative over several orders of magnitude (10^{-5}–10^{-11} M) of the concentration of the peptides alone and of various molar ratios of their combination within the binary ACL-system. The same uniformly negative results are obtained for spasmogenic activity (smooth muscle of guinea pig) and chemotactic activity for neutrophils of CCT. Methods and results on which these data are based, are given in Wissler *et al.* [33–40, 45] and Stecher *et al.* [28]. Reprinted in comprehensively altered form from Wissler *et al.* [38] with kind permission of the publisher (Karger, Basel).

[1] CA–N, CA–M, CA–E = chemotactic activity for neutrophils, macrophages, eosinophils, respectively. nd = not done.

Properties of Chemotactic Activities for Neutrophils and Eosinophils,
Displayed by the Leucotactic Binary Peptide System

Concerning *chemotactic activity for neutrophils* displayed by the *combination* of CAT and CCT, the magnitude of this activity is strongly dependent
on the absolute concentrations *and* the molar ratio of the two peptide
components. Figure 3 and 5 disclose that maximum chemotactic activity
for neutrophils of the peptide combination is only displayed within a molar
ratio range of CCT:CAT of 1:100 to 8:1. Increasing or decreasing molar
ratios leads to a decreased observable chemotactic activity for neutrophils.
Hence, a CAT preparation which is contaminated with 1% CCT displays
maximum chemotactic activity for neutrophils, whereas the same cell type is
almost non-responsive chemotactically to a CCT preparation which is
contaminated even with 5% CAT. Additionally, neutrophils are chemotactically non-responsive to ratios CCT:CAT>50:1, as judged from the
Boyden assay device [17, 38–40].

A further interesting point is that the magnitude of chemotactic response
for neutrophils to CAT:CCT combinations in the range of the ratios CCT:
CAT of 1:100 to 8:1 (ratio range of *responsiveness* of neutrophil chemotaxis)
is strongly *dependent* on the absolute concentration of the two peptides. In
contrast, the *non-responsiveness* of neutrophils in terms of chemotaxis is
independent of the absolute concentration of the two peptides, provided that
the molar ratio CCT:CAT is>50:1. Similar results have been obtained for
different species in the homologous and heterologous system [28, 38, 39].
Table III summarizes the experiments.

Just as for neutrophils, *chemotactic activity for eosinophils* is displayed
by the peptide combination of CAT and CCT, but not for macrophages.
The magnitude of this activity depends, too, on the absolute concentration
and on the molar ratio of CAT and CCT (fig. 4, 6). However, since the two
peptides *per se* attract eosinophils *in vitro* (see above, table II; fig. 4, 6),
there exists no distinct range of molar ratios of the peptide combination in
which eosinophils are non-responsive in terms of chemotaxis, in contrast
to the neutrophil response (fig. 3, 5) [39, 40]. Additionally, the influence of
alterations of the molar ratio of the two peptides on the magnitude of
chemotactic activity for eosinophils is less pronounced than in neutrophil
chemotaxis (fig. 3–6). The regulatory implications of these results in terms
of cell-specific and cell-selective attraction of leucocytes are subject of the
next chapter.

In summary, these findings clearly demonstrate the critical point of

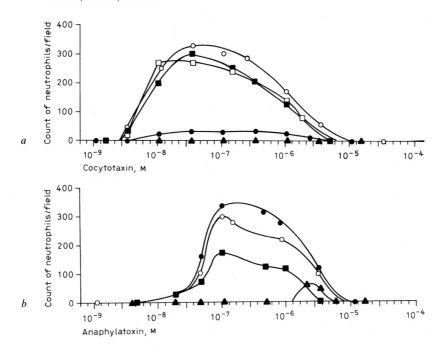

Fig. 3. Dependence of chemotactic activity for neutrophil leucocytes on the concentration of chemotactically active preparations. For technical reasons, in the examples given below, peptides from pig serum activated with yeast, and peritoneal neutrophils from rabbits have been used. This is permissible in view of the results presented in table III and in STECHER *et al.* [28]. Chemotactic activity is presented as neutrophil counts/microscopic field. The concentration of peptides is based on a molecular weight of 9,500 daltons for CAT [37, 39] and 8,500 daltons for CCT [33, 39], determined spectrophotometrically on a dry weight basis [33, 37, 38]. For implications using other protein determination methods on the concentration scale, especially staining methods, see WISSLER [37] and WISSLER *et al*]. 38]. *a.* The binary ACL-system as chemotactically active peptide preparation (compare to figure 5) with varying concentrations of CCT in combination with constant doses of CAT. Constant concentrations of CAT used: $\blacktriangle = 0$ M, $\bullet = 3.50 \cdot 10^{-8}$ M, $\bigcirc = 1.12 \cdot 10^{-7}$ M, $\blacksquare = 2.80 \cdot 10^{-7}$ M, $\square = 9.80 \cdot 10^{-7}$ M. *b.* The binary ACL-system as chemotactically active peptide preparation (compare to figure 5) with varying concentrations of CAT in combination with constant doses of CCT. Constant concentrations of CCT used: $\blacktriangle = 0$ M, $\bullet = 5.90 \cdot 10^{-8}$ M, $\bigcirc = 1.18 \cdot 10^{-7}$ M, $\blacksquare = 1.19 \cdot 10^{-6}$ M.

Reprinted from WISSLER *et al.* [40] with kind permission of the publisher (Pergamon Press, Oxford). There, too, technical details are given on which these results are based.

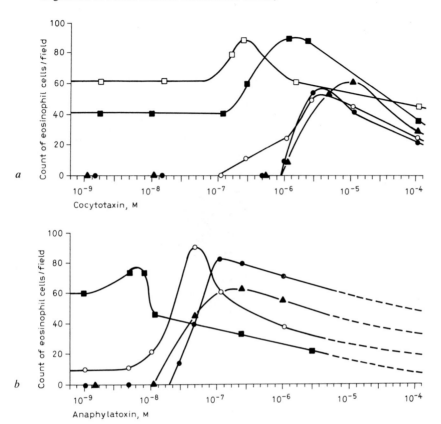

Fig. 4. Dependence of chemotactic activity for eosinophil leucocytes on the concentration of chemotactically active preparations. For technical reasons, in the examples given, peptides from pig serum, activated with yeast, and peritoneal eosinophils from guinea pigs have been used (for species specificity see legend to figure 3). Chemotactic activity is presented as eosinophil counts/microscopic field. The molecular weights of the peptides on which the given concentrations are based, and methods of their determination are specified in figure 3. Other methods concerning these figures are given by WISSLER *et al.* [40], from there, too, figure 4a and b are reprinted with kind permission of the publisher (Pergamon Press, Oxford). Details on the chemotactic assay for eosinophil cells are reported in STECHER *et al.* [28]. *a.* The binary ACL-system as chemotactically active peptide preparation (compare to figure 6) with varying concentrations of CCT in combination with constant doses of CAT. Constant concentrations of CAT used: $\blacktriangle = 0$ M, $\bullet = 1.40 \cdot 10^{-9}$ M, $\circ = 1.40 \cdot 10^{-8}$ M, $\blacksquare = 7.00 \cdot 10^{-8}$ M, $\square = 4.20 \cdot 10^{-7}$ M. *b.* The binary ACL-system as chemotactically active peptide preparation (compare to figure 6) with varying concentrations of CAT in combination with constant doses of CCT. Constant concentrations of CCT used: $\blacktriangle = 0$ M, $\bullet = 1.18 \cdot 10^{-7}$ M, $\circ = 1.18 \cdot 10^{-6}$ M, $\blacksquare = 1.18 \cdot 10^{-5}$ M.

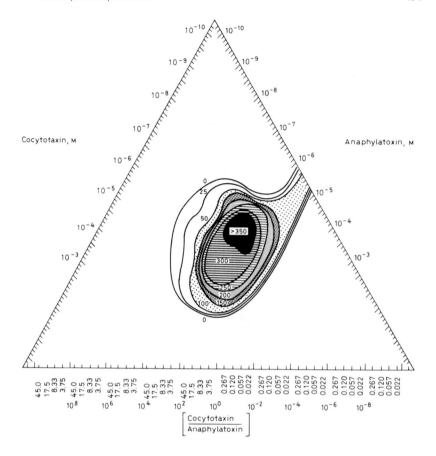

Fig. 5. Biological function *in vitro* of the ACL-system: Homogeneity of CAT and chemotaxis of neutrophils. The activity phase diagram of the binary ACL-system represents chemotactic activity for neutrophils as acquired biological activity of CAT in isochemotaxis curves, as analogously given two-dimensionally in figure 3a and b with a less complete experimental set. Parameters from which chemotactic activity is dependent are the absolute concentrations (M) of CAT and CCT, and the molar ratio of the two peptides. Isochemotaxis curves represent constant levels of chemotactic activity as cell counts/ microscopic field. Only within the area limited by the zero-isochemotaxis curve, the peptide system displays chemotactic activity for neutrophils. Maximum chemotactic activity is observed within the concentration and molar ratio range, given by the black area. Non-responsiveness of neutrophils to the binary ACL-system in terms of chemotaxis is observed in the concentration and molar ratio ranges outside of the zero-isochemotaxis

molecular homogeneity of a peptide preparation, especially when considering chemotactic activity for neutrophils [38, 40]. CAT may easily *acquire* a *biological activity* which, physiologically and physicochemically, is not an intrinsic property of the CAT peptide. This principle evaluated for CAT can be shown to be extendable to other non-mediatory proteins [32, 35, 43, 48]. Additionally, the aforementioned intrinsic, mediatory activity (spasmogenicity on smooth muscle cells) of the CAT peptide is not altered qualitatively, or abolished by contamination with CCT, and therefore, is displayed irrespective of the molar ratio of the two peptides above the threshold. However, as in the case of eosinophil chemotaxis, an influence may be exerted by changes of the molar ratio of the two peptides on the magnitude of the intrinsic activity (spasmogenicity) of CAT, depending on the tissue in which this activity is displayed [35, 36]. Furthermore, a non-responsiveness phase, determined by the molar ratio of CAT and CCT, is not existing for the intrinsic, spasmogenic activity of CAT. However, compared to eosinophil chemotaxis, where two active principles are affecting cells (fig. 6) within the mediatory action of this binary peptide system, only one component (CAT) of the binary peptide system exerts spasmogenicity. We have tentatively proposed the name *'anaphylatoxin-related leucotactic binary peptide system'* or *'anaphylatoxin-cocytotaxin leucotactic binary peptide system' (ACL-system)* [38, 39] to comprise the multitude of biological activities displayed regulatorily (next chapter) by the combinations of CAT and CCT, as well as in part by the peptides themselves [39]. This definition takes into account the possibility that in sera of species other than pig, rat and guinea pig, or in the sera of these species under varied conditions of plasma or serum activation, different activity principles in terms of leucotaxis and spasmogenicity might be effective [36, 39].

---

curve. In contrast to responsiveness in terms of chemotaxis for neutrophils, the non-responsiveness of these cells is *independent* of the absolute concentration of the two peptide components, provided that the molar ratio CCT:CAT>50:1. Note that significant chemotactic activity for neutrophils is observed only when the native CAT is contaminated with CCT. Methods for assaying chemotaxis are reported in WISSLER et al. [38]. Peptides from pig serum, activated with yeast, and peritoneal neutrophils from rabbits have been used for technical reasons. This is permissible in view of the species independence of the results obtained (table III) [28]. For molecular weights on which the concentration values are based and protein determination methods, see legend figure 3 [33, 37]. The activity phase diagram has been prepared according to the recommendations of PEREL'MAN [14] on geometric presentation of multicomponent systems. Reprinted from WISSLER et al. [39] with kind permission of the publisher (Karger, Basel).

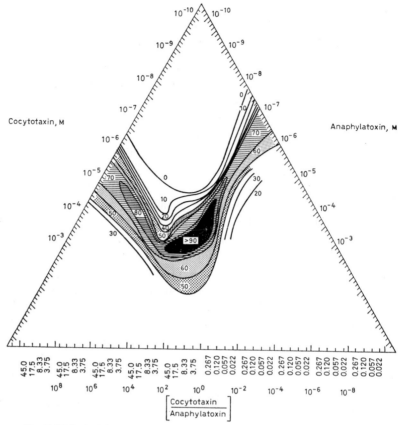

Fig. 6. Biological function *in vitro* of the ACL-system: Chemo-attraction of eosino-phils. The activity phase diagram of the binary ACL-system represents chemotactic activity for eosinophil leucocytes in isochemotaxis curves (compare to figure 5), as analogously given two-dimensionally in figure 4a and b with a less complete experimental set. Para-meters from which chemotactic activity is dependent, are the absolute concentrations (M) of CAT and CCT and the molar ratio of the two components of the binary ACL-system. Isochemotaxis curves represent constant levels of chemotactic activity as cell counts/ microscopic field. Chemotactic activity for eosinophil cells is displayed by the binary ACL-system in different magnitudes in the areas limited by the correspondent isochemo-taxis curve. Maximum activity is observed within the concentration and molar ratio range, shown by the black area. Notably, in contrast to the behaviour of neutrophil cells (fig. 5), eosinophils show responsiveness within the whole phase diagram below the zero-isochemo-taxis curve. For technical reasons, peptides from pig serum, contact-activated with yeast and guinea pig peritoneal eosinophil cells have been used. This is permissible in view of the species independence of the results obtained (table III) [28]. For molecular weights on which the given concentration values of the peptides are based, and on protein determina-tion methods, see legends to figure 3 [33, 37]. Methods and assay for chemotaxis are reported by WISSLER *et al.* [38, 40] and STECHER *et al.* [28]. For presentation of the phase diagram, see figure 5. Reprinted from WISSLER *et al.* [39] with kind permission of the publisher (Karger, Basel).

Mapping of
Biological Activities of the Anaphylatoxin-Related
Leucotactic Binary Peptide System

Phase diagrams of the binary ACL-system for chemotactic activity for neutrophils and eosinophils are presented in figure 5 and 6.

These figures have been used [39] for mapping, in part, the biological activities of the binary ACL-system. Figure 7 represents a partial activity map of the binary ACL-system, i.e. for chemotactic activity for neutrophils and eosinophils obtained by superimposing of figures 5 and 6. In order to simplify the presentation, spasmogenic activity of CAT alone and its variance in terms of its magnitude and its dependence on the absolute concentration and the molar ratio of the peptides has been omitted (in part, e.g. for ileum contraction, phase limit of spasmogenic activity mostly follows the iso-concentration [threshold] of CAT). Figure 7 demonstrates that the binary ACL-system can display multiple biological activities either specifically or selectively, depending on only two variable parameters, namely the absolute concentration of CAT (or CCT) *and* the molar ratio of CAT and CCT, through the existence of incongruent, but partially overlapping activity phases. Together with the spasmogenic activity of CAT in terms of contraction of smooth muscle, the following activities can be displayed regulatorily and, in part, be delineated from figure 7: (a) chemotactic attraction of neutrophils, predominantly; (b) chemotactic attraction of eosinophils, predominantly; (c) spasmogenic activity of CAT, e.g. in terms of contraction of ileum; (d) activities (a) and (b) together, elicited in the overlap range of the corresponding activity phases and leading to selective chemotactic attraction of granulocytes (since macrophages are not subject to chemotactic attraction by this system); (e) activities (a) and (c) together, elicited in the overlap range of the corresponding activity phases (not given in figure 7) [35]; (f) activities (b) and (c) together, elicited in the overlap range of the corresponding activity phases (not given in figure 7) [35], and (g) activities (a), (b) and (c) together, elicited in the overlap range of the corresponding phases (not included in figure 7) [35].

Since the subtle balance of the concentration and of the molar ratio of CAT and CCT may decide the biological effects displayed by the binary ACL-system, and since macrophages do not respond chemotactically to it, three regulatory functions of the ACL-system concerning its chemotactic activity for granulocytes can be delineated from the activity map, represented by figure 7 [39, 40]:

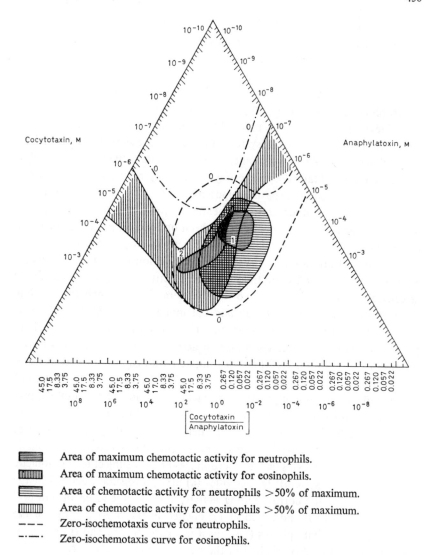

Area of maximum chemotactic activity for neutrophils.

Area of maximum chemotactic activity for eosinophils.

Area of chemotactic activity for neutrophils >50% of maximum.

Area of chemotactic activity for eosinophils >50% of maximum.

— — — Zero-isochemotaxis curve for neutrophils.

—·—· Zero-isochemotaxis curve for eosinophils.

Fig. 7. Biological function *in vitro* of the ACL-system: Regulated cell-specific and cell-selective chemo-attraction of neutrophil and eosinophil leucocytes. Comparison and demonstration of partial overlap of the activity phases of chemotactic activity for neutrophil and eosinophil leucocytes, displayed by the binary ACL-system with its components CAT and CCT. The diagram is obtained by superimposition of the correspondent activity phase diagrams, given in figure 5 and 6. There, too, the technical details and methods are given on which the results are based. Note that areas of maximum activity as well as the whole activity phase of each cell type are far from being congruent. Therefore, the binary

(a) Group specificity of chemotaxis of granulocytes (neutrophils and eosinophils) through the responsiveness-nonresponsiveness principle.

(b) Cell selectivity within the granulocyte group by preferential attraction of either granulocyte type, characterized by differences in the magnitude of activities and incongruence of the maximum activity phases through variation of the absolute concentration and of the molar ratio of the peptide components.

(c) Cell specificity of attraction by cell discrimination within the granulocyte group through the existence of a non–responsiveness phase for either cell type.

The detailed position in the activity map of further candidates for biological activities of the ACL-system, displayed specifically or selectively, is under investigation [35, 48, 49]. These are spasmogenic activity of CAT in terms of enhancement of capillary permeability, histamine liberation (mast cell degranulation, histamine levels in blood) and shock production, as well as chemo-attractive trapping of eosinophils. At a first glance, inspection of the map of chemotactic activities (fig. 7) as well as of the other above-

---

ACL-system can attract *in vitro* either specifically eosinophils *or* neutrophils alone. In the area of overlapping activity phases it can attract both cell types together, the magnitude of selectivity being dependent only on the absolute concentrations of CAT and CCT and their molar ratio. In preliminary experiments, satisfactory results have been obtained concerning the cell selectivity and specificity of attraction *in vivo* by application of the peptide combinations at positions 1 and 2 in the phase diagram. Thus, in the skin test in rabbits, peptide combination 2 preferably attracted eosinophils, and with combination 1 neutrophils have been predominant [39]. Since the peptide components of the ACL-system interact in displaying chemotaxis [34, 35, 38, 43, 44], but do not form a *stable* complex in free solution [38], kinetics and thermodynamics of their formation and of their inactivation may decide the balance of their concentration and molar ratio and, thus, the specificity and selectivity of cell attraction *in vivo*. In summary, since macrophages do not respond to the ACL-system in terms of chemotaxis, three regulatory functions of the ACL-system in terms of chemotaxis may be delineated from the diagram [40]: (1) Group specificity for granulocytes by cell discrimination through the responsiveness–non-responsiveness principle. (2) Cell selectivity within the granulocytes by preferential attraction of one cell type through variation of the absolute concentrations and of the molar ratio of the peptide components characterized by differences in the magnitudes of activity and incongruence of the phases of maximum activity. (3) Cell specificity by cell discrimination within the granulocyte group through the existence of a non-responsiveness phase for either cell type.

Reprinted from WISSLER *et al.* [39] with kind permission of the publisher (Karger, Basel). For presentation of the phase diagram, see figure 5. For further activity mapping, see text.

mentioned biological effects [35] of the ACL-system may suggest that most of the iso-activity positions, e.g. the maxima or zero positions of phases of different activities, seem to lie closely together. However, as the scale on the axis indicates, iso-activity positions of phases of different biological effects of the ACL-system may differ by up to several orders of magnitude in terms of the absolute concentrations of the two peptides or of their molar ratio. Hence, if the principles evaluated *in vitro* should be effective *in vivo* too, small differences in the thermodynamics and kinetics of formation and inactivation processes of CAT and CCT are powerful means in determining and overregulating the activity displayed by the ACL-system. Time-dependent shift of the peptide ratio may contribute to a broad spectrum of biological effects to be apparent in sequence *or* coincidently. Doubtless, *in vivo*, any situation is much more complex than *in vitro* assays can ever disclose. Nevertheless, encouraging results have been obtained in preliminary *in vivo* assays concerning predominant cell attraction by application of combinations of CAT and CCT of different molar ratios, as mentioned in the legend of figure 7.

Summary

The physicochemical and biological characteristics of the leucotactic binary peptide system with classical anaphylatoxin and cocytotaxin as peptide components of contact reaction-activated mammalian sera (pig, rat, guinea pig) are reviewed. The regulatory principles by which the peptide system governs cell selectivity of chemotactic attraction are emplasized.

Acknowledgement

This work has been supported by the 'Deutsche Forschungsgemeinschaft', Grant No. Wi 406/1a and Wi 406/1b, by the 'Schweizerischer Nationalfonds zur Förderung der wissenschaftlichen Forschung', Grant No. 3.8750-72 and by Hoffmann La Roche AG, Basel.

References

1   COCHRANE, C. G.: The Arthus reaction; in ZWEIFACH, GRANT and McCLUSKEY The inflammatory process, pp. 613–648 (Academic Press, New York 1965).
2   STETSON, C. A., jr.: Similarities in the mechanisms determining the Arthus and Shwartzman phenomena. J. exp. Med. *94:* 347–357 (1951).

3   BENACERRAF, B.: Hypersensibilité retardée; dans BORDET Immunologie. Coll. Méd.-
    Chirurg. Rev. Period. pp. 1069–1078 (Flammarion, Paris 1972).
4   BENACERRAF, B.: Delayed hypersensitivity; in ZWEIFACH, GRANT and McCLUSKEY
    The inflammatory process, pp. 577–586 (Academic Press, New York 1965).
5   TURK, J. L.: Delayed hypersensitivity; in NEUENBERGER and TATUM Frontiers in
    biology. North-Holland Research Monographs, vol. 4 (North-Holland, Amsterdam
    1967).
6   LEE, L. and STETSON, C. A., jr.: The local and generalized Shwartzman phenomena;
    in ZWEIFACH, GRANT and McCLUSKEY The inflammatory process, pp. 791–817
    (Academic Press, New York 1965).
7   SPECTOR, W. G.: Histology of allergic inflammation. Brit. med. Bull. 23: 35–38
    (1967).
8   SCHILD, H. O. and WILLOUGHBY, D. A.: Possible pharmacological mediators of
    delayed hypersensitivity. Brit. med. Bull. 23: 46–51 (1967).
9   McCLUSKEY, R. T.; BENACERRAF, B., and McCLUSKEY, J. W.: Studies on the
    specificity of the cellular infiltrate in delayed hypersensitivity reactions. J. Immunol.
    90: 466–477 (1963).
10 FREED, J. J. and LEBOWITZ, M. M.: The association of a class of saltatory movements
    with microtubules in cultured cells. J. Cell Biol. 45: 334–354 (1970).
11 REBUCK, J. W.; BRENNAN, M. J.; HALL, J. A., and BARTH, C. L.: Leucocytic trophism
    for inflammation and autochthonous and homologous tumor cells. J. reticuloendot.
    Soc. 1: 450–463 (1964).
12 KASSEL, R. L.; OLD, L. J.; CARSWELL, E. A.; FIORE, N. C., and HARDY, W. D., jr.:
    Serum-mediated leukemia cell destruction in AKR mice. Role of complement in the
    phenomenon. J. exp. Med. 138: 925–938 (1973).
13 PARISH, W. E.: Substances that attract eosinophils in vitro and in vivo and that elicit
    blood eosinophilia; in SORKIN Chemotaxis, its biology and biochemistry (Karger,
    Basel 1974).
14 PEREL'MAN, F. M. (ed.): Phase diagrams of multicomponent systems. Geometric
    methods, A special research report. Translated from Russian by D. A. PATERSON
    (Consultants Bureau Enterprises, New York 1966).
15 WINN, H. J.; BALDAMUS, C. A.; JOOSTE, S. V., and RUSSELL, P. S.: Acute destruction
    by humoral antibody of rat skin grafted to mice. The role of complement and poly-
    morphonuclear leukocytes. J. exp. Med. 137: 893–910 (1973).
16 HIRSCH, J. G.: Neutrophil and eosinophil leucocytes; in ZWEIFACH, GRANT and
    McCLUSKEY The inflammatory process, pp. 245–280 (Academic Press, New York
    1965).
17 BOYDEN, S. V.: The chemotactic effect of mixtures of antibody and antigen on poly-
    morphonuclear leucocytes. J. exp. Med. 115: 453–466 (1962).
18 DAVIS, D. J.: NIAID initiatives in allergy research. J. Allergy 49: 323–328 (1972).
19 ODLAND, G. and ROSS, R.: Human wound repair. I. Epidermal regeneration. J. Cell
    Biol. 39: 135–151 (1968).
20 SCHILLING, J. A.: Wound healing. Physiol. Rev. 48: 374–423 (1968).
21 CLINE, M. J.: Leukocyte function in inflammation. The ingestion, killing, and diges-
    tion of microorganisms. Ser. Haemat. 3/2: 3–16 (1970).
22 EBERT, R. H.: The experimental approach to inflammation; in ZWEIFACH, GRANT and
    McCLUSKEY The inflammatory process, pp. 1–33 (Academic Press, New York 1965).

23   ROBINEAUX, R. et BONA, C.: Phagocytose et pinocytose; dans BORDET Immunologie. Coll. Méd.-Chirurg. Rev. Period., pp. 155–207 (Flammarion, Paris 1972).

24   KUNZE, W. P. and GROPP, A.: Einschlüsse von Granulozyten in Krebszellen. Z. Krebsforsch. *73:* 218–222 (1970).

25   COCHRANE, C. G.: Immunologic tissue injury mediated by neutrophilic leucocytes. Adv. Immunol. *9:* 97–162 (1968).

26   FLOREY, H.: Chemotaxis, phagocytosis and the formation of abscesses; in FLOREY General pathology, pp. 98–127 (Lloyd-Luke, London 1962).

27   GIERTZ, H. und HAHN, F.: Makromoleculare Histaminliberatoren; in EICHLER and FARAH Handbook of experimental pharmacology, vol. 18/1, pp. 481–568 (Springer, Berlin 1966).

28   STECHER, V. J.; WISSLER, J. H., and SORKIN, E.: Chemistry and biology of the ana-phylatoxin-related serum peptide system. IV. Species specificity of the leucotactic activity (manuscript to be submitted).

29   PARISH, W. E.: Eosinophilia. III. The anaphylactic release from isolated human basophils of a substance that selectively attracts eosinophils. Clin. Allergy *2:* 381–390 (1972).

30   BERNAUER, W.; HAHN, F.; WISSLER, J. H.; NIMPTSCH, P. und FILIPOWSKI, P.: Unter-suchungen mit klassischem und kristallisiertem Anaphylatoxin an isolierten Herz-präparaten. Arch. Pharmakol. *269:* 413 (1971).

31   BERNAUER, W.; HAHN, F.; NIMPTSCH, P., and WISSLER, J. H.: Studies on heart anaphylaxis. V. Cross-desensitization between antigen, anaphylatoxin, and compound 48/80 in the guinea pig papillary muscle. Int. Arch. Allergy *42:* 136–151 (1972).

32   WISSLER, J. H. and SORKIN, E.: Mechanism of cellular recognition and chemotactic activity for neutrophil leucocytes. Abstr. Commun. 9th Int. Congr. Biochem., Stockholm 1973, p. 318.

33   WISSLER, J. H.: Chemistry and biology of the anaphylatoxin-related serum peptide system. II. Purification, crystallization and properties of a new basic peptide, co-cytotaxin, from rat serum. Europ. J. Immunol. *2:* 84–89 (1972).

34   WISSLER, J. H. and SORKIN, E.: Nature and mechanism of cellular recognition and regulation of leucocyte migration. Rep. Workshop Chemotaxis of Leucocytes, Paris 1973. Nouv. Rev. franç. Hémat. (in press).

35   WISSLER, J. H.: Cellular recognition, information analysis, and regulation of motor responses of leucocytes: Studies with natural and artificial model reaction systems. (manuscript to be submitted).

36   WISSLER, J. H.: A new biologically active peptide system related to classical ana-phylatoxin. Experientia *27:* 1147–1148 (1971).

37   WISSLER, J. H.: Chemistry and biology of the anaphylatoxin-related serum peptide system. I. Purification, crystallization, and properties of classical anaphylatoxin from rat serum. Europ. J. Immunol. *2:* 73–83 (1972).

38   WISSLER, J. H.; STECHER, V. J., and SORKIN, E.: Chemistry and biology of the ana-phylatoxin-related serum peptide system. III. Evaluation of leucotactic activity as a property of a new peptide system with classical anaphylatoxin and cocytotaxin as components. Europ. J. Immunol. *2:* 90–96 (1972).

39   WISSLER, J. H.; STECHER, V. J., and SORKIN, E.: Biochemistry and biology of a leucotactic binary peptide system related to anaphylatoxin. Int. Arch. Allergy *42:* 722–747 (1972).

40 WISSLER, J. H.; STECHER, V. J., and SORKIN, E.: Regulation of chemotaxis of leucocytes by the anaphylatoxin-related peptide system; in PEETERS Proc. XXth Coll.
Protides of the Biological Fluids, pp. 411–416 (Pergamon Press, Oxford 1973).

41 SORKIN, E.; STECHER, V. J., and BOREL, J. F.: Chemotaxis of leucocytes and inflammation. Ser. Haemat. 3/1: 131–162 (1970).

42 SORKIN, E.; STECHER, V. J., and WISSLER, J. H.: The anaphylatoxin-related leukotactic
binary peptide system; in BRAUN and UNGAR Non-specific factors influencing host
resistance. A Reexamination, pp. 196–204 (Karger, Basel 1973).

43 WISSLER, J. H.; STECHER, V. J.; SORKIN, E., and JUNGI, T.: Mechanism of leucotaxis.
Abstr. Commun. 4. Tag. Ges. Immunologie, Bern 1972, p. 61.

44 WISSLER, J. H.; STECHER, V. J., and SORKIN, E.: Secondary structural properties of
anaphylatoxin preparations and chemotactic activity for neutrophils. J. Immunol.
111: 314–315 (1973).

45 WISSLER, J. H. A new leucotactic peptide system; in SORKIN and WARD Chemotaxis
of leucocytes. Immunologic and nonimmunologic factors. 1st. Int. Congr. Immunology, Washington. Progr. Immunol., vol. 1, pp. 1411–1415 (Academic Press,
New York 1971).

46 WISSLER, J.; GIERTZ, H.; ERNENPUTSCH, I.; WALLENFELS, K., and HAHN, F.: Purification and crystallization of hog serum anaphylatoxin. Abstr. Commun. 6th FEBS-
Meeting, Madrid 1969, Abstr. 229.

47 WISSLER, J.; GIERTZ, H.; HAHN, P.; WALLENFELS, K., and ERNENPUTSCH, I.: Crystallized anaphylatoxin of hog serum. Abstr. Commun. 4th Int. Congr. Pharmacology, Basel 1969, p. 217.

48 WISSLER, J. H.: Modes of cellular recognition governing migratory responses of
leucocytes. Abstr. Commun. 9th FEBS-Meeting, Budapest 1974 p. 300.

49 WISSLER, J. H.; SORKIN, E.; JUNGI, T. W.; STECHER, V. J., and ARCON, A.: Mechanisms regulating leucocyte accumulation. Cellular recognition and migratory
behaviour of phagocytes; in WILLOUGHBY, VELO, GIROUD, ARRIGONI-MARTELLI and
ROSENTHALE Proc. Int. Meeting Future Trends Inflammation, Verona 1973 (Picini,
Padova 1974, in press).

50 JENSEN, J. A.: Anaphylatoxin (s), in INGRAM, D.G. (Ed.) Biological activities of
complement, pp. 136–157 (Karger, Basel 1972).

51 LIEFLÄNDER, M.; DIELENBERG, D.; SCHMIDT, G., and VOGT, W.: Structural elements of anaphylatoxin obtained by contact activation of hog serum. Z. Physiol.
Chem. 353: 385–392 (1972).

Author's address: Dr. JOSEF H. WISSLER, Schweiz. Forschungsinstitut, Medizinische
Abteilung, CH-7270 Davos-Platz (Switzerland)

Subject Index

Actin in cells 192
Activity mapping of serum peptide
  system 457
Actomyosin system, inhibition by AMP
  209
Albumin, recognition of denatured
  425
Amoebae 96–110
AMP, cyclic
– effect on amoebae 96–110
Anaphylatoxin, classical (CAT) 233,
  442–463
– biological properties 448–449
– chemotactic activity for phagocytes
  451–460
– combination with cocytotaxin 451
– molecular homogeneity 443
– molecular weight 449
– physico-chemical properties 448
– purification 444
– spasmogenic activity 443
Anaphylactic mast cells, eosinophil
  attraction 253
Anaphylactic substances eliciting
  eosinophilia 241
– attracting eosinophils 241
Anaphylatoxin-related leucotactic
  peptide system 451–460
Arthus reaction 9
Assays of chemotaxis 1, 13, 55, 99, 115,
  126, 146, 161

Bacteria, chemotactic factor from,
  382–408
Basophils, release of eosinophil
  attracting substance 247
Blood coagulation and chemotaxis 362
Blood eosinophilia 233, 256
Boyden chamber 2
– modifications 117, 127, 146

Casein 423, 433
Cell aggregation in Dictoystelium 96
Cell derived chemotactic factors 273,
  241, 247
Cellular polarity 205
Cell specificity 8
Chediak-Higashi Syndrome 151, 345, 347
Chemoreception in bacteria 12–20
Chemoreceptors, number of 15–18
– mode of action 17
– nature of – in *E. coli* 13
– nature of – in leucocytes 429
Chemotaxis
– and mobility 184
– and inflammation 296
– and random mobility 338–349
  – clinical correlates 344
– cell specificity of 8
– chambers 117, 128
– criteria for chemotactic factor 297
– detection of a gradient 79–93
– defects of 151, 291, 345–347, 350